Lecture Notes in
Control and
Information Sciences

Edited by M. Thoma

IFIP

69

Stochastic Differential Systems

Filtering and Control

Proceedings of the
IFIP-WG 7/1 Working Conference
Marseille-Luminy, France, March 12-17, 1984

Edited by M. Metivier and E. Pardoux

Springer-Verlag
Berlin Heidelberg New York Tokyo

Series Editors
M. Thoma

Advisory Board
A. V. Balakrishnan · L. D. Davisson · A. G. J. MacFarlane
H. Kwakernaak · J. L. Massey · Ya Z. Tsypkin · A. J. Viterbi

Editors
Prof. Michel Metivier
Département de Mathématiques Appliquées
Ecole Polytechnique
91128 Palaiseau Cedex (France)

Prof. Etienne Pardoux
U. E. R. de Mathématiques
Université de Provence
3, place Victor Hugo
13331 Marseille Cedex 03 (France)

ISBN 3-540-15176-1 Springer-Verlag Berlin Heidelberg New York Tokyo
ISBN 0-387-15176-1 Springer-Verlag New York Heidelberg Berlin Tokyo

Library of Congress Cataloging in Publication Data
Main entry under title:
Stochastic differential systems.
(Lecture notes in control and information sciences ; 69)
1. Stochastic differential equations--Congresses.
2. Filters (Mathematics)--Congresses.
3. Control theory--Congresses.
I. Métivier, Michel
II. Pardoux, E. (Etienne)
III. IFIP WG 7.1.
IV. Series.
QA274.23.S85 1985 519.2 85-2710

Offsetprinting: Mercedes-Druck, Berlin
Binding: Lüderitz und Bauer, Berlin
2061/3020-543210

PREFACE

The fourth IFIP working conference on stochastic differential systems took place in the "Centre International de Rencontres Mathématiques" at Marseille-Luminy, France, with a special support from the "Centre National de la Recherche Scientifique" and the "Société Mathématiques de France".

The meeting was intended to focus on topics in full development in the field of Stochastic differential systems. It was announced that a particular emphasis would be put on infinite dimensional problems, especially those connected with the modeling and control of distributed systems and non linear filtering. Contributions to the study of asymptotic properties, of small perturbations, fluctuation theory... were explicitly asked. This orientation of the meeting can be seen in the content of these proceedings and explain the relatively small amount of talks devoted to classical finite dimensional stochastic equations.

We tried to group together the contributions dealing with connected subjects. For this reason the proceedings are divided into five parts :

1. *Stochastic partial differential equations and infinite dimensional martingale problems.*
2. *Fluctuations and asymptotic analysis of finite and infinite dimensional systems.*
3. *Stochastic equations. Diffusions.*
4. *Filtering*
5. *Control theory.*

We hope that the proceedings reflect truly the high quality of the lectures and discussions and seize the opportunity to thank all the participants and contributors who made this meeting a success.

The Editors.

LIST OF PARTICIPANTS

J. AZEMA
laboratoire de Probabilités, tour 56
Université de Paris 6
4, place Jussieu
75230 PARIS CEDEX 05

A.V. BALAKRISHNAN
Systeme Science Department
University of California
4532 Boelter Hall
LOS ANGELES, Ca 90024 (U.S.A.)

S. BELLIZZI
LMA-CNRS
31 Ch. J. Aignier
13000 MARSEILLE

M. BENAROUS
Ecole Normale Supérieure
45 rue d'Ulm
75020 PARIS CEDEX 20

A. BENASSI
Laboratoire de Probabilités, tour 56
Université de Paris 6
4, place Jussieu
75230 PARIS CEDEX 05

A. BENSOUSSAN
INRIA
Domaine de Voluceau
Rocquencourt
B.P. 105
78153 LE CHESNAY CEDEX

G. BLANKENSHIP
Electrical Engineering Department
University of Maryland
College Park - M.D. 20742 (U.S.A.)

R. BOUC
LMA-CNRS
31, Ch. J. Aignier
13000 MARSEILLE

F. CAMPILLO
LMA-CNRS
31, Ch. J. Aignier
13000 MARSEILLE

M. CHALEYAT-MAUREL
Laboratoire de Probabilités, Tour 56
Université de Paris 6
4, place Jussieu
75230 PARIS CEDEX 05

A. CHOJNOWSKA-MICHALIK
Institute of Mathematics
Lodz University
ul. Stefana Banacha 22
90238 ŁODZ (Poland)

J.M.C. CLARK
Department of Electrical Engineering
Imperial College of Science & Technology
Exhibition Road
LONDON SW7 2BT (England)

N. CUTLAND
Department of Pure Mathematics
University of Hull
HULL HU6 7RX (England)

G. DA PRATO
Scuola Normale Superiore
Piazza dei Cavalieri
56100 PISA (Italia)

D. DAWSON
Department of Mathematics & Statistics
Carleton University
OTTAWA K1S 5B6 (Canada)

G. DEL GROSSO
Istituto Matematico "Guido Castelnuovo"
Universita di Roma
ROME (Italia)

L. ELIE
Département de Mathématiques
Université de Paris 7
2, place Jussieu
75321 PARIS CEDEX 05

N. EL KAROUI
Laboratoire de Probabilités, tour 56
Université de Paris 6
4, place Jussieu
75230 PARIS CEDEX 05

R.J. ELLIOTT
Department of Pure Mathematics
The University of Hull
Cottingham Road
HULL - HU6 7RX (England)

H.J. ENGELBERT
Sektion Mathematik
Friedrich Schiller Universität Jena
Universitatshochhaus, 17 O.G
D.D.R. - 6900 JENA

W.H. FLEMING
Division of Applied Mathematics
Brown University
PROVIDENCE, Rhode Island 02912 (U.S.A.)

H. FÖLLMER
E.T.H. Zentrum
Mathematik Department
CH - 8092 ZURICH (Switzerland)

J.P. FOUQUE
Laboratoire de Probabilités, tour 56
Université de Paris 6
4, place Jussieu
75230 PARIS CEDEX 05

A. GERARDI
Istituto Matematico "Guido Castelnuovo"
Universita di Roma "La Sapienza"
ROME (Italia)

L. GOROSTIZA
Centro de Investigacion y Estudios
Avanzados
Departemento de Matematicas
Apartado Postal 14-740
MEXICO 07000 - D.F. (Mexico)

B. GRIGELIONIS
Institute of Mathematics & Cybernetics
POZELOS 54, Vilnius - Lithuania (U.S.S.R.)

D. HADJIEV
Institute of Mathematics
8, rue Acad. G. Bontchev.
1113 - SOFIA (Bulgaria)

U. HAUSSMANN
University of British Columbia
VANCOUVER, B.C. V6T 1Y4 (Canada)

A. ICHIKAWA
Faculty of Engineering
Shizuoka University
HAMAMATSU 432 (Japan)

G. JONA LASINIO
Istituto di Fisica "G. Marconi"
Piazzale A. Moro 2
0185 ROME (Italia)

G. KALLIANPUR
Department of Statistics
University of North Carolina
321 Phillips Hall 039 A
CHAPEL HILL, NC 27514 (U.S.A.)

H. KOREZLIOGLU
E.N.S.T.
46, rue Barrault
75013 PARIS

T. KURTZ
Department of Mathematics
University of Wisconsin
480 Lincoln Drive
MADISON, WI. 53706 (U.S.A.)

F. LE GLAND
INRIA
Route des Lucioles
Sophia-Antipolis
06560 VALBONNE

P. LEYLAND
LMA-CNRS
31, Ch. J. Aignier
13000 MARSEILLE

P.L. LIONS
CEREMADE
Université de Paris IX-Dauphine
Place Mal de Lattre de Tassigny
75775 PARIS CEDEX 16

G. MAZZIOTTO
C.N.E.T.
38-40 rue du Général Leclerc
92131 ISSY LES MOULINEAUX

D. McDONALD
Département de Mathématiques
Faculté des Sciences et de Génie
Université d'Ottawa
585 King Edward
OTTAWA - KIN 984 Ontario (Canada)

J.L. MENALDI
INRIA
Domaine de Voluceau
Rocquencourt - BP 105
78153 LE CHESNAY CEDEX

M. METIVIER
Département de Mathématiques Appliquées
Ecole Polytechnique
91128 PALAISEAU CEDEX

M. MUSIELA
Laboratoire T.I.M. 3
Institut IMAG
B.P. 68
38402 ST MARTIN D'HERES CEDEX

J. NEVEU
Laboratoire de Probabilités, tour 56
Université de Paris 6
4, place Jussieu
75230 PARIS CEDEX 05

D. OCONE
Mathematics Department
Rutgers University
NEW BRUNSWICK, N.J. 08903 (U.S.A.)

J.Y. OUVRARD
Institut IMAG
B.P. 68
38402 ST MARTIN D'HERES CEDEX

G. PAPANICOLAOU
C I M S
New York University
251 Mercer Street
NEW YORK - NY 10012 (U.S.A.)

E. PARDOUX
U.E.R. de Mathématiques
Université de Provence
3, place Victor Hugo
13331 MARSEILLE CEDEX 03

J. PICARD
INRIA
Route des Lucioles
Sophia Antipolis
06560 VALBONNE

E. PLATEN
Institut fur Mathematik
Akademie der Wissenschaften der DDR
Mohrenstrasse 39
DDR-1086 BERLIN

L. STETTNER
Institut de Mathématiques
Académie des Sciences Polonaise
Sniadeckich 8
00950 VARSOVIE (Poland)

H.J. SUSSMANN
Department of Mathematics
Rutgers University
NEW BRUNSWICK, New Jersey 08903 (U.S.A.)

J. SZPIRGLAS
C N E T /PAA/TIM/MTI
38-40 av. Général Leclerc
92131 ISSY LES MOULINEAUX

D. TALAY
INRIA
Route des Lucioles
Sophia Antipolis
06560 VALBONNE

S. USTUNEL
Laboratoire de Probabilités
Tour 56 - Université Paris VI
4, place Jussieu
75230 PARIS CEDEX 05

S.R.S. VARADHAN
Courant Institute of Math Sciences
New York University
251 Mercer Street
NEW YORK, NY 10012 (U.S.A.)

V. WIHSTUTZ
Forschungsschwerpunkt Dynamische
Systeme
Universitat Bremen
Bibliothekstrasse
2800 BREMEN 33 (R.F.A.)

V. YURINSKII
Mathematical Institute
Siberian Branch
Academy of Sciences of U.S.S.R.
NOVOSIBIRSK 630090 (U.S.S.R.)

CONTENTS

1. STOCHASTIC PARTIAL DIFFERENTIAL EQUATIONS AND INFINITE DIMENSIONAL MARTINGALE PROBLEMS

HYPOELLIPTICITE DES EQUATIONS AUX DERIVEES PARTIELLES STOCHASTIQUES A COEFFICIENTS ALEATOIRES

M. Chaleyat-Maurel
Université Paris VI
Laboratoire de Probabilités
Tour 56
4 Place Jussieu
75230 PARIS CEDEX 05 - France -

I - EQUATIONS AUX DERIVEES PARTIELLES STOCHASTIQUES

Soit $(\Omega,\mathcal{F},(\mathcal{F}_t)_{t\geq 0},P)$ un espace de probabilité filtré usuel et $(\beta_t)_{t\geq 0}$ un $(\mathcal{F}_t\text{-}P)$ mouvement brownien sur Ω à valeurs dans $\mathbb{R}^p$.

Considérons une équation aux dérivées partielles stochastiques (EDPS) à coefficients aléatoires construite sur β, écrite au sens faible pour une semi-martingale u_t à valeurs dans les mesures de Radon sur $\mathbb{R}^n$:

$$(\mathrm{I})\quad \begin{cases} du_t f = u_t\, \mathcal{M}_o(t,\omega,.)f\, dt + u_t\, \mathcal{M}_i(t,\omega,.)f\, d\beta_t^i \\ u_o \text{ fixé ; p.s. } \forall f\in C_c^\infty(\mathbb{R}^n) \end{cases}$$

où $\mathcal{M}_o(t,\omega,.)$ est un opérateur différentiel d'ordre 2 sur $\mathbb{R}^n$ dont les coefficients sont des applications progressivement mesurables sur $\mathbb{R}_+\times\Omega\times\mathbb{R}^n$, et $\mathcal{M}_i(t,\omega,.)$, $1\leq i\leq p$ sont des opérateurs différentiels d'ordre 1 à coefficients vérifiant les mêmes propriétés que ci-dessus.

d désigne la différentielle de Stratonovitch.

On suppose que u_t, $\mathcal{M}_i$, $\mathcal{M}_o$ vérifient de plus des conditions qui donnent un sens à l'équation.

Ce type d'équation apparaît naturellement en filtrage (équation de Zakaï) ; nous le verrons dans le paragraphe II.

On s'intéresse ici à l'hypoellipticité de cette équation i.e. :

Soit $h_t(x)$ une semi-martingale à valeurs dans $C^\infty(\mathbb{R}^n)$ telle que $\forall f\in C_c^\infty(\mathbb{R}^n)$,

$$u_t(f)-u_o(f) - \int_o^t u_s(\mathcal{M}_o f)ds - \int_o^t u_s(\mathcal{M}_i f)d\beta_s^i = \int_{\mathbb{R}^n} f(x)\, h_t(x)dx,$$

on cherche des conditions sur $\mathcal{M}_o$ et $\mathcal{M}_i$ $(1\leq i\leq p)$ pour que p.s., $\forall t>0$, u_t admette une densité C^∞.

Cette équation est associée à l'opérateur de la chaleur formel :

$$\frac{\partial}{\partial t} - \mathcal{M}_o^* - \mathcal{M}_i^* \,\text{"}\frac{d\beta_t^i}{dt}\text{"}$$

où $\mathcal{M}_i^*$ désigne l'adjoint de $\mathcal{M}_i$ $0 \leq i \leq p$.

Ces opérateurs présentent deux difficultés qui interdisent d'appliquer les théorèmes classiques d'hypoellipticité des équations aux dérivées partielles : tout d'abord la dépendance en temps qui peut être irrégulière et ensuite l'intervention du hasard à travers $d\beta$ qui ne permet pas de raisonner trajectoires par trajectoires.

E. Pardoux [11] a démontré la régularité des solutions de (I) dans le cas où les coefficients sont C^∞, bornés, déterministes et l'opérateur $\mathcal{M}_o$ est elliptique. Il est clair que sa méthode s'applique (toujours sous l'hypothèse d'ellipticité) à des coefficients aléatoires.

Si l'on désire des résultats dans le cas dégénéré avec des coefficients aléatoires, il faut supposer un peu plus sur la dépendance en (t,ω) des coefficients, par exemple que les $\mathcal{M}_i$ $(0 \leq i \leq p)$ dépendent du hasard à travers une semi-martingale construite sur β et on pourra ainsi effectuer un calcul stochastique. D'autre part, si on veut des hypothèses géométriques, il faut donner une forme intrinsèque "somme de carrés" comme dans le célèbre théorème d'Hörmander (cf. F. Trèves [13]).

Introduisons les champs de vecteurs qui vont donner une forme géométrique à l'opérateur.

Soit $C_b^\infty = C_b^\infty(\mathbb{R}^n, \mathbb{R}^r)$ l'espace des fonctions continûment différentiables bornées ainsi que toutes leurs dérivées.

On suppose donnés :

. $V_o, V_1, \dots V_m, \tilde{V}_1, \dots \tilde{V}_p$ $m+p+1$ champs de vecteurs sur $\mathbb{R}^n \times \mathbb{R}^r$ tels que :

$$V_i(x,y) = \sum_{j=1}^n V_i^j(x,y) \frac{\partial}{\partial x_j} \qquad 0 \leq i \leq m$$

$$\tilde{V}_i(x,y) = \sum_{j=1}^n \tilde{V}_i^j(x,y) \frac{\partial}{\partial x_j} \qquad 1 \leq i \leq p$$

où les V_i^j et les $\tilde{V}_i^j$ sont dans C_b^∞.

. $g_o(x,y), g_1(x,y), \dots g_p(x,y)$ $p+1$ fonctions de C_b^∞.

La classe de mesures aléatoires sur laquelle nous travaillons est la même que dans [2], à savoir : $\mathcal{U}$ est la classe des variables aléatoires à valeurs dans les mesures de Radon sur $\mathbb{R}^+ \times \mathbb{R}^n$ de la forme $\mu_t(\omega, dx)dt$ telles que :

(i) $\mu_\cdot(\cdot,dx)$ est $\mathcal{F}_t$-progressivement mesurable.

(ii) l'application $t \to \mu_t$ envoie les ensembles bornés de $\mathbb{R}^{++}$ dans les bornés de l'espace des mesures de Radon aléatoires que l'on norme par :

$$\|\mu(\omega)\| = E(\|\mu\|^q_{\mathcal{M}(\mathbb{R}^n)})^{1/q}$$

avec $q=\sup(n+1,4)$ et où $\mathcal{M}(\mathbb{R}^n)$ désigne les mesures de Radon sur $\mathbb{R}^n$.

Enfin, soient $h_o,h_1,\ldots h_p$ $p+1$ fonctions sur $\mathbb{R}^+ \times \mathbb{R}^n \times \Omega$ mesurables par rapport à la tribu produit, C^∞ en $x \in \mathbb{R}^n$ et telles que si on note $h(t,x,\omega) = \int_o^t h_o(s,x,\omega)ds + \int_o^t h_i(s,x,\omega)\, d\beta_s^i$, toutes les dérivées en x de h_i et h soient dans $L^q(\mathbb{R}^+ \times \mathbb{R}^n \times \Omega)$.

ⓐ. Coefficients dépendant d'une semi-martingale

Soient $\eta_o,\eta_1,\ldots\eta_p$ $p+1$ fonctions de $\mathbb{R}^+ \times \Omega$, bornées à valeurs dans $\mathbb{R}^r$ telles que :

. $\forall 0 \le i \le p$, η_i est $\mathcal{B}(\mathbb{R}^+) \otimes \mathcal{F}_t$-mesurable

. $\forall 0 \le i \le p$, η_i est dans $L^2(\mathbb{R} \times \Omega)$.

On notera η la matrice des vecteurs colonnes η_i et soit $(\eta_t)_{t \ge 0}$ le semi-martingale

$$\eta_t = \int_o^t \eta_o(s,\omega)ds + \int_o^t \eta(s,\omega)\, d\beta_s.$$

C'est cette semi-martingale qui donne la dépendance en (t,ω) de l'opérateur, i.e. :

$$(1) \quad \begin{cases} \mathcal{M}_o(t,\omega,.) = (\sum_{i=1}^m V_i^2 + V_o + g_o)\,(.,\eta_t(\omega)) \\ \mathcal{M}_i(t,\omega,.) = (\tilde{V}_i+g_i)\,(.,\eta_t(\omega)) \qquad 1 \le i \le p \end{cases}$$

Bien que les coefficients ne soient que continus par rapport à t, on a un résultat d'hypoellipticité avec une hypothèse d'Hörmander restreinte où l'algèbre de Lie est la suivante :

Soit $\mathcal{S}$ l'algèbre de Lie à coefficients dans C_b^∞ engendrée par $V_1,\ldots V_m$ et si $N \in \mathbb{N}$, on note $\mathcal{S}_N$ le sous espace de $\mathcal{S}$ formé par les crochets de longueur inférieure ou égale à N.

Le théorème de régularité s'énonce ainsi :

Théorème 1

(H_1) Supposons que $\forall K$ compact de $\mathbb{R}^n$, il existe $N \in \mathbb{N}$ tel que $\mathcal{S}_N(x,y) = \mathbb{R}^n$ pour tout $(x,y) \in K \times \mathbb{R}^r$.

Soit $u = \mu_t\, dt$ dans $\mathcal{G}$ vérifiant (I) où les $\mathcal{M}_i$ $(0 \le i \le p)$ sont donnés par (1) ; alors p.s. $\forall t > 0$, μ_t admet une densité C^∞.

Remarque : c'est une hypothèse restreinte (au sens de J.M. Bismut) car elle ne fait intervenir que les sommes de carrés.

Formellement, le drift s'écrit $\tilde{V}_o + \tilde{V}_i \frac{"d\beta^i"}{dt}$, il se splite en $p+1$ composantes indépendantes et on va voir qu'en supposant un peu plus sur la dépendance en (t,ω) des coefficients (i.e. qu'elle se fasse à travers une diffusion construite sur β) on va pouvoir travailler sous une hypothèse d'Hörmander générale faisant intervenir les $\tilde{V}_i$.

ⓑ. Coefficients dépendant d'une diffusion

Soient $A_o, A_1, \ldots A_p$ $p+1$ champs de vecteurs sur $\mathbb{R}^r$ tels que :

$$A_i = \sum_{j=1}^{r} A_i^j(y) \frac{\partial}{\partial y_j} \qquad 0 \leq i \leq p$$

où les A_i^j sont dans $C_b^\infty(\mathbb{R}^r)$.

La dépendance des coefficients de la EDPS par rapport au passé du mouvement brownien β se fait par l'intermédiaire d'une diffusion ξ_t définie par l'équation différentielle stochastique suivante :

$$\begin{cases} d\xi_t = A_o(\xi_t)dt + A_i(\xi_t)d\beta_t^i \\ \xi_o \text{ fixé.} \end{cases}$$

Remarque : les hypothèses sur les A_i entrainent l'existence et l'unicité de la diffusion ξ_t sur $\mathbb{R}_+$ tout entier.

On prend donc :

$$(2) \quad \begin{aligned} \mathcal{M}_o(t,\omega,.) &= \left(\sum_{i=1}^{m} V_i^2 + V_o + g_o\right)(.,\xi_t(\omega)) \\ \mathcal{M}_i(t,\omega,.) &= (\tilde{V}_i + g_i)(.,\xi_t(\omega)) \qquad 1 \leq i \leq p. \end{aligned}$$

Introduisons enfin l'algèbre de Lie qui intervient dans notre hypothèse d'hypoellipticité.

Soit $\mathcal{N}$ l'algèbre de Lie à coefficients dans C_b^∞ engendrée par $V_1, \ldots V_m$ et les crochets des $V_1, \ldots V_m, \tilde{V}_1 + A_1, \ldots \tilde{V}_p + A_p$ où apparaît au moins un V_i.

Si $N \in \mathbb{N}$, on note $\mathcal{N}_N$ le sous espace de $\mathcal{N}$ formé par les crochets de longueur inférieure ou égale à N.

Le théorème de régularité des solutions de (I) s'énonce ainsi :

Théorème 2

$(H)_2$ Supposons que $\forall K$ compact de $\mathbb{R}^n$, il existe $N \in \mathbb{N}$ tel que $\mathcal{N}_N(x,y) = \mathbb{R}^n$ pour tout $(x,y) \in K \times \mathbb{R}^r$.

Soit $u = \mu_t\, dt$ dans $\mathcal{G}$ vérifiant (I), où les coefficients sont donnés par (2) ; alors p.s. $\forall t > 0$, μ_t admet une densité C^∞.

Remarque : cette hypothèse n'est pas fortuite. Si on considère μ_t comme une diffusion dont le drift est $W = \tilde{V}_o(x,\xi_t(\omega)) + \tilde{V}_i(x,\xi_t(\omega)) \dfrac{"d\beta_t^i"}{dt}$, le calcul formel suivant montre qu'on a inclut le drift dans l'hypothèse d'hypoellipticité :

$$\left[\frac{\partial}{\partial t} + W, V_i\right] = \left[\tilde{V}_o + A_o, V_i\right] + \left[\tilde{V}_j + A_j, V_i\right] \frac{"d\beta_t^j"}{dt}.$$

Dans le cas ⓐ on ne peut faire un calcul analogue car il n'y a pas de champs de vecteurs sur une variété de dimension finie naturellement associés à une semi-martingale.

Ces résultats sont le prolongement d'un travail commun avec D. Michel [2] dans le cas où $\xi_t = \beta_t$ et où figurent toutes les idées.

La méthode consiste à adapter au cadre probabiliste la technique de J.J. Kohn [5] pour démontrer le théorème d'hypoellipticité d'Hörmander.

Nous ne donnons pas la démonstration de ces deux théorèmes qui découle aisément de [2] et que l'on peut trouver dans [3].

Signalons le travail de N.V. Krylov et B. Rozovsky [6] qui démontrent dans le cas où $\xi_t = \beta_t$ et où les V_i sont elliptiques la régularité des solutions de I ; tandis que H. Kunita [7] obtient un résultat d'hypoellipticité sous $(H)_2$ mais pour des coefficients déterministes.

Ces deux théorèmes ont une application directe au filtrage : le premier redonne le résultat bien connu de J.M. Bismut et D. Michel [1] mais sous des hypothèses plus restrictives ; le deuxième étend à un cas hypoelliptique les résultats obtenus par D. Michel [10] sous des hypothèses d'uniforme ellipticité en utilisant le calcul de Malliavin.

II - APPLICATIONS AU FILTRAGE

Rappelons quelles sont les situations de filtrage données par des équations différentielles stochastiques qu'on sait traiter et indiquons les résultats de régularité dans chacun des cas.

II.1. D. Michel dans [10] considère le système suivant :

$$\begin{cases} dx_t = a(t,x_t,z)dt + b(t,x_t,z)\delta W_t \\ dz_t = A(t,x_t,z)dt + B(t,z)\delta W_t \\ x_o, z_o \quad \text{fixés} \end{cases}$$

$W_1,\dots W_n$ est un mouvement brownien $n=m+p$ dimensionnel, x est m dimensionnel et z est p dimensionnel et δ désigne la différentielle de Ito.

Si on suppose BB^* uniformément elliptique, on montre que l'on peut décomposer le signal en $dx_t = \tilde{a}\,dt + D\,\delta z_t + f\,\delta W_t$ où z et $F\,\delta W$ deviennent indépendants ce qui conduit au système suivant :

$$\begin{cases} dx_t = \tilde{a}(t,x_t,z)dt + \tilde{b}_1(t,x_t,z)\,\delta\tilde{W}_t^1 + \tilde{b}_2(t,x_t,z)\delta\tilde{W}_t^2 \\ dz_t = (BB^*)^{1/2}\,(t,z)\delta\,\tilde{W}_t^2 + A(t,x_t,z)dt \\ x_o, z_o \quad \text{fixés.} \end{cases}$$

Les lois conditionnelles du filtrage sont les π_t définies par :

$$\pi_t f = E(f(x_t)/\mathcal{B}_t^z)$$

avec f mesurable bornée et $\mathcal{B}_t^z = \sigma(z_s,\ s \leq t)$.

En utilisant le calcul de Malliavin, D. Michel [10] a obtenu le résultat suivant :

Si $\tilde{a}, \tilde{b}_1, \tilde{b}_2, B, A$ sont de classe C^∞ en x et sont continues bornées par rapport à l'ensemble des variables ainsi que leurs dérivées successives et si on suppose la matrice $\tilde{b}_1$ uniformément elliptique, p.s. $\forall t > 0$, π_t admet une densité C^∞.

II.2.- Si on suppose que les coefficients dépendent du passé de l'observation par l'intermédiaire d'une semi-martingale ν_t, il est naturel d'utiliser la notation champ de vecteurs et le problème se pose ainsi :

Soit $\mathcal{W}$ (resp. $\tilde{\mathcal{W}}$) l'espace $\mathcal{C}(\mathbb{R}^+, \mathbb{R}^m)$ (resp. $\mathcal{C}(\mathbb{R}^+, \mathbb{R}^p)$; on note ω(resp. $\tilde{\omega}$) un point de $\mathcal{W}$(resp. $\tilde{\mathcal{W}}$) et soit Q(resp. $\tilde{Q}$) la mesure de Wiener sur $\mathcal{W}$ (resp. $\tilde{\mathcal{W}}$) avec $Q(\omega_o = 0) = 1$ (resp. $\tilde{Q}(\tilde{\omega}_o = 0) = 1$).

Si x est un processus stochastique sur $\mathcal{W} \times \tilde{\mathcal{W}}$, on note $\mathcal{B}_t^x$ la tribu $\sigma(x_s,\ s \leq t)$.

. Soient $X_o, X_1,\dots X_m, \tilde{X}_1,\dots \tilde{X}_p$ $m+p+1$ champs de vecteurs définis sur $\mathbb{R}^n \times \mathbb{R}^r$ tels que :

$$X_i(x,y) = \sum_{j=1}^{n} X_i^j(x,y) \frac{\partial}{\partial x_j} \qquad 0 \leq i \leq m$$

$$\tilde{X}_i(x,y) = \sum_{j=1}^{n} \tilde{X}_i^j(x,y) \frac{\partial}{\partial x_j} \qquad 1 \leq i \leq p$$

où les X_i^j et $\tilde{X}_i^j$ sont dans C_b^∞.

. Soient $Z_1(y),\ldots Z_p(y)$ p champs de vecteurs sur $\mathbb{R}^r$ à coefficients dans $C_b^\infty(\mathbb{R}^r)$.

. Soient $\ell_1(x,y),\ldots \ell_p(x,y)$ p fonctions sur $\mathbb{R}^n \times \mathbb{R}^r$ dans C_b^∞ (on notera $\ell = (\ell_1,\ldots,\ell_p)$).

. Soient $\nu_o(s,\omega),\ldots \nu_p(s,\omega)$ $p+1$ fonctions progressivement mesurables bornées à valeurs dans $\mathbb{R}^r$ telles que ν_i $0 \leq i \leq p$ soit dans $L^2(\mathbb{R}^+ \times \Omega)$ (on notera ν la matrice des vecteurs colonnes ν_i)

$$(3) \qquad \begin{cases} dx_t = X_o(x_t,\nu_t)dt + X_i(x_t,\nu_t)\delta w_t^i + \tilde{X}_i(x_t,\nu_t)(\delta \tilde{w}_t^i + \ell_i(x_t,\nu_t)dt) \\ dz_t = Z_i(\nu_t)(\delta \tilde{w}_t^i + \ell_i(x_t,\nu_t)dt) \\ \nu_t = \int_o^t \nu_o(s,\omega)ds + \int_o^t \nu(s,\omega)\delta z_s \\ x_o,\ z_o \text{ fixés.} \end{cases}$$

Remarques sur le système

On introduit les ℓ_i dans l'équation de x_t pour simplifier la transformation de Girsanov mais cela n'enlève rien à la généralité du problème en effet $(BB^*)^{1/2}$ étant inversible, on prend ℓ tel que $A=(BB^*)^{1/2}\,\ell$ et on pose $(BB^*)^{1/2} = Z$. De même on prend $a = X_o + X\,\ell$.

La forme de l'équation n'étant plus invariante par difféomorphisme, nous écrivons le système dans le calcul d'Ito (remarquons que l'écriture du système dans le calcul de Stratonovitch entrainerait des termes parasites dans les coefficients des équations du filtrage, i.e. la multiplication par les fonctions progressivement mesurables bornées $\eta_i(s,\omega)$ ce qui conduirait à un théorème d'hypoellipticité pour les EDPS voisin du théorème 1 mais sans beaucoup d'intérêt).

On utilise la transformation de Girsanov pour se ramener à une équation aux dérivées partielles stochastiques, l'équation de Zakaï que l'on peut alors traiter par les méthodes du paragraphe I.

On définit :

$$L_t = \exp\{\int_o^t \ell_i(x_s,\nu_s)\delta \tilde{w}_s^i + \frac{1}{2}\sum_{i=1}^{p}\int_o^t \ell_i^2(x_s,\nu_s)ds\}$$

et Q_o la mesure de probabilité sur $\mathcal{W} \times \tilde{\mathcal{W}}$, absolument continue par rapport à $Q \otimes \tilde{Q}$ telle que

$$\frac{dQ_o}{dQ \otimes d\tilde{Q}}\Big/\mathcal{B}_t^{w,\tilde{w}} = L_t^{-1}.$$

Le théorème de Girsanov entraîne alors que :

$\tilde{w}'_t = \tilde{w}_t + \int_o^t \ell(x_s,\nu_s)ds$ est un Q_o-mouvement brownien.

On introduit le filtre non normalisé ρ_t i.e. si f est dans $C_b^2(\mathbb{R}^n)$,

$$\rho_t f = E_o(f(x_t)L_t/\mathcal{B}_t^z)$$

ρ_t est alors relié à π_t par la formule de Kallianpur-Striebel-Kuschner :

$$\pi_t f = \frac{\rho_t\, f}{\rho_t\, 1}.$$

On démontre alors facilement en utilisant la méthode de E. Pardoux ([11]) que ρ_t vérifie l'équation de Zakaï :

$$\rho_t f = \rho_o f + \int_o^t \rho_s\, \Delta_s f\, ds + \int_o^t \rho_s\, B_i f\, \delta z_s^i$$

avec $B_i = (\tilde{X}_i + \ell_i)(.,\nu_s)$ et $\Delta = \{\frac{1}{2}\sum_{i=1}^m X_i^2 + \sum_{i=1}^p \tilde{X}_i^2 + X_o + \ell_i\, \tilde{X}^i\}\,(.,\nu_s)$.

Cette équation écrite au sens faible, est du type I avec le cas ⓐ ; le théorème 1 donne immédiatement un résultat de régularité sous une hypothèse d'hypoellipticité faisant intervenir l'algèbre de Lie suivante :

Soit $\mathcal{L}(X_1, X_2,\ldots X_m)$ l'algèbre de Lie à coefficients dans C_b^∞ engendrée par $X_1,\ldots X_m$ et si $N \in \mathbb{N}$ on note $\mathcal{L}_N(X_1, X_2,\ldots X_m)$ le sous espace de $\mathcal{L}(X_1,\ldots X_m)$ formé par les crochets de longueur inférieure ou égale à N.

Théorème

Supposons que $\forall K$ compact de $\mathbb{R}^n$, il existe $N \in \mathbb{N}$ tel que $\mathcal{L}_N(X_1,\ldots X_m)(x,y) = \mathbb{R}^n$ pour tout $(x,y) \in K \times \mathbb{R}^r$.

Alors, p.s. $\forall t > 0$, π_t admet une densité C^∞.

Remarques :

a). Ce théorème étend donc au cas hypoelliptique le résultat de D. Michel [10] mais seulement dans le cas où la dépendance des coefficients par rapport au passé de z se fait par l'intermédiaire d'une semi-martingale construite sur z, alors que la méthode de D. Michel basée sur le calcul de Malliavin autorise une dépendance quelconque des coefficients par rapport au passé de z.

b). Ceci inclut le cas où les coefficients du système (3) dépendent également du temps ; en effet il suffit d'ajouter une composante t à la semi-martingale ν_t.

II.3. Lorsque la semi-martingale qui décrit le passé de z est une diffusion y_t construite sur z, l'étude d'un système de filtrage donné par une diffusion donne le résultat car il suffit de considérer le signal comme une composante séparée de cette diffusion y_t.

Montrons comment le théorème 2 redonne le résultat de régularité de J.M. Bismut et D. Michel [1] mais sous une hypothèse plus contraignante.

Pour ces résultats qui sont de nature géométrique nous adoptons la formulation Stratonovich.

Considérons les champs de vecteurs $X_o, X_1, \dots X_m$, $\tilde{X}_1, \dots \tilde{X}_p$, $Z_1, \dots Z_p$ définis en II.2. et vérifiant les mêmes hypothèses ainsi que les p fonctions $\ell_1, \dots \ell_p$.

On considère de plus p+1 champs de vecteurs sur $\mathbb{R}^s$ à coefficients dans $C_b^\infty(\mathbb{R}^s) = Y_o, Y_1, \dots Y_p$.

Le système de filtrage est alors le suivant :

$$\begin{cases} dx_t = X_o(x_t, y_t, z_t)dt + X_i(x_t, y_t, z_t)dw_t^i \\ \qquad + \tilde{X}_i(x_t, y_t, z_t)(d\tilde{w}_t^i + \ell_i(x_t, z_t)dt) \\ dz_t = Z_i(z_t)(d\tilde{w}_t^i + \ell_i(x_t, z_t)dt) \\ dy_t = Y_o(y_t)dt + Y_i(y_t)dz_t \\ x_o, z_o, y_o \text{ donnés.} \end{cases}$$

Le signal étant x, ceci se ramène au problème du filtrage de la diffusion $(x_t, y_t) = \nu_t$ par l'observation z.

En posant $U_o = \begin{pmatrix} X_o \\ Y_o \end{pmatrix}$, $U_i = \begin{pmatrix} X_i \\ 0 \end{pmatrix}$ $1 \leq i \leq m$ et $(\tilde{U}_i) = \begin{pmatrix} \tilde{X}_i \\ Y_i Z_i \end{pmatrix}$ $1 \leq i \leq p$, le système devient

$$\begin{cases} dv_t = U_o(v_t,z_t)dt + U_i(v_t,z_t)dw_t^i + \tilde{U}_i(v_t,z_t)(d\tilde{w}_t^i + \ell_i(v_t,z_t)dt) \\ dz_t = Z_i(z_t)(d\tilde{w}_t^i + \ell_i(v_t,z_t)dt) \\ v_o, z_o \quad \text{donnés} \end{cases}$$

où les ℓ_i ne dépendent en fait que de x_t et z_t.

On opère la transformation de Girsanov de la même façon que dans le paragraphe précédent, on introduit le filtre non normalisé ρ_t qui satisfait à une équation de Zakaï du type (I). On retrouve la régularité de ρ_t sous une hypothèse d'hypo-ellipticité concernant l'algèbre de Lie suivante :

Soit $\mathcal{R}$ l'algèbre de Lie à coefficients dans C_b^∞ engendrée par $U_1,\dots U_m$ et les crochets de $U_1,\dots U_m$, $\tilde{U}_1+Z_1,\dots\tilde{U}_p+Z_p$ où apparaît au moins un U_i, et si $N \in \mathbb{N}$, on note $\mathcal{R}_N$ le sous espace de $\mathcal{R}$ formé par les crochets de longueur inférieure ou égale à N.

Le théorème 2.1. conduit alors au résultat suivant :

Théorème

Supposons que $\forall K$ compact de $\mathbb{R}^n \times \mathbb{R}^s$, il existe $N \in \mathbb{N}$ tel que $\mathcal{R}_N(x,y,z) = \mathbb{R}^n$ quel que soit $(x,y,z) \in K \times \mathbb{R}^p$.

Alors, p.s. $\forall t > 0$, ρ_t admet une densité C^∞.

On voit que si la dépendance d'un système de filtrage par rapport au passé de l'observation se fait par l'intermédiaire d'une diffusion construite sur z, le problème est en fait un cas markovien classique.

Signalons un article de S. Kusuoka - D.W. Stroock [8] qui redémontre le résultat de régularité de J.M. Bismut - D. Michel [2] en utilisant leur propre formalisme du Calcul de Malliavin.

BIBLIOGRAPHIE

[1] J.M. BISMUT, D. MICHEL : Diffusions conditionnelles. J. Funct. Anal. Part I, 44, 174-211, 1981. Part II, 4, 274-292, 1982.

[2] M. CHALEYAT-MAUREL, D. MICHEL : Hypoellipticity theorems and conditional laws. Z.f.W. 65, 579-597, 1984.

[3] M. CHALEYAT-MAUREL : Thèse de Doctorat d'Etat. Paris VI, 1984.

[4] L. HÖRMANDER : Hypoelliptic second order différential equations. Acta Math. 117, 147-171, 1967.

[5] J.J. KOHN : Pseudo differential operators and hypoellipticity. Proc. Symp. Pure Math. 23, 61-69, 1973.

[6] N.V. KRYLOV - B.L. ROZOVSKY : On the Cauchy problem for linear stochastic partial differential equations. Izv. Akad. Nauk. SSSR 41, 1329-1347, 1977.

[7] H. KUNITA : Stochastic partial differential equations connected with non-linear filtering. CIME, Lect. Notes in Math. n° 972, Springer, 1982.

[8] S. KUSUOKA, D.W. STROOCK : The partial Malliavin Calculus and its applications to non linear filtering. A paraître.

[9] R.S. LIPTSER, A.N. SHIRYAYEV : Statistics of Random Processes,Applications of Mathematics n° 5, Springer-Verlag, Berlin/Heidelberg/New-York, 1977.

[10] D. MICHEL : Régularité des lois conditionnelles en théorie du filtrage non-linéaire et calcul des variations stochastiques. Journal of functional analysis, Vol. 41, n° 1, 8-36, 1981.

[11] E. PARDOUX : Stochastic partial differential equations and filtering of diffusion processes. Stochastic 3, 127-167, 1979.

[12] E. PARDOUX : Equations of non-linear filtering, and application to stochastic control with partial observation. CIME Loc. cit.

[13] F. TREVES : Introduction to pseudo-differential and Fourier integral operators. Vol. 1, New York, Plenum Press, 1981.

[14] S. USTUNEL : Hypoellipticity of the stochastic partial differential operators. A paraître.

STATIONARY DISTRIBUTIONS FOR ∞-DIMENSIONAL LINEAR EQUATIONS WITH GENERAL NOISE

Anna Chojnowska-Michalik
Institute of Mathematics
Łódź University
90-238 Łódź, Poland

0. Introduction and notations

We consider stationary measures for the linear stochastic infinite dimensional equation, formally written as:

$$(*) \qquad dX_t = AX_t dt + dZ_t, \qquad t \geq 0$$

In this equation $\underline{A}$ denotes the infinitesimal generator of a strongly continuous semigroup $\underline{T_t}$, $t \geq 0$, of linear bounded operators acting on real separable Hilbert space H and $\underline{Z_t}$, $t \geq 0$, is an H-valued process with homogeneous independent increments defined on fixed probability space $(\Omega, \mathfrak{M}, P)$, continuous in probability, cadlag, with $Z_o \equiv 0$.
A process Z_t with these properties will be shortly called P.H.I.I.

The mild solution of (*) is given by the explicit formula (section 2):

$$(0.1) \qquad X_t = T_t X_o + \int_o^t T_{t-s} dZ_s,$$

where the stochastic integral is defined as Stieltjes integral in the sense of convergence in probablity (section 1).

The main results of this paper are contained in section 3, where the problem of existence of stationary distributions for processes (0.1) is investigated. For $H = R^n$ this problem has been studied in [10], in [9] (under additional assumption that T_t is stable) and recently in [4] and [5]. Here we obtain some generalization of these results. Namely, under certain assumption on the semigroup T_t (which is automatically satisfied in R^n) we get sufficient and necessary conditions (for the existence of stationary distribution) similar as in [10](th. 3.11) or as in [9],[4],[5] (remark 3.16). However, in general, the main condition in [9],[4],[5]: $E \log (1+|Z_1|) < +\infty$ has not to be necessary in infinite dimensional case, as example 3.15 shows.

Let us note that our methods of proofs are different. First, we prove that in this problem the noise Z_t can be reduced to Z_t without gaussian component (th. 3.5).

On the other hand, we also generalize some infinite dimensional results for $Z_t = W_t$ from [11].

0.1. Notation

$L(H,K)$ - the space of linear bold operators from H to K

$\langle\cdot,\cdot\rangle$ is the scalar product in H

$|\cdot|$ is the norm in H

$\|\cdot\|$ is the norm in $L(H,K)$

$\mathcal{L}(X)$ denotes the probability distribution of random variable (r.v.) X

$\hat{\mu}$ is the characteristic functional of probability measure μ

$\Longrightarrow$ denotes weak convergence of measures

$\mathcal{B}(H)$ is the Borel σ-field on H

$B(\varepsilon) = \{x \in H : |x| \leq \varepsilon\}$, $B'(\varepsilon) = H \setminus B(\varepsilon)$

χ_D is the indicator function of set D

$t \wedge s = \min (t,s)$

0.2. Representation of the process Z_t

Since Z_t is P.H.I.I., it can be represented in the form ([3])

$$Z_t = at + W_t + \xi_t, \tag{0.2}$$

where $a \in H$, W_t is an H-valued Wiener process, ξ_t is the jump process and W_t and ξ_t are independent.

The characteristic functional of Z_t has the form:

$$E (\exp i \langle y,Z_t\rangle) = \exp t\psi(y), \quad t \geq 0, \quad y \in H, \tag{0.3}$$

where

$$\psi(y) = i \langle a,y\rangle - \frac{1}{2} \langle Ry,y\rangle + \int_H (e^{i\langle x,y\rangle} - 1 - i\langle x,y\rangle \chi_{B(1)}(x))M(dx) \tag{0.4}$$

In the formula (0.4):

$\underline{R}$ is the covariance operator of the Wiener process W_t (i.e. $\mathrm{Cov}\, W_t = tR$), therefore R is self-adjoint, non-negative nuclear operator;

$\underline{M}$ is the Lévy spectral measure (of the process ξ_t) i.e. σ-finite measure on $\mathcal{B}(H)$ s.t. $M(\{0\}) = 0$ and $\int_H (|x|^2 \wedge 1) M(dx) < +\infty$.

We shall call $\underline{[a,R,M]}$ the Lévy characterization of r.v. Z_1. (Then Z_t has the Lévy characterization $[ta,tR,tM]$).

1. Stochastic integral of Stieltjes type

In [1] we define stochastic integral with respect to P.H.I.I. for a class of deterministic integrands containing strongly continuous semigroups of operators, using as a tool infinitely divisible measures instead of semimartingales. This approach gives immediately some important properties of the integral, for instance the explicit formula for the characteristic function.

This definition is a generalization of the definition for finite dimensional case given in [7] and [9].

Let us remark that it follows from (0.2) that one needs to define only the integral w.r.t. the jump process.

If φ is a step function the integral $\int_{]u,t]} \varphi(s)dZ_s$ is defined as a sum. Next we show that we can pass to the limit in probability in the space $L^\infty_{[u,t]}(H,K)$, where

$$L^\infty_{[u,t]}(H,K) := \{\varphi : [u,t] \to L(H,K) \text{ s.t.}$$

$$\text{(i) } \forall\, h \in H, \quad \varphi h : [u,t] \to K \text{ is measurable}$$

$$\text{(ii) } \sup_{s\in[u,t]} \|\varphi(s)\| < +\infty\}$$

THEOREM 1.1 ([1])

If $\varphi \in L^\infty_{[u,t]}(H,K)$, then the integral $\int_{]u,t]} \varphi(s)dZ_s$ is well defined K-valued r.v., which has ∞-divisible distribution with the Lévy representation

$$(1.1) \qquad [\int_u^t \varphi(s)a\,ds + \int_u^t \int_H \varphi(s)x[\chi_{B(1)}(\varphi(s)x) - \chi_{B(1)}(x)]M(dx)ds,$$

$$\int_u^t \varphi(s)R\varphi^*(s)ds, \quad \int_u^t \tilde{M}(\varphi(s)^{-1}dx)ds]$$

where

$$\tilde{M}(\varphi^{-1}dx) := \begin{cases} M(\varphi^{-1}dx)|_{\mathcal{B}(K \setminus \{0\})} \\ \tilde{M}(\varphi^{-1}(\{0\})) := 0. \end{cases}$$

In particular the integrals $\int_{]u,t]} T_{t-s}dZ_s$ and $\int_{]u,t]} T_s dZ_s$ are well defined.

Definition 1.2

We define $\int_u^{+\infty} \varphi(s)dZ_s$ as the limit in probability of $\int_u^t \varphi(s)dZ_s$, as $t \nearrow \infty$.

2. Mild solution of the equation (*)

Let X_o be H-valued r.v. independent of $(Z_t)_{t>0}$ and consider the equation (*) with the initial condition X_o.

Definition 2.1

A process X_t is a mild solution of (*) iff

$$\forall y \in \mathcal{D}(A^{*2}) \quad \forall t \quad \text{wp } 1$$

$$<X_t, y> = <X_o, y> + \int_o^t <X_s, A^*y>ds + <Z_t, y>.$$

THEOREM 2.2 ([1])

The equation (*) has a unique mild solution X_t. X_t is given by the formula (0.1).

3. Stationary distributions for (*)

In this section we shall use the following notations:

If μ is a measure on $\mathcal{B}(H)$, $T_t\mu$ means the measure s.t. $(T_t\mu)(B) = \mu(T_t^{-1}(B))$ for $B \in \mathcal{B}(H)$.

$$\nu_t := \mathcal{L}\Big(\int_{]0,t]} T_{t-s}dZ_s\Big), \qquad \nu_t^1 := \mathcal{L}\Big(\int_{]0,t]} T_{t-s}dW_s\Big),$$

$$\nu_t^2 := \mathcal{L}\Big(\int_{]0,t]} T_{t-s}d\xi_s\Big), \qquad b_t := \int_o^t T_s\, a ds.$$

The representation (0.2) implies:

$$\nu_t = \nu_t^1 * \nu_t^2 * \delta_{b_t}$$

With the notations above a probability measure μ is stationary for (*) (i.e. (0.1)) iff

$$\mu = T_t\mu * \nu_t, \qquad \text{for any} \quad t \geq 0.$$

LEMMA 3.1

ν_t is weakly convergent as $t \nearrow \infty$ iff $\int_0^\infty T_s dZ_s$ exists. In this case $\nu = w - \lim_{t\nearrow\infty} \nu_t$ is the distribution of $\int_0^\infty T_s dZ_s$.

Proof. It is enough to notice that $\nu_t = L(\int_0^t T_s dZ_s)$.

PROPOSITION 3.2 ([1])

If $\int_0^\infty T_s dZ_s$ exists, then $L(\int_0^\infty T_s dZ_s)$ is a stationary distribution for (*).

In certain situations the last condition is also necessary for the existence of stationary measure. Namely, we have

PROPOSITION 3.3

If the semigroup T_t is stable (i.e. $\forall_{x\in H}\ T_t x \to 0$, as $t \nearrow \infty$), then: there exists a stationary for (*) distribution μ iff $\int_0^\infty T_s dZ_s$ exists. In this case $\mu = L(\int_0^\infty T_s dZ_s)$

COROLLARY 3.4. Uniqueness

If T_t is stable, then there may exist at most one stationary distribution for (*).

Proof of Proposition 3.3. Since T_t is stable, $T_t\mu \to \delta_o$, as $t \nearrow \infty$, for any probability measure μ. Now let μ be a stationary distribution for (*) and let $t_n \nearrow \infty$. Then $\mu = T_{t_n}\mu * \nu_{t_n}$. Therefore Th. 2.1, p.58 and Lemma 2.1, p.153 in [8] imply the convergence $\nu_{t_n} \Rightarrow \mu$. Hence, by lemma 3.1, the proposition follows.

The theorem below gives some representation of stationary measure and shows that the existence problem may be "reduced" in a certain sense.

THEOREM 3.5 ([1])

I) There exists a stationary distribution μ for $(*)$ iff the following conditions are satisfied:

(C1) There exists $\nu^1 = w - \lim_{t \nearrow \infty} \nu_t^1$

($\widetilde{C2}$) There exists a stationary μ^2 for $(*)$ with $Z_t = at + \xi_t$.

II) If (C1), (C2) are satisfied, then any stationary distribution μ is of the form

$$\mu = \nu^1 * \mu^2.$$

COROLLARY 3.6 (compare [11])

I) The following conditions are equivalent:

(C1)

(C1') $\int_0^\infty T_s dW_s$ exists

(C1") $\int_0^\infty \operatorname{tr}(T_s R T_s^*) ds < +\infty$

(C1"') There exists a stationary distribution μ for

$$dX_t = AX_t dt + dW_t$$

II) If any of these conditions is satisfied, then the stationary measure μ has the form $\mu = \mu^2 * \nu^1$, where μ^2 is a stationary probability measure for the deterministic system $dX_t = AX_t dt$ (i.e. $\mu^2 = T_t \mu^2$).

Then the main difficulty is to consider the existence of stationary measure problem for $(*)$ with Z_t without gaussian part i.e. ($\widetilde{C2}$).

PROPOSITION 3.7 ([1])

I) ($\widetilde{C2}$) holds if the following conditions are satisfied:

(C2) $\int_0^\infty \int_H (|T_t x|^2 \wedge 1) M(dx) dt < +\infty$

(C3) There exists

$$\lim_{t \nearrow \infty} \int_0^t \left[\int_H T_s x (\chi_{B(1)}(T_s x) - \chi_{B(1)}(x)) M(dx) \right] ds$$

(C4) There exists a stationary probablity measure α for deterministic system:

$$dX_t = AX_t dt + a dt \qquad (\text{i.e.} \quad \alpha = T_t\alpha * \delta_{b_t})$$

II) If (C2), (C3), (C4) are satisfied, then any stationary measure μ^2 in $(\widetilde{C2})$ is of the form:

$$\mu^2 = \alpha * \nu^2, \quad \text{where} \quad \nu^2 = w - \lim_{t\to\infty} \nu_t^2$$

Remark 3.8

(C4) is satisfied if $(\widetilde{\widetilde{C4}})$: there exists $b \in \mathcal{D}(A)$ s.t. $a = Ab$.
Then $\alpha = \delta_{-b} * \beta$, where β is an invariant probability measure for $dX_t = AX_t dt$.
If range A is closed (so, in particular, in R^n), then $(C4) \Leftrightarrow (\widetilde{\widetilde{C4}})$ (see [10]).

3.1. Space of values of the process ξ_t and W_t

Let $H_2 := \{x \in H : \int_0^\infty |T_t x|^2 \, dt < +\infty\}$, let $\gamma(m,S)$ means the gaussian measure on H with the expectation m and the covariance operator S.

PROPOSITION 3.9

If there exists stationary measure for (*), then:

i) $M(H \setminus H_2) = 0$ (see [10] for R^n)

ii) the measure $\gamma(0,R)$ is concentrated on the subspace $\overline{H}_2$ (see[4] for R^n but the proof is rather complicated).

Proof. i) follows from (C2), which is necessary for $(\widetilde{C2})$ (see [1]).
ii) from the identity:

$$\mathrm{tr} T_t R T_t^* = \| T_t R^{1/2} \|^2_{H-S} \quad (\text{where } \| \ \|_{H-S} \text{ denotes the}$$

Hilbert-Schmidt norm) and (C1'') we get

$$\int_0^\infty \sum_j < T_t R^{1/2} h_j , T_t R^{1/2} h_j > dt < + \infty \quad \text{for any orthonor-}$$

mal basis $\{h_j\}$ in H. Now let $y \in$ range $(R) \subset$ range $(R^{1/2})$, $y \neq 0$. Then for some $x \in H$, $y = R^{1/2}x$. Let $\{h_j\}$ be an orthonormal basis in H s.t. $h_1 = \frac{x}{|x|}$.

Therefore we obtain

$$\frac{1}{|x|^2} \int_0^\infty \langle T_t R^{1/2} x, T_t R^{1/2} x \rangle dt \leq \int_0^\infty \sum_j \langle T_t R^{1/2} h_j, T_t R^{1/2} h_j \rangle dt < +\infty,$$

which means that $y \in H_2$.

Hence range $(R) \subset H_2$, from which ii) follows.

<u>COROLLARY 3.9'</u>

If there exists a stationary distribution for $(*)$, then the processes ξ_t and W_t take their values in the subspace $\overline{H}_2$.

3.2. Necessary and sufficient conditions for the existence of stationary distribution

<u>THEOREM 3.10</u> ([1])

If the Lévy measure M is symmetric, then the conditions (C1''), (C2), (C4) are necessary and sufficient for the existence of stationary measure for $(*)$.

<u>THEOREM 3.11</u>

If the subspace H_2 is closed, then:

I) The conditions (C1''), (C2), (C3) and (C4) are necessary and sufficient for the existence of stationary distribution for $(*)$.

II) Any stationary measure μ for $(*)$ is of the form:

$$\mu = \alpha * \nu^1 * \nu^2,$$

where α and ν^2 are defined in prop. 3.7, ν^1 is defined in the condition (C1) and, moreover, the distributions ν^1 and ν^2 are concentrated on the space H_2.

<u>Proof</u>. II) follows from th. 3.5, prop. 3.7 and cor. 3.9'.

I) By th. 3.5, cor. 3.6 and prop. 3.7 we need to prove that (C2), (C3) and (C4) are necessary for $(\widetilde{C2})$. Since H_2 is a closed subspace of H, H is isomorphic with the Banach space $\mathfrak{X} = H_2 \times H/H_2$.

(Let $h \to (h',h'')$ means the canonical isomorphism).

Moreover, $T_t(H_2) \subset H_2$ and $T_t^{-1}(H_2) \subset H_2$.

From the first inclusion, the closedness of H_2 and cor. 3.9' it follows that the process $\int_0^t T_{t-s} d\xi_s$ takes its values in H_2 for all $t \geq 0$. Therefore, our process (0.1) (with $Z_t = at + \xi_t$) is isomorphic with the following process in $\mathcal{X}$:

(3.1) $$X_t' = T_t X_0' + \int_0^t T_{t-s} a' ds + \int_0^t T_{t-s} d\xi_s$$

(3.2) $$X_t'' = \tilde{T}_t X_0'' + \int_0^t \tilde{T}_{t-s} a'' ds.$$

Now, the existence of stationary distribution for (0.1) implies the existence of stationary measure for (3.1). But $T_{t|H_2}$ is exponentially stable semigroup, i.e. $\|T_t\| \leq Ce^{-\lambda t}$ for some $C \geq 1$, $\lambda > 0$ for all $t \geq 0$ (from [2]). Then from prop. 3.3 the integral $\int_0^\infty T_s(d\xi_s + a'ds)$ exists, which, in turn, implies existence of $\int_0^\infty T_s d\xi_s$ (again by exponential stability). This is equivalent to (C2) and (C3) ([1]). Finally, we deduce (C4) similarly as in the proof of th. 3.5.

As an immediate consequence of th. 3.11, cor. 3.9' and Datko [2] result, we obtain:

<u>COROLLARY 3.12</u>

In the th. 3.11 we can replace the condition (C1'') by the following one:

($\widetilde{C1}$) W_t is H_2-valued Wiener process.

<u>Remark</u>. Th. 3.11 implies Zabczyk [10] result for $H = R^n$.

3.3. Stationary distributions, when T_t is an exponentially stable semigroup

In this case in R^n any stationary measure for (*) is an operator-selfdecomposable measure and conversely ([9],[4],[5]). This is some motivation to consider model (*) with exponentially stable semigroups.

The following simple condition (J) is sufficient and necessary for the existence of stationary measure for (*) in this situation in R^n ([9],[4],[5]) (as in th. 3.11 the general situation in R^n can be reduced to this case)

$$(J) \qquad \int_{R^n} \log^+ |x| \, M(dx) < +\infty.$$

The condition above has firstly been obtained by Jurek [6] in the characterization of operator-selfdecomposable measure in Banach space and it is equivalent to

$$(J') \qquad E(\log(|\xi_1| + 1)) < +\infty \qquad \text{(see [6])}.$$

Now we consider the following question:
Is (J) also sufficient and necessary in H for the existence of stationary measure for (*) with exponentially stable semigroup?

THEOREM 3.13 (J) is sufficient ([1])

Let $(T_t)_{t\geq 0}$ be exp. stable semigroup on H, M be the Lévy measure of process Z_t s.t.

$$(J) \qquad \int_H \log^+ |x| M(dx) < +\infty.$$

Then there exists a stationary measure for (*).

PROPOSITION 3.14 ([1])

If $(T_t)_{t\geq 0}$ is exp. stable semigroup on H and, moreover, $(T_t)_{t\in R}$ is a group, then the condition (J) is also necessary for the existence of stationary distribution for (*).

It seems that the group property of T_t is important for (J) to be necessary. Namely, consider the following example.

EXAMPLE 3.15

Let $H = L^2[0,1]$ and
$$\underset{\substack{t\geq 0 \\ s\in[0,1]}}{(T_t x)(s)} = \begin{cases} x(t+s), & \text{for } t+s \leq 1 \\ 0, & \text{otherwise.} \end{cases}$$

Then $T_t x = 0$ for $t \geq 1$, and $\|T_t\| \leq 1$ for $0 \leq t < 1$. Therefore T_t is exp. stable semigroup (for instance $\|T_t\| \leq e \cdot e^{-t}$).

Let us notice that for this semigroup T_t

$$\int_0^\infty T_t dZ_t = \int_{]0,1]} T_t dZ_t,$$

so, this integral exists without any additional condition on the Lévy measure M. Then, by prop. 3.3 the process (0.1) has a unique stationary distribution $\mu = L(\int_{]0,1]} T_t dZ_t)$.

Remark 3.16

From th. 3.13 and prop. 3.14 we get th. 5.2 in [9]. Cor. 3.12, th. 3.13 and prop. 3.14 imply th. 4.1 in [4].

REFERENCES

[1] A. Chojnowska-Michalik: On processes of Ornstein-Uhlenbeck type in Hilbert space (to appear)

[2] R. Datko: Extending a theorem of A.M. Liapunov to Hilbert space, J. Math. Anal. Appl., 32 (1970), pp.610-616

[3] I.I. Gikhman and A.V. Skorochod: Theory of random processes, vol. 2, Moscow (1973), Nauka (in Russian)

[4] J.B. Gravereaux: Probabilités de Lévy sur R^d et équations differentielles stochastiques linéaires, Sém. Probab. de Rennes (1982)

[5] J. Jacod: Grossissement de filtration et processus d'Ornstein--Uhlenbeck generalisé", preprint, 1983

[6] Z.J. Jurek: An integral representation of operator-selfdecomposable random variables, Bull. Acad. Pol. Sci., Vol. XXX, No 7-8, (1982), pp.385-393

[7] E. Lukacs: Stochastic convergences, Second edition, Academic Press, New York (1975)

[8] K.R. Parthasarathy: Probability measures on metric spaces, Acad. Press, New York and London (1967)

[9] K. Sato and M. Yamazato: Stationary processes of Ornstein-Uhlenbeck type, Proc. IV Japan-Russian Symp. on Probability 1982, Lect. Notes in Math.

[10] J. Zabczyk: Stationary distributions for linear equations driven by general noise, Report No 73, (August 1982) Universität Bremen (published in Bull. Acad. Pol. Sci. 1983)

[11] J. Zabczyk: Linear stochastic systems in Hilbert spaces; structural properties and limit behaviour, Preprint Nr 236, Institute of Mathematics, Polish Academy of Sciences (March 1981)

NON-LINEAR EVOLUTION EQUATIONS AND FUNCTIONNALS OF MEASURE-VALUED BRANCHING PROCESSES

by Nicole EL KAROUI

I- INTRODUCTION

We are concerned by the Branching motion of mass distributions (MB-process), arising as "the high density limit" of infinite particle branching Markov process.

The concept of branching motion of mass distributions, (or continuous state branching process) was introducted by M. Jirina, ([JI]), and was studied by J. Lamperti ([LA]), S.Watanabe, ([WA.1], [WA.2]), between 1960 and 1968.

These processes have received much attention in recent years, (1977-1980), and have been studied extensively: for example see D.Dawson, ([DA.1], [DA.2], [DA.3], [DA.4]), A. Bose and D.Dawson, ([BO.DA]), Holley-Stroock, ([HO.ST]), T.Kurtz, ([KU]), D.Dawson and K.Hochberg, ([DA.HO]). D. Dawson and K.Hochberg have studied the local structure of state X_t of the process, for t fixed, in ([DA.2]), and ([DA.HO]).

In this paper, we shall study the laws of the weighted occupation times: $\int_0^t X_s(f)\, ds$ for $f \geq 0$

and more generally the laws of the measure-valued linear functional of X, X_μ, definited by : $X_\mu(f) = \int \mu(dt)\, X_t(f)$, where μ is a positive Radon measure on $(0,\infty)$.

The probability distribution of X_μ is uniquely determined by the Laplace transform : $L_\mu(f)(m) = E_m \exp-(X_\mu, f)$ for $f \in C_K$

When the measure on R^+ is the Lebesgue measure, denoted by λ, the physical meaning of the functional X_λ is clear: $X_\lambda(B)$ is a estimation of the amount of the individuals in the set B, weighted by the amount of time each spent in B.

This case was studied by I.Iscoe, ([IS.1], [IS.2]), using a semi-group approximation theorem, when X is the limit of infinity particle system, governed by a symétric stable semi-group on R^d.

Our main tool is the stochastic calculus: Branching motions of mass distribution (MB-processes) can be characterized by a martingal problem. This was proved by S.Roelly-Coppoletta in ([RO]), when the trajectories of X are supposed to be continuous. In this paper, we will consider only this case. The general case is studied in ([EK.RC]).

Using the martingal characterization, we prove the existence of a function, F(t,x), which converge to 0 if x goes to infinity, and such that :

$$\exp\left(-(X_t, F(t,.)) - \int_{(o,t]} \mu(ds) \langle X_s, f\rangle\right) \quad \text{is a martingal.}$$

This function F is a mild solution of non-linear evolution equation.

When there is no particle motion , X_t is a Branching continuous process on R^+. The law of the variables X_μ was described in the same way by J.Pitman and M.Yor, ([PI.YO]), using the solutions of the Sturm-Liouville equation $y'' + y.\mu = 0$

After a short review of concepts and notations, we give the martingal problem and its equivalent formulations. Next, a theorem of Girsanov and Feynman-Kac type is proved. We deduce the Laplace transform of $\langle X_\mu, f\rangle$ from solution of non-linear evolution equation. The general properties of this equations are given in the last section.

II- THE MARTINGAL PROBLEM

Let E be a locally compact space, with denumberable basis. It is the space of state of particle motion.

$C_0(E)$ will denote the Banach space of continuous functions on E, which vanish at infinity.

$M_f(E)$ is the space of finite positive measures on $\underline{\underline{B}}(E)$,the Borel σ-algebra of E . $M_f(E)$ carries the weak topology.

Integration will be denoted often by the pairing $\langle\ ,.\rangle$ ($\langle\mu, f\rangle = \int \mu(dx)\, f(x)$).

DEFINITION 1 : A Branching motion of mass distribution is a measure valued Feller process, (Ω, $\underline{\underline{F}}$, $\underline{\underline{F}}_t$, X_t , ζ, P_m ; $m \in M_f(E)$) for which the branching property is satisfied:

$$\text{(BP.1)} \qquad \Pi_t(m_1,.) * \Pi_t(m_2,.) = \Pi_t(m_1 + m_2 ,.)$$

where Π_t denote the semigroup of MB-process.

The probability distribution of the measure-valued random variable X_t is characterized by the Laplace transform :

$$L_t(f) = \Pi_t(\exp - \langle .,f\rangle) \qquad f \in Co(E)$$

The branching property is equivalent to :

$$\text{(BP.2)} \qquad L_t(f)(m) = \exp - \langle m, U_t f\rangle$$

and by the Markov property, U_t is a non linear semi-group on $C_0(E)$

In this paper, we consider the particular case, where $U_t f$ is a mild solution of non-linear evolution equation:

$$\text{(E1)} \quad \left| \begin{array}{l} \frac{du(t)}{dt} = A\,u(t) - \gamma\, u^2(t) \\ u(o) = f \end{array} \right.$$

i.e (E1') $\quad U_t f = S_t f - \gamma \int_o^t S_{t-s}(U_s^2 f)\, ds$

S_t is a strongly continuous semi-group on $C_o(E)$, which governe the particle motion;its infinitesimal generator is denoted by A with domain $D(A)$. $D(A)$ is dense in $C_o(E)$, and A is a closed linear operator. ((PA.p5))
This equations are studied in the last section.

For simplicity, we suppose $S_t 1 = 1$

REMARK: By the hypothesis $S_t 1 = 1$, for all $k \in R^+$, $U_t(k)$ is constant in x, and satisfy the differential Riccati equation (on R^+)

$$\frac{dv}{dt} = - \gamma v^2 \qquad v(o) = k$$

i.e. $\quad U_t(k) = k/1+ \gamma kt$

The process $x_t = \langle X_t, 1\rangle$ is a continuous Branching process on R^+, which has moments of all orders. ((WA. 1) , (PI.YO)) .

In ((RO.CO)), S.Roelly-Coppoletta studies the MB-process, which the Laplace transform is associated with the propagator of the evolution equation (E1).
Using tightness result for distribution probability on the set $D(R^+, M_f(E))$ of functions from R^+ into $M_f(E)$ which are "càdlàg" , and the martingal problem mentionned below (Theorem 2) , she shows that there exist a sequence of branching diffusion renormalized processes, which laws on $D(R^+, M_f)$ converge weakly to probability distribution of this MB-process.

THEOREM 2: THE MARTINGAL PROBLEM

Let $(\Omega, \underline{\underline{F}}, \underline{\underline{F}}_t, P)$ be a probability space, satisfying the usual conditions.

Let X. be a càdlàg process, taking value in $M_f(E)$.

The following propositions are equivalent:

1) For all f in $C_o(E)^+$, and T>0
$\exp - \langle X_t, U_{T-t} f\rangle$ is a P-martingal on $(0,T)$

2) For all f in $D(A)^+$
$\exp\left(-\langle X_t, f\rangle + \langle X_o, f\rangle + \int_o^t \langle X_s, Af - \gamma f^2\rangle\, ds\right) = H_t f$
is a P- local martingal

3) For all f in $D(A)$,

$$<X_t,f> - <X_o,f> - \int_0^t <X_s,Af> ds = M_t f$$ is a P-local martingal, with increasing process $2\gamma \int_0^t < X_s, f^2> ds$.

What is more, the process X has continuous trajectories.

4) For all G ,twice continuously differentiable on $\mathbb{R}$, and all f in $D(A)$

$$G(<X_t,f>) - G(<X_o,f>) - \int_0^t G'(<X_s,f>) <X_s,Af> - \gamma G''(<X_s,f>) <X_s,f^2> ds$$

is a P-local martingal

A probability P_m , which sastisfies 1) ,2) ,3) , or 4), and $P_m(X_0 = m) = 1$ is a solution of martingal problem $\mathcal{P}_m$.

This theorem is proved in ([RO-CO]) .

COROLLARY 3: The branching process $x_t = <X_t,1>$ on $(0, \infty)$ is a continuous martingal with increasing process:

$$d\langle\langle x \rangle\rangle_t = 2\gamma x_t \, dt$$

It is the square of Bessel process, and satisfies to stochastic equation :

$$dx_t = \sqrt{2\gamma x_t} \, dB_t \qquad \text{where } B \text{ is a Brownian motion .}$$

The existence and uniqueness of solution to martingal problem is proved in ([RO-CO]).

THEOREM 4 : On the canonical space $D(\mathbb{R}^+,M_f)$, there exist a unique probability P_m , solution of the martingal problem $\mathcal{P}_m$.

P_m has it support in $C(\mathbb{R}^+,M_f)$.

REMARK: For P and Q two probabilities on $\mathbb{D}(\mathbb{R}^+, M_f)$, $P \oplus Q$ will denote the distribution of $X + Y$, with (X_t) and (Y_t) two independant processes, respectively P and Q distribued.

By Theorem 2 and 4 , the branching property can be extented to law of process as follows :

$$\text{(BP.3)} \qquad P_m \oplus P_{m'} = P_{m+m'} \qquad \text{for } m, m' \text{ in } M_f \ .$$

A Theorem of Girsanov and Feynman-Kac type

We are going to describe how the law of X is transformed by murder and by absolute continuous probability change.

It is clear, that if $H_t g$,(Theorem 2) is a martingal, the Branching property (BP.3) is still true for the probability Q_m définited by

$$Q_m = H_t g \cdot P_m \qquad \text{on } \underline{\underline{F}}_t$$

THEOREM 5 : Let $(\Omega, \underline{\underline{F}}, \underline{\underline{F}}_t, X_t, \zeta, P_m ; m\in M_f)$ be the MB.process associated with the propagator of evolution equation (E1). ζ is infinite P_m a.s.

Let g be a continuous function in D(A), such that:

$$H_t g = \exp\left(-\langle X_t, g\rangle + \langle X_0, g\rangle + \int_0^t \langle X_s, Ag - \gamma g^2\rangle ds\right)$$

is a P_m- martingal , for all $m\in M_f$.

Let c be a non-negative continuous function, and $C_t = \int_0^t \langle X_s, c\rangle ds$.

We will denote by Q_m the probability define on $\underline{\underline{F}}_t \cap \{t<\zeta\}$ by :

$$Q_m = H_t g \cdot \exp - C_t \cdot P_m$$

The process $(\Omega, \underline{\underline{F}}, \underline{\underline{F}}_t, X_t, \zeta, Q_m; m\in M_f)$ is a continuous MB-process with life time ζ , and for all f in $C_0(E)^+$,

$$Q_m(1_{\{t<\zeta\}} \exp - \langle X_t, f\rangle) = \exp - \langle m, V_t f\rangle$$

where $V_t f$ is the mild solution of evolution equation :

$$\text{(E2)} \quad \begin{cases} \dfrac{dv(t)}{dt} = Av(t) - \gamma v^2(t) - 2\gamma g v(t) + c \\ v(o) = f \end{cases}$$

PROOF: We have noticed that the process $(\Omega, \underline{\underline{F}}, \underline{\underline{F}}_t, X_t, \zeta, Q_m)$ is still a MB-process. We only have to calculate its Laplace transform.

By the theorem 2, for all f in D(A), $H_t(f+g)$ is a P_m-local martingal.

We can write $H_t(f+g)$ as:

$$H_t(f+g) = \exp\left(-\langle X_t, f\rangle + \langle X_0, f\rangle + \int_0^t \langle X_s, Bf\rangle ds\right) H_t g \exp -C_t$$

where B denote the operator on D(A) :

$$Bf = A(f+g) - \gamma(f+g)^2 - (A(g) - \gamma g^2) + c$$

Then,it is clear that :

$$1_{\{t<\zeta\}} \exp\left(-\langle X_t, f\rangle + \langle X_0, f\rangle + \int_0^t \langle X_s, Bf\rangle ds\right)$$ is a Q_m- local martingal. This property can be extented to time-depending function f in D(d/dt +A) where the generator A is replaced by $\hat{A} = d/dt + A$ and B by $\hat{B} = d/dt + B$.

Particulary, if $V_t f$ is a strong non-negative solution of evolution equation:

$\frac{dv(t)}{dt} = Bv(t)$, $v(o) = f$, $\hat{B}V_{T-t}f = 0$ and

$$1_{\{t<\zeta\}} \exp\left(-\langle X_t, V_{T-t}f\rangle + \langle X_0, V_T f\rangle\right)$$ is a local bounded martingal with respect to Q_m .

Q.E.D.

More generally, we can prove exactly by the same way, that:

THEOREM 6: Let μ be a radon positive measure on (o,∞).

For $c(t) \geq 0$, continuous w.r.x, let $F(T,t)$ be a non-negative solution, right continuous w.r.t of P.D.E. of Riccati type :

$$(E3) \quad \begin{cases} d_sF(T,s) + AF(T,s)\,ds = \gamma F^2(T,s)\,ds + c(s)\,\mu(ds) & \text{(in weak sens)} \\ F(T,0) = 0 \end{cases}$$

Then
$$E_m(\exp - \int_{(0,T(} \mu(dt) \langle X_t, c(t)\rangle\;) = \exp - \langle m, F(T,0)\rangle$$

PROOF: By extension of Theorem 2 , if $G(s)$ is a continuous function w.r.x and the repartition function w.r. s . . of transition measure on (o,∞), $G(ds)$, then

$$\langle X_t, G(t)\rangle - \langle X_o, G(o)\rangle - \int_{(o,t)} \langle X_s, AG(s)\rangle ds + \langle X_s, G(ds)\rangle$$

is a P_m- continuous local martingal, with increasing process $2\gamma \langle X_s, G^2(s)\rangle ds$

The exponential formula of Theorem 2 is again true for $G(t)$.

The remainder of proof is just as in Theorem 5.

REMARK 1: If $\mu = dt$, then $F(T,t) = V_{T-t}(0)$ satisfies (E3)

REMARK 2: If $G(t^-)$ denotes the leftcontinuous function $F(T-t)$, and if μ_T is the image measure of μ, by the transformation $u \to T-u$ on $(0,T)$, the right-regularized function of $G(t^-)$, $G(t)$, is solution of :

$$(E4) \quad \begin{cases} d_sG(s) = (AG(s) - \gamma G^2(s))ds + c(T-s)\,\mu_T(ds) \\ G(0) = 0 \end{cases}$$

We can give a description of the asymptotic behavior of $V_t(0)$, solution of evolution equation (E2), when g is the null function . ([IS.2])

THEOREM 7: We suppose that $\int_0^\infty S_t(c)\,dt$ is bounded by K.
$V_t(0)$ increases to a continuous function V^* , and

$$V_t(V^*) = V^* \qquad \text{for all } t > 0$$

Moreover, V^* is a mild solution of :

$$Au - \gamma u^2 + c = 0$$

PROOF: That $V_t(0)$ increases to a function V^* can be seen from:

$$E_{\delta_x}(\exp -C_t\;) = \exp - V_t(0)(x)$$

The application $\omega \to C_t(\omega)$ is continuous, and the probabilities P_m are weakly continuous. ([RO-CO])

Therefore, the function V^* is continuous .

By the Markov property, $E_m(\exp - C_\infty) = E_m(\exp - (C_t + \langle X_t, V^*\rangle))$

and by the Theorem 5
$$V_t(V^*) = V^*$$

In the last section, we will prove that $V_t(0)$ is positive and bounded by K.

Let S_t^α be $e^{-\alpha t}S_t$

Then, $V_t f = S_t^\alpha f + \int_0^t S_{t-s}^\alpha(\alpha V_s f + c - \gamma V_s^2 f)\, ds$ (Lemma 10)

and

$$V^* = S_t^\alpha V^* + \int_0^t S_{t-s}^\alpha(\alpha V^* + c - \gamma V^{*2})\, ds$$

If $t \to \infty$, V^* is bounded and $S_t^\alpha 1$ goes to 0

$$V^* = \int_0^\infty S_t^\alpha(\alpha V^* + c - \gamma V^{*2})\, dt$$

Q.E.D.

III- NON-LINEAR EVOLUTION EQUATIONS

We have seen many evolution equations of Riccati type.

For example :

(E1) $\frac{du(t)}{dt} = Au(t) - \gamma u^2(t)$, (E2) $\frac{dv(t)}{dt} = Av(t) + c - 2\gamma gv(t) - \gamma v^2(t)$

(E5) $d_s G(s) = (AG(s) - \gamma G^2(s) + g\, G(s))\, ds + c(s)\, \mu(ds)$

This first two are classical type, and we systematicaly refere to ([PA]) to solve them.

The last is a little more general, but can be treated in the same way.

DEFINTIONS 8: A function $u: [0,T) \to C_0(E)$ is a strong solution of

(1) $\frac{du(t)}{dt} = Au(t) + k(t,u(t))$ $u(0) = f$

if u is continuous on $[0,T)$, continuously differentiable on $(0,T)$, $u(t)$ is in $D(A)$ and (1) is satisfied on $[0,T[$.

ii) A function $u : [0,T) \to C_0(E)$ is a mild solution of (1)

if u is continuous on $[0,T)$ and if :

(2) $u(t) = S_t f + \int_0^t S_{t-s} k(s,u(s))\, ds$

iii) A mild solution for (E5) is a right-continuous function G on $[0,T)$ that

(3) $G(t) = S_t f + \int_{[0,t)} S_{t-s} c(s)\, \mu(ds) + \int_{(0,t)} S_{t-s}(gG(s) - \gamma G^2(s))\, ds$

(cf ([PA p.105]))

THEOREM 9: ([PA p.186])

We suppose that k only satisfies a local lipschitz condition in u, uniformly in t on $(0,T)$, that is

for every $T > 0$ and $c \geq 0$ $\|k(t,u) - k(t,v)\| \leq L(T,c)\, \|u-v\|$ for u, v with $\|u\| \leq c$, $\|v\| \leq c$.

i) For f in $C_0(E)$, there is a $T_{max} \leq \infty$ such that :

(1) has a unique mild solution on $(0,T_{max})$

Moreover, if $T_{max} < \infty$, $\lim_{t\to T_{max}} \|u(t)\| = \infty$

ii) For $f \in D(A)$, the mild solution is a strong solution

The evolution equation (E1) and (E2) have a unique mild solution on $(0,T_{max})$, because

$$k(t,u) = c(t) - 2\gamma g u(t) - \gamma u^2(t)$$

is locally lipschitz.

Exactly by the same way, we can prove that (E5) has a unique mild solution on $(0,T_{max})$.

We are going to prove that T_{max} is infinite, in the particular case, which interest us, by establishing that all mild solution of (E5) is positive , and therefore its norm is smaller than

$$\|S_t f + \int_{(0,t)} S_{t-s} c(s)\, \mu(ds)\|$$

LEMME 10: Let $(W, \underline{\underline{G}}, \underline{\underline{G}}_t, Y_t, Q_x ; x \in E)$ be a realization of the Feller semi-group S_t .

We suppose that $h(t)$ is a continuous function such that $:\sup_{t \leq t_o} \|h(t)\| < \infty$

$\Phi(t)$ is a mild right continuous solution of linear evolution equation:

$$(10.1) \qquad d_s u(s) = (Au(s) - h(s)\,u(s))\,ds + c(s)\,\mu(ds) \qquad \text{on } (0,T_{max})$$
$$u(o) = f$$

if and only if :

$$(10.2) \quad \Phi(t) = Q.(f(Y_t)\, M_o^t + \int_{(0,t)} M_s^t\, c(s)(Y_{t-s})\, \mu(ds))$$

where $M_s^t = \exp - \int_s^t h(r)(Y_{t-r})\, dr$

PROOF: Let's denote by $W(t)$ the last term in (10.2)

Using the integration by parts formula, we can prove that:

$$M_s^t = 1 - \int_s^t \exp - \int_s^u h(r)(Y_{t-r})dr\; h(u)(Y_{t-u})\, du$$
$$= 1 - \int_s^t h(u)(Y_{t-u})\, M_s^u \circ \theta_{t-u} \quad du$$

and that :

$$W(t) = Q.\,[f(Y_t) + \int_{(o,t)} c(s)(Y_{t-s}) \quad \mu(ds)$$
$$- \int_o^t h(u)(Y_{t-u})\, du \quad (f(Y_t)\, M_o^u \circ \theta_{t-u} + \int_{(o,u)} M_s^u \circ \theta_{t-u} c(s)(Y_{t-s}) \mu(ds))]$$

And by Markov property,

$$W(t) = S_t f - \int_{(o,t)} S_{t-s}\,(h(s)W(s))ds + \int_{(o,t)} S_{t-s} c(s)\, \mu(ds)$$

Conversively, let $\Phi(t)$ be a solution of (10.1) . Let P_n be the polynom

$$P_n(x) = \sum_{k=o,n} (-1)^k \frac{x^k}{k!}$$

We can prove by iteration, that if $N_u^t = \int_u^t h(s)(Y_{t-s})\, ds$

$$\Phi(t) = Q.\left(f(Y_t)P_n(N_o^t) + \int_{(o,t]} c(s)(Y_{t-s})P_n(N_s^t)\mu(ds) - \int_o^t h(s)(Y_{t-s})\,\Phi(s)(Y_{t-s})(-1)^n \frac{(N_s^t)^n}{n!}\, ds\right)$$

It is enough to replace Φ by its value in this relation. As $||h(s)||$ is bounded we can let n go to infinity, and obtain $\Phi(t)$ as mentionned in (10.2) .

COROLLARY 11: <u>If</u> f <u>and</u> c <u>are non-negative, a mild solution of (E5) satisfies the integral equation</u> :

$$G(t,x) = Q_x\left(f(Y_t)\exp-\int_o^t (\gamma G - g)(s,Y_{t-s})\, ds + \int_{(o,t]} c(s,Y_{t-s})\mu(ds)\exp-\int_s^t (\gamma G - g)(r,Y_{t-r})\, dr\right)$$

<u>and therefore</u> G <u>is non negative</u> .

REFERENCES

(BO.DA) A.BOSE and D.DAWSON: A class od measure-valued Markov processes Lect.Notes in Mat n°695 - Springer Verlag (1978)

(DA.1) D.DAWSON : Stochastic evolution equation and related measure processes J.Multivariate Ana. 5 .pp. 1-52 (1975)

(DA.2) D.DAWSON: The critical measure diffusion process. Z f. Warsch. 40 pp 125-145 (1977)

(DA.3) D.DAWSON : Geostochastic calculus Canad.J.Statistic 6 .pp. 143-168. (1978)

(DA.4) D.DAWSON : Stochastic measure diffusion processes Canad.Math.Bull.22.pp.129-138. (1979)

(DA.HO) D.DAWSON and K.J.HOCHBERG : The carrying dimension of a stochastic measure diffusion. Annals of Probability vol.7.n°4. (1979)

(EK.RC) N.EL KAROUI and S.ROELLY-COPPOLETTA : Study of a general class of measure-valued branching processes by stochastic calculus (à paraitre) (1984)

(HO.ST) R.HOLLEY and D.STROOCK : Generalized Ornstein-Uhlenbeck processes and infinite particle branching Brownian motions. Publ.Res.inst.Math.Sci. 14.pp. 741-788. (1978)

(IS.1) I.ISCOE : On the support of measure-valued critical branching Brownian motion. (à paraitre) (1984)

(IS.2) I.ISCOE : A weighted occupation time for a class of measure-valued branching processes . (preprint) (1984)

(JI) M.JIRINA : Stochastic branching processes with continuous state space. Czechoslovak.Math.Jour. 8 (1958)

(KU) T.KURTZ : Approximation of population processes Soc.for Ind.andAppl.Math. (1981)

(LA) J.LAMPERTI : Continuous state branching processes Bull.Amer.Math.Soc. 136 (1967)

(PA) A.PAZY : Semigroups of linear operators and applications to partial differential equations Applied Math.Sciences.Volumes 44 . Springer-Verlag (1983)

(PI.YO) J.PITMAN and M.YOR: A decomposition of Bessel bridges. Z.für Wahrsch. 59 (1982)

(RO.CO) S.ROELLY-COPPOLETTA : Un critère de convergence pour les lois de processus à valeurs mesures. Application aux processus de branchement. (Preprint) (1984)

(WA.1) S.WATANABE : On two dimensional Markov processes with branching property. Trans.Amer.Math.Soc. (1969)

(WA.2) S.WATANABE : A limit theorem of branching processes and continuous state branching processes. Jour.Math.Kyoto.Univ. 8. (1968)

Nicole EL KAROUI
Ecole Normale Supérieure
31, Avenue LOMBART
92260- FONTENAY-AUX-ROSES

DNA DISRIBUTION AS A MEASURE VALUED PROCESS

A. Gerardi and G. Nappo
Dipartimento di matematica
Università degli Studi "La Sapienza"
Roma, I 00187

1. Introduction.

The purpose of this paper is to study a cell population growth in relation to the amount of DNA contained in each cell.

There is a rich literature on this subject (see e.g. [1] for a list of references). This problem has been studied with two different basic model types: discrete and continuous. In discrete models the cell cycle is divided into compartments corresponding to a subdivision of the range of DNA content; continuous models are derived by hydrodynamic tecniques (continuity-type equation). They are not usually the limit of discrete models except in a formal sense. Our approach is to derive a continuous model as the limit of a sequence of discrete ones in the sense of weak convergence.

Since the quantity of interest is the number of cells at a certain DNA level, it seems natural to describe the population growth by a measure valued process. This kind of approach has been already used by several authors (see e.g. [2], [3], [4]).

We assume that each cell moves independently of each other accordingly to the following model:

- each cell stays in G (gap period) for an exponential time of parameter λ_1. In the gap period there is no DNA synthesis.
- then the cell starts to produce DNA with a synthesis velocity $v(x)$ (synthesis period). This period ends when its DNA amount doubles.
- from now on, after another gap period, there is the mitotic period. On the whole they last an eponential time of parameter λ_2 .
- at the end of mitosis the cell has two possibilities: to die or to produce two identical daughters which start anew the same cycle.

So each cell moves in $X = G \cup [1,2]$, and the process describing the population growth has values in M, the space of non negative Radon measures on X, endowed with the weak topology.

In order to construct discrete models, we fix k>0 and set

$$X_k = G \cup \{ x_i^k = 1 + \frac{i}{k} , i = 0,1,\ldots,k \}$$

Let M_k M be the set of atomic measures with atoms in X_k; for every $f \in C_b(M_k)$ define

$$\tilde{L}_k f(\mu) := \sum_{i=0}^{k-1} \{f(\mu+\delta_{x_{i+1}^k} - \delta_{x_i^k}) - f(\mu) \} kv(x_i^k) \mu(x_i^k) +$$

$$+ \alpha\lambda_2\mu(2) \{f(\mu+2\delta_G-\delta_2)-f(\mu)\} + (1-\alpha)\lambda_2\mu(2) \{f(\mu-\delta_2) - f(\mu)\} +$$

$$+ \lambda_1\mu(G) \{f(\mu+\delta_1-\delta_G) - f(\mu)\} .$$

where α is the probability of reproducing itself.

This operator is the generator of a pure jump Markov process $\tilde{\mu}_k$ which represents the behviour of the population if we assume that DNA increases by a quantity 1/k. We consider now the process $\mu_k := \tilde{\mu}_k/\varepsilon(k)$, where $\lim_{k\to\infty}\varepsilon(k) = +\infty$. We will prove the weak convergence of $\{\mu_k\}$ to a process μ , which is defined by a generator L of the following form:

$$Lf(\mu) := \nabla F(\langle\mu,\phi\rangle)\langle\mu,B\phi\rangle \tag{1.1}$$

- on functions $f(\mu) = F(\langle\mu,\phi\rangle) \in C_b(M)$, under suitable assumptions on F,ϕ.
- B is defined by the relations:

$$\begin{aligned} B\phi(G) &= \lambda_1\{\phi(1) - \phi(G)\} \\ B\phi(x) &= v(x)\phi'(x) \qquad x \in [1,2) \\ B\phi(2) &= \lambda_2\{2\alpha\phi(G) - \phi(2)\} \end{aligned} \tag{1.2}$$

In order to prove this convergence we need the following facts:

- the martingale problem (MGP) associated with L has a solution.
- the solution is unique, hence it defines a Markov process μ.
- a suitable convergence (as will be specified later) of the generators L_k of μ_k to L is a sufficient condition of weak convergence in $D(R^+;M)$. For particular initial distributions one can get also convergence in probability.

There are various approaches to verify the first two points. The here are related to measure valued processes in a natural way. Mor qualitative properties of the limit process are investigated.

2. Existence.

We set the MGP fo (L,π) on the class S of functions $f \in C_b(M)$ $f(\mu) = F(\langle\mu,\phi\rangle)$ with

i) $F \in C^1((R^+)^d)$ with compact support for some d.

ii) $\phi = (\phi_1,\ldots,\phi_d)$, $\phi_i \in C_b(X) \cap C^1[1,2]$.

iii) $\phi_i \geq 0$, $\Sigma_i \phi_i > 0$.

iv) $B\phi(\cdot) \in C_b(X^d)$.

We recall that this means to look for a probability measure P or

- the distribution of $\mu(0)$ under P is π,
- for every $f \in S$, $f(\mu_t) - \int_0^t (Lf)(\mu_s)ds$ is a P-local martingale canonical process.

The solution to this MGP is constructed by considering the solu (one point compactification of M) and then by proving that it does

THEOREM 1. Let $L'f(\mu) := L(f|_M - f(\Delta))(\mu), \mu \in M$, $L'f(\Delta) := 0$, f and such that $f|_M - f(\Delta) \in S$. For every initial distribution on M', solution of the MGP for L' in $D(R^+;M')$.

Proof. As one can find in [6], thm IV.5.4, a sufficient condit is the maximum principle for L. To this end we observe that if μ_0 i for f, then, $\langle\mu_0,\phi_i\rangle > 0$ implies $\partial_i F(\langle\mu_0,\phi\rangle) = 0$ and $\langle\mu_0, \phi_i\rangle = 0$ $\partial_i F(\langle\mu_0,\phi\rangle) \leq 0$ and $\langle\mu_0,B\phi_i\rangle \geq 0$, so $Lf(\mu_0) \leq 0$.

THEOREM 2. Every solution of the MGP for L' in $D(R^+;M')$ such t verifies $P(\mu \in D(R^+;M)) = 1$, so it is a solution to the MGP for

Proof. If T is the time of explosion , one can prove (see [6], for every t , $P(T > t) = 1$ if there exists a sequence of functions

i) $f_n \to I_M$ (indicator function of M in M')

ii) $Lf_n \to 0$

ded pointwise convergence). In our case we can choose $f_n(\mu) = F_n(\mu(X))$, $f_n(\Delta) = 0$

$F_n \in C^1(R^+)$ and $F_n(y) \uparrow 1$ for every y, supp $F_n \subseteq [0,n]$, $|F_n'(y)| \leq O(1/n)$.

iqueness.

ually uniqueness for MGP is more difficult to prove than existence, but the wing theorem ([7]) allows to tranform uniqueness problems into existence ones.

EOREM 3. Let A_i be linear operators on $C_b(E_i)$, where E_i are separable metric s (i = 1,2). Assume that for every solution x_1, x_2 of the MGPs (A_1,π), (A_2,δ_y) , whenever $y \in E_2$ and π has compact support, we have

$$(x_1(t),y) = \int Ef(x,x_2(t)) \exp(\int_0^t h(x_2(s))ds)\ \pi(dx) \tag{3.1}$$

f and h are continuous functions.

E_1 is complete and $\{f(\cdot,y)\}$ is a separating family , then existence for the $_2,\delta_y)$ implies uniqueness for the MGP (A_1,π) for every π (not necessarily with t support).

s result allows us to prove the following

OREM 4. The MGP (L,π) has a unique solution for every probability measure π.

of. First of all we observe that the Cauchy Problem $\phi_t = B\phi_t$, $\phi_0 = \phi$ has ue solution ϕ_t in the set $E \subset C_b(X)$ of functions satisfying ii,iii,iv So we can define on $C_b(E)$ the Markov semigroup $T_t f(\phi) := f(\phi_t)$, whose tor we denote by A. Obviously, ϕ_t is the solution of the MGP (A,δ_ϕ). we will prove (3.1) for L and A with $h = 0$, $f(\mu,\phi) = \exp(-\langle\mu,\phi\rangle)$ and this tes the proof. To this end it is enough to see that

$$\cdot,\phi)(\mu) = Af(\mu,\cdot)(\phi) := g(\mu,\phi) \tag{3.2}$$

$$\text{up}_{s,t\leq N} |g(\mu_s,\phi_t)|\ ) < \infty \qquad \text{for every } N > 0 \tag{3.3}$$

as one can easily deduce from [7] . The relation (3.2) is obvious by definition, while the relation (3,3) depends on the following facts:

$$|g(\mu_s,\phi_t)| \leq \sup_{t\leq T}||B_t|| \sup_{s\leq N} \mu_s(X)$$

$$\mu_s(X) \leq \text{const.}<\mu_s,\psi> \quad , \quad \psi = 1 \text{ on } [1,2] \text{ and } \quad \psi \in E.$$

Further $<\mu_t,\psi>$ is a super (sub) martingale when $\alpha \geq \frac{1}{2}$ ($\alpha \leq \frac{1}{2}$).

4, Qualitative analysis,

First we observe that the semigroup U_t associated to μ_t operates in a deterministic way

$$U_t F(<\cdot,\phi>)(\mu) = F(<\mu,\phi_t>) \tag{4.1}$$

for every $f(\mu) = F(<\mu,\phi>) \in S$, as one can easily prove by the same argument used in thm 4. More precisely, the process μ_t is uniquely defined by the conditions

$$<\mu_t,\phi> = <\mu,\phi_t> \tag{4.2}$$

for every $t \geq 0$, $\mu \in M$, $\phi \in E$. Moreover the subset of $\mu \in M$, with density $m(x)$ in [1,2) is invariant, In this subset we look for an equilibrium measure $\hat{\mu}$ in the sense that $\hat{\mu}_t = c(t)\hat{\mu}$, It turns out that $c(t) = \exp \beta t$; where β satisfies $<\mu,B\phi> = \beta<\mu,\phi>$ for every $\phi \in E$, and so we have the following

THEOREM 5, For every $C>0$ there exists one and only one equilibrium measure $\hat{\mu}$, in the subset under consideration, such that $\hat{\mu}(X) = C$,

Proof, From the previous observation it follows that

$$\hat{m}(x) = (\overline{C}/v(x)) \exp(-\beta\int_1^x \frac{ds}{v(x)})$$

$$\hat{\mu}(G) = \overline{C}/\lambda_1 \qquad \hat{\mu}(2) = \frac{\overline{C}}{\lambda_2}\frac{\lambda_1+\beta}{2\alpha\lambda_1} \tag{4.3}$$

where the parameter β is uniquely defined by the conditions

$$\lim_{x\to 2-} \hat{m}(x) = \overline{C}\,\frac{(\lambda_1+\beta)(\lambda_2+\beta)}{v(2)\;2\alpha\lambda_1\lambda_2}$$

$$\lambda_1 + \beta \geq 0 \tag{4.4}$$

Note that conditions (4.4) become the equation given in [5] when $\alpha = 1$, in the limit $\lambda_1, \lambda_2 \to \infty$.

For a discussion of further qualitative properties of the process μ, in a particular case, one can see [8] and the references there quoted.

5. Convergence.

Finally we study the convergence of the sequence $\{\mu_t^k\}$. By using a sufficient condition given in [9] we can prove the weak convergence of the sequence to the process described in the previous sections. Further, since the limit process is a deterministic one, we can deduce the convergence in probability, when the initial conditions converge in probability to the equilibrium measure.

THEOREM 6. The sequence $\{\mu_t^k\}$ converges weakly to the process μ_t in $D(R^+;M)$ whenever $\{\mu_0^k\}$ converges weakly to μ_0 in M. If $\mu_0 = \hat{\mu}$, then

$$P(\sup_{[0,N]} |\langle\mu_t^k,\phi\rangle - \langle\hat{\mu}_t,\phi\rangle| > \varepsilon) \to 0 \tag{5.1}$$

for every $\varepsilon>0, N>0$.

Proof. Note that, from (4.1) we can deduce that $U_t S \subseteq S$. Then, since S is dense in $C_0(M)$, the set of functions vanishing at infinity (the statement in [2] immediately extends to the present case) S is a core for L. Now to have weak convergence, it is sufficient to see that (here π_n is the restriction to M_n)

$$\lim_{n\to\infty} \sup_{\mu\in M_n} |L_n(\pi_n f)(\mu) - \pi_n(Lf)(\mu)| = 0 \tag{5.2}$$

for every $f \in S$ ([9]). But, taking into account the definition of S, we see that there exists a number $r>0$ such that

$$\sup_{\mu\in M_n} |L_n(\pi_n f)(\mu) - \pi_n(Lf)(\mu)| = \sup_{\mu\in M_n; \|\mu\|\leq r} |L_n(\pi_n f)(\mu) - \pi_n(Lf)(\mu)|$$

and now one obtains (5.2) by a direct computation.

As far as (5.1) is concerned, we observe that $\mu_0 = \hat{\mu}$ implies the weak convergence of $\{\mu_t^k \exp(-\beta t)\}$ to $\hat{\mu}$ in $D(R^+;M)$, that is in probability. Now the fact that $\hat{\mu}$ is constant implies that

$$\sup_{[0,N]} d(\mu_t^k \exp(-\beta t),\hat{\mu}) \to 0$$

in probability, when d is a translation invariant bounded metric in M, which in turn implies (5.1).

The next natural step would be to study fluctuations around the equilibrium point. It turns out that the natural environment for this problem is that of distribution valued processes. In this direction, we have already reported some preliminary results in [10].

References.

[1] Bertuzzi, A. - Gandolfi, A. - Giovenco, M.A. Mathematical Models of Cell Cycle with a View to Tumor Studies. Math. Biosci. 53, 159 (1981)

[2] Fleming, W.H. - Viot, M. Some measure-valued Markov Processes in Population Genetics Theory Indiana Univ. Math. J., 28, n.5 (1979)

[3] Dawson, D. - Ivanoff, G. Branching Diffusions and Random Measures in "Branching Processes" 5, (Joffe, A. and Ney, P. eds), Dekker N.Y. (1978)

[4] Kurtz, T.G. Approximation of Population Processes in "N.S.F. Regional Conference Montana" (1979)

[5] Rubinow, S.I. A Maturity-time Representation for all Populations Biophys. J. , 8 (1968)

[6] Ethier, S.N. - Kurtz, T.G. "Markov Processes: Characterization and Convergence" to appear.

[7] Dawson, D. - Kurtz, T.G. Applications of Duality to Measure-valued Diffusion Processes in Lect. Notes in Contr. Inf. Sci., 42, Springer Verlag (1983)

[8] Bertuzzi, A. - Gandolfi, A. - Germani, A. - Vitelli, R. A general Expression for sequential DNA-fluorescence Histograms J. Theor. Biol., 102, (1983)

[9] Kurtz, T.G. Semigroups of Conditioned Shifts and Approximations of Markov Processes Ann. Prob. 3 , n.4, (1975)

[10] Gerardi, A. - Nappo, G. Martingale Approach for Modeling DNA Synthesis Int. Report, Dept. of Math. Univ. of Rome, (1984)

WEAK SOLUTIONS OF STOCHASTIC EVOLUTION EQUATIONS

B. Grigelionis, R. Mikulevicius
Institute of Mathematics and Cybernetics
Lithuanian Academy of Sciences
University of Vilnius

INTRODUCTION.

Let $V \subset H \equiv H' \subset V'$ be a rigged Hilbert space, V be a separable reflexive Banach space, which is a dense subset of H, an imbedding $V \longrightarrow H$ be compact, $(\Omega, \mathcal{F}, \mathbb{F}, P)$ be a filtered probability space (see [1] for the standard terminology and notations).

We shall consider the following stochastic evolution equations :

$$X_t = X_0 + \int_0^t A(s, X_s)\, dN_s + \sum_{j=1}^{\infty} \int_0^t B^j(s, X_{s-})\, dM_s^j, \quad t \geq 0 \qquad (1)$$

where $X_0 \in H$, N is a continuous increasing non-random function, $A : R_+ \times \Omega \times V \longrightarrow V'$ is $(\mathcal{B}(V'), \mathcal{P}(\mathbb{F}) \times \mathcal{B}(V))$ measurable, $\int_0^t |A(s, X_s)|_{V'}\, dN_s < \infty$ p a.e. for each $t > 0$, $M^j \in M^2_{loc}(P, \mathbb{F})$, $\langle M^j \rangle_t = \int_0^t Q_s^j\, dN_s$, $t \geq 0$, $\langle M^j, M^k \rangle_t \equiv 0$, as $j \neq k$, the $(P, \mathbb{F})$-compensator of the jump measure of $M = (M^j, j \geq 1)$ is $\Pi([0,t] \times \Gamma) = \int_0^t \pi(s, \Gamma)\, dN_s$ $t \geq 0$, $\Gamma \in \mathcal{B}(R^{\infty} \setminus \{0\})$, $B^j : R_+ \times \Omega \times V \longrightarrow H$ is $(\mathcal{B}(H), P(\mathbb{F}) \otimes \mathcal{B}(V))$-measurable, $j \geq 1$, $\sum_{j=1}^{\infty} \int_0^t |B^j(s, X_s)|_H^2\, Q_s^j\, dN_s < \infty$ P-a.e. for each $t > 0$.

Denote $\mathcal{X} = D_{[0,\infty]}(H_w) \cap L^2_{loc,w}(V, N) \cap L^2_{loc}(H, N)$, endowed with the suprenum toplogy, where $D_{[0,\infty]}(H_w)$ is the Skorokhod space with $\mathcal{J}_1$-topology of càdlág functions $x(.) : R_+ \longrightarrow H$, considering the weak topology in H, $L^2_{loc,w}(V, N) = \{x(.) :$ $\int_0^t |X(s)|_H^2\, dN_s < \infty,\ \forall\, t > 0\}$ with the topology of weak convergence $L^2_{loc}(H, N) =$ $\{x(.) : \int_0^t |X(s)|_H^2\, dN_s < \infty,\ \forall\, t \geq 0\}$ with the topology of strong convergence. Let $\mathbb{D} = \{\mathcal{D}_t, t \geq 0\}$ be the canonical filtration on $\mathcal{X}$, $\mathcal{D} = \bigvee_{t>0} \mathcal{D}_t$, $\overline{\Omega} = \Omega \times \mathcal{X}$, $\overline{\mathcal{F}} = \mathcal{F} \otimes \mathcal{D}$, $\overline{\mathcal{F}}_t = \bigcap_{s>t} \mathcal{F}_s \otimes \mathcal{D}_s$, $t \geq 0$, $\overline{\mathbb{F}} = \{\overline{\mathcal{F}}_t, t \geq 0\}$, $X_t(\omega, X) = x(t)$, $t \geq 0$, $(\omega, X) \in \overline{\Omega}$.

Following [2], a probability measure $\overline{P}$ on $(\overline{\Omega}, \overline{\mathcal{F}})$ is called a weak solution of the stochastic evolution equation (1), if $\overline{P}_{|\Omega} = P$, $M^j \in M^2_{loc}(\overline{P}, \overline{F})$, $\langle M^j \rangle_t = \int_0^t Q^j_s \, dN_s$, $t \geq 0$, $j \geq 1$, $\langle M^j, M^k \rangle_t \equiv 0$, as $j \neq k$, the $(P, \mathbb{F})$-compensator of the jump measure of $M = (M^j, j \geq 1)$ is $\int_0^t \pi(s, \Gamma) \, dN_s$, $t \geq 0$, $\Gamma \in \mathcal{B}(R^\infty \setminus \{0\})$, and X satisfies (1), considering the natural extensions to $\overline{\Omega}$ of functions, defined only on Ω and denoting $\overline{P}_{|\Omega}$ the Ω-marginal of $\overline{P}$.

The aim of this paper is the investigation of conditions for existence, stability and uniqueness of weak solutions of the stochastic evolution equation (1), developping some ideas from the papers [2] - [6]. Remark, that a construction of stochastic integrals with respect to the Fréchet Valued local martingales is considered in [5].

§ 1 - MARTINGALE CHARACTERIZATION OF WEAK SOLUTIONS.

Let $\overline{P}$ be a weak solution to (1), $\tilde{\mathfrak{X}} = H \times R^\infty$, $\tilde{X} = \{\tilde{X}_t = (X_t - X_0, M_t), t \geq 0\}$ $\tilde{P}$ be a jump measure of $\tilde{X}$, $\tilde{\Pi}$ be the $(\overline{P}, \overline{\mathbb{F}})$-compensator of $\tilde{P}$. From the equations (1) we find that

$$\tilde{\Pi}([0,t] \times \tilde{\Gamma}) = \int_0^t \tilde{\pi}(s, \tilde{\Gamma}) \, dN_s, \ t \geq 0, \ \tilde{\Gamma} \in \mathcal{B}(\tilde{\mathfrak{X}} \setminus \{0\}),$$

where

$$\tilde{\pi}(s, \tilde{\Gamma}) = \int_{R^\infty \setminus \{0\}} \chi_{\tilde{\Gamma}} \left(\sum_{j=1}^{\infty} B^j(s, X_s) z_j, z \right) \pi(s, dz).$$

Applying the Ito's formula to the one dimensional semimartingale

$$\langle \tilde{X}' \mid \tilde{X}_t \rangle = (X_t - X_0, h)_H + \sum_{j=1}^{\infty} M^j_t z_j, \quad \tilde{X}' = (h, z) \in \tilde{\mathfrak{X}}',$$

we obtain, that

$$M^{\tilde{X}'}_t = e^{i \langle \tilde{X}' \mid \tilde{X}_t \rangle} - \int_0^t e^{i \langle \tilde{X}' \mid \tilde{X}_s \rangle} \varphi_{\tilde{X}'}(s, \tilde{X}_s) \, dN_s, \ t \geq 0, \tag{2}$$

is a $(\bar{P}, \bar{\mathbb{F}})$-local martingale, where

$$\varphi_{\tilde{X}'}(s, \tilde{X}_s) = \int_{\tilde{\mathcal{X}} - \{0\}} [e^{i < \tilde{X}' | \tilde{y} >} - 1 - i < \tilde{X}' | \tilde{y} >] \tilde{\pi}(s, d\tilde{y}) +$$

$$+ i < A(s, X_s), h > - \frac{1}{2} \sum_{j=1}^{\infty} [(B^j(s, X_s), h)_H +$$

$$+ 2 (B^j(s, X_s), h)_H X_j + X_j^2] Q_s^j .$$

For any Hilbert space $\hat{H}$ and $(t, \omega) \in R_+ \times \Omega$ denote $L_t^2(\hat{H}^\infty)$ a Hilbert space of sequences $\hat{h} = (\hat{h}_1, \hat{h}_2, \dots)$, $\hat{h}_j \in \hat{H}$, $j \geq 1$, such that the norm

$$|\hat{h}|_t = (\sum_{j=1}^{\infty} |\hat{h}_j|_{\hat{H}}^2 Q_t^j)^{\frac{1}{2}} < \infty .$$

According to the definition of weak solution we have that $dN\, d\bar{P}$-a.e. $B(t, X_t) = (B^j(t, X_t),\ j \geq 1) \in L_t^2(H^\infty)$.

Theorem 1 - (cf. [2]). A probability measure $\bar{P}$ on $(\bar{\Omega}, \bar{\mathcal{F}})$ is a weak solution of the stochastic evolution equation (1) if and only if $\bar{P}_{|\Omega} = P$, for each $t > 0$ $\bar{P}$-a.e.

$$\int_0^t |A(s, X_s)|_{V'} dN_s < \infty , \quad \int_0^t |B(s, X_s)|_s^2 dN_s < \infty$$

and for each $\tilde{X}' \in \tilde{\mathcal{X}}'$ $M^{\tilde{X}'}$ is a $(\bar{P}, \bar{\mathbb{F}})$-local martingale.

Proof of necessity part of the theorem follows directly from the definition of the weak solution and the above statement. Sufficiency part can be proved easily in an analoguous way as in theorem 1 in [7].

§ 2 - STABILITY OF WEAK SOLUTIONS.

Let $\{\bar{P}^n, n \geq 1\}$ be a sequence of weak solutions corresponding to the sequences of the coefficients $\{A_{n,1}, n \geq 1\}$ and $B_n, n \geq 1\}$, i.e. $\bar{P}^n_{|\Omega} = P$, $M^j \in \mathcal{M}^2_{loc}(\bar{P}^n, \bar{\mathbb{F}})$, $< M^j >_t = \int_0^t Q_s^j dN_s$, $t \geq 0$, $j \geq 1$, $< M^j, M^k >_t \equiv 0$, as $j \neq k$, the $(\bar{P}^n, \bar{\mathbb{F}})$-compensators of the jump measure of $M = (M^j, j \geq 1)$ is $\int_0^t \pi(s, \Gamma)\, dN_s$, $t \geq 0$,

$\Gamma \in \mathcal{B}(R^{\infty} \setminus \{0\})$ and X satisfies the following stochastic evolution equations :

$$X_t = X_0 + \int_0^t A_{n,1}(s, X_s)\, dN_s + \sum_{j=1}^{\infty} \int_0^t B_n^j(s, X_{s-})\, dM_s^j \quad , t \geq 0 , n \geq 1 ,$$

Assuming that $A_{n,1}$ and B_n satisfy the measurability and integrability assumptions analoguous to A and B.

Call $M_{mc}(\bar{\Omega})$ the space of all finite positive measures μ on $\bar{\mathcal{F}}$, endowed with the weakest topology, under which the mappings $\mu \longrightarrow \mu(g) = \int_{\bar{\Omega}} g(\omega, x)\, \mu(d\omega, dx)$ are continuous for all bounded measurable functions g on $(\bar{\Omega}, \bar{\mathcal{F}})$, such that $g(\omega, .)$ are continuous in $\mathcal{X}$ for each $\omega \in \Omega$, we shall find conditions, for $\bar{P}^n \Longrightarrow \bar{P}$, as $n \longrightarrow \infty$, in the space $M_{mc}(\bar{\Omega})$. The properties of this space are investigated in [8], [9].

Calling $< v', v >$, $v' \in V'$, $v \in V$, a canonical bilinear form, we shall use the following assumptions (cf. [3] - [5]) :

I - There exist mappings $A_{n,2} : R_+ \times \Omega \times V \longrightarrow V'$, $n \geq 1$, which are $(\mathcal{B}(V'), \mathcal{P}(\mathbb{F}) \times \mathcal{B}(V))$-measurable and such that $dN\, d\bar{P}^n$-a.e.

$A_{n,1}(s, X_{s-}) = A_{n,2}(s, X_{s-})$;

II - For each $v \in V$, $t \geq 0$

$$2 < A_{n,2}(t, v), v > + |B_n(t, v)|_t^2 + \varepsilon |v|_V^2 \leq g_t (1 + |v|_H^2) \quad ,$$

where $\varepsilon > 0$, $g \in \mathcal{P}^+(\mathbb{F})$ and P-a.e. for all $t > 0$

$$\int_0^t g_s\, dN_s < \infty \quad ;$$

III - For each $v \in V$, $t \geq 0$

$$|A_{n,j}(t,v)|_{V'}^2 \leq g_t (1 + |v|_H^2) + R |v|_V^2 \quad , R > 0 , j = 1, 2 ;$$

IV - For some dense subset $\tilde{V}$ of V and each $v_1 \in V$, $v_2 \in \tilde{V}$, dN dP-a.e.

$$< A(t, v_1), v_2 > = < f(t, v_2), v_1 > + g(t, v_1, v_2) ,$$

where for each $v_2 \in \tilde{V}$ the mappings $v_1 \longrightarrow g(t, v_1, v_2)$ are continuous in H, $g(t, v_1, v_2)$ is $\mathcal{P}(\mathbb{F}) \times \mathcal{B}(V)$-measurable, P-a.e. for all $t > 0$

$$\int_0^t |f(s, v_2)|^2_{V'} \, dN_s < \infty ,$$

and the mappings $\bar{v}_1 \longrightarrow ((B^j(t, \bar{v}_1), v_2)_H, j \geq 1) \equiv B(t, \bar{v}_1) . v_2$ from H into $L^2_t(R^\infty)$ are continuous $dN \, dP$-a.e. ;

V - For each $v_2 \in \tilde{V}$ and bounded subsets $W \subset V$, as $n \longrightarrow \infty$,

$$\sup_{v_1 \in W} |\langle A_{n,1}(s, v_1), v_2 \rangle - \langle A(s, v_1), v_2 \rangle| \longrightarrow 0 ,$$

$$\sup_{v_1 \in W} |B_n(t, v_1) . v_2 - B(t, v_1) . v_2|_t \longrightarrow 0$$

in measure $dN \, dP$.

Theorem 2 - if the assumptions I - V are fulfiled, then the sequence $\{ \bar{P}^n, n \geq 1 \}$ is relatively compact in the space $M_{mc}(\bar{\Omega})$ and each limiting point is the weak solution of equation (1).

Proof - Denote by $M_c(\mathcal{X})$ the space of all finite positive measures on $\mathcal{B}(\mathcal{X})$ with the topology of weak convergence. According to [8], [9], the relative compactness of the $\mathcal{X}$-marginals $\{ \bar{P}^n_{|\mathcal{X}}, n \geq 1 \}$ in $M_c(\mathcal{X})$ implies the relative compactness of the sequence $\{ \bar{P}^n, n \geq 1 \}$ in $M_{mc}(\bar{\Omega})$. Since the topological space $\mathcal{X}$ is completely regular, the tightness criterion of Prokhorov holds (see [10]). In order to apply this criterion we shall describe the structure of compact subsets in $\mathcal{X}$. Denote by J_1 the J_1-topology of Skorokhod in $D_{[0,\infty]}(H_W)$, J_2 the topology of weak convergence in $L^2_{loc,w}(V, N)$, J_3 the topology of strong convergence in $L^2_{loc}(H, N)$ and $J = \sup(J_1, J_2, J_3)$, let $\Delta_c^{[0,T]}(x)$ be the "modulus of continuity" in the space $D_{[0,T]}(V')$ with respect to the Skorokhod metric d.

We shall need the following statement.

Lemma - A subset $K \subset \mathcal{X}$ is relatively compact in the J topology, if for each $T > 0$

$$\sup_{x \in K} \sup_{s \leq T} |x(s)|_H < \infty , \quad \sup_{x \in K} \int_0^T |x(t)|^2_V \, dN_t < \infty$$

and

$$\lim_{c \to 0} \sup_{x \in K} \Delta_c^{[0,T]}(x) = 0 \tag{3}$$

Proof of lemma - We can assume, that K is a closed set in the topology T. Fix $T > 0$ and set $R = \sup_{x \in K} \sup_{s \leq T} |X(s)|_H$. The topology, induced on the set $\{h : |h|_H \leq R\}$ by the weak topology on H is metrisable, because it is enough to take the metric

$$\rho(h, h') = \sum_{k=1}^{\infty} (|(h_k, h)_H - (h_k, h')_H| \wedge 1) 2^{-k}$$ for some trongly dense sequence $\{h_k, k \geq 1\}$ in $\{h : |h|_H \leq R\}$. Therefore the topology, induced on the set K by the topology J_1, is also metrisable. The same is true for the topologies on K, induced by J_2 and J_3. From (3) it follows that for each sequence $\{x_n, n \geq 1\}$ in K $\sup_{n \geq 1} \Delta_c^{[0,T]}(x_n) \longrightarrow 0$, as $c \longrightarrow 0$.

Because the sequences $\{(x_n(.), h_k)_H, k \geq 1\}$ are bounded, there exists a subsequence $\{x_{n_k}, k \geq 1\}$ and $x \in D_{[0,T]}(H_W) \cap \{x : \sup_{s \leq T} |x(s)|_H \leq R\}$ such that $x_{n_k} \longrightarrow x$ in $D_{[0,T]}(H_W)$. Analoguously for each sequence $\{x_n, n \geq 1\}$ in K there exists a subsequence $\{x_{n_k}, k \geq 1\}$ and $x \in L^2_{[0,T],W}(V, N) \cap \{x : \int_0^T |x(s)|_V^2 \, dN_s \leq \leq \sup_{y \in K} |\int_0^T |y(s)|_V^2 \, dN_s\}$ such that $x_{n_k} \longrightarrow x$ in $L^2_{[0,T],W}(V, N)$.

From the compactness of the imbedding $V \longrightarrow H$ we have, that for each $\eta > 0$ there exists $c(\eta) > 0$ satisfying the inequality :

$$|v|_H \leq \eta |v|_V + c(\eta) |v|_{V'}, \; v \in V. \tag{4}$$

Let $\{x_n, n \geq 1\}$ be a sequence in K. Then dN-a.e. $\underline{\lim}_{n \to \infty} |x_n(.)|_V < \infty$, as otherwise from the Fatou lemma we shall have, that

$$\underline{\lim}_{n \to \infty} \int_0^T |x_n(t)| \, dN_t = \infty ,$$

contradicting to our assumption. Let $S = \{t \leq T : \underline{\lim}_{n \to \infty} |x_n(t)|_V < \infty\}$ and $I = \{t_k, k \geq 1\} \subset S$ be such, that $\overline{I} = \operatorname{supp}(dN)$. Using the compactness of the imbedding and the definition of I, we can take a subsequence $\{x_{n_k}, k \geq 1\}$ such that for each $t_i \in I$ there

exists $v_i \in V'$ and $|x_{n_k}(t_i) - v_i| \longrightarrow 0$, as $k \longrightarrow \infty$. From (4) it follows, that the func-

tion $\bar{x}(t_i) = v_i$, $t_i \in I$, can be uniquelly extended as a càdlág function x on I. Let $x(t) = \bar{x}(t)$

for $t \in \bar{I}$, $\bar{x}(\inf(s \in \bar{I}, s \geq t))$ for $t \notin I$. Now if $t \in \bar{I}$, t is a continuity point of $x(.)$ and

there exists a sequences $t_n \uparrow t$, $t'_n \downarrow t$, $t_n < t$, $t'_n > t$, t_n, $t'_n \in I$, then using (4) we

find, that $|x_{n_k}(t) - x(t)|_{V'} \longrightarrow 0$, as $k \longrightarrow \infty$, i.e. $x_{n_k} \longrightarrow x$ in $L^2_{[0,T]}(V', N)$. To

conclude the proof of lemma it is enough to note that, if $x_n \longrightarrow x$ in J-topology, $x_n \longrightarrow x'$ in J_2

topology and $x_n \longrightarrow x''$ in J_3 topology, as $n \longrightarrow \infty$, then $x = x' = x''$.

Setting $S_k = \inf\{t: \int_0^t g_s \, dN_s > k\} \wedge T$, $T > 0$, we shall have obviously that for

each $T > 0$

$$\lim_{k \to \infty} \overline{\lim_{n \to \infty}} \, \overline{P}^n \{S_k < T\} = 0$$

Applying the Ito's formula for $|X_t|_H^2$, $t \geq 0$, from the assumption II and lemma 2 in $[5]$,

we obtain that for each $T > 0$, $k \geq 1$

$$\sup_{n \geq 1} \overline{E}^n (\sup_{s \leq S_k} |X_s|) < \infty$$

and

$$\lim_{\ell \to \infty} \overline{\lim_{n \to \infty}} \, \overline{P}^n \{\int_0^T |X_s|_V^2 \, dN_s > \ell\} = 0$$

It remains to prove, that for each $\varepsilon > 0$ and $T > 0$

$$\lim_{c \to 0} \overline{\lim_{n \to \infty}} \, \overline{P}^n \{\Delta_c^{[0,T]}(X) > \varepsilon\} = 0 \, . \tag{5}$$

According to the know results (see $[11]$) the equality (5) is implied by the Aldous-Rebolledo condition : for each sequences $T_n \in T(\overline{\mathbb{F}})$, $T_n \leq T$, $\delta_n \downarrow 0$ and $\varepsilon > 0$

$$\overline{\lim_{n \to \infty}} \, \overline{P}^n \{|X_{T_n + \delta_n} - X_{T_n}|_{V'} > \varepsilon\} = 0 \, . \tag{6}$$

As far as

$$|X_{T_n+\delta_n} - X_{T_n}|_{V'} \leqslant \int_{T_n}^{T_n+\delta_n} |A_{n,1}(s, X_s)|_{V'}\, dN_s +$$

$$+ \left| \int_{T_n}^{T_n+\delta_n} \sum_{j=1}^{\infty} B_n^j(s, X_{s-})\, dM_s^j \right|_{H'} \qquad (7)$$

using III and the know inequality from [12], after the standard estimations we shall obtain, that for each $\varepsilon > 0$, $\delta > 0$

$$\overline{\lim_{n\to\infty}} \; \bar{P}^n \{ \int_{T_n}^{T_n+\delta_n} |A_{n,1}(s, X_s)|_{V'}\, dN_s > \delta \} = 0 , \qquad (8)$$

$$P^n \{ \left| \int_{T_n}^{T_n+\delta_n} \sum_{j=1}^{\infty} B_n^j(s, X_{s-})\, dM_s^j \right|_H > \delta \} \leqslant \frac{\varepsilon}{\delta^2} +$$

$$+ P^n \{ \int_{T_n}^{T_n+\delta_n} |B_n(s, X_s)|_s^2\, dN_s > \varepsilon \} \quad (9)$$

From II and III we have that

$$|B_n(s, v)|_t^2 \leqslant 2\, |v|_V\, |A_{n,2}(t, v)|_{V'} + g\,(1 + |v|_H^2) .$$

Because for each $\gamma > 0$ and $m \geqslant 1$

$$\int_{T_n}^{T_n+\delta_n} |X_s|_V\, |A_{n,1}(s, X_s)|_{V'}\, dN_s \leqslant \int_{T_n}^{T_n+\delta_n} |X_s|_V\, \chi_{\{|X_s|_V > m\}} \times$$

$$\times |A_{n,1}(s, X_s)|_{V'}\, dN_s + m \int_{T_n}^{T_n+\delta_n} |A_{n,1}(s, X_s)|_{V'}\, dN_s \leqslant$$

$$\leqslant \gamma^2 \int_{T_n}^{T_n+\delta_n} |A_{n,1}(s, X_s)|_{V'}^2\, dN_s + \frac{1}{m\gamma^2} \int_{T_n}^{T_n+\delta_n} |X_s|_V^2\, dN_s +$$

$$+ m \int_{T_n}^{T_n + \delta_n} | A_{n,1} (s , X_s) |_{V'} \, dN_s \tag{10}$$

Now the equality (6) follows easily from (7) - (10). The relative compactness of the sequence $\{ \bar{P}^n , n \geq 1 \}$ follows from the lemma.

Denote

$$M_t^{n,\tilde{x}'} = e^{i < \tilde{x}' | \tilde{X}_t >} - \int_0^t e^{i < \tilde{x}' | \tilde{X}_s >} \varphi^n_{\tilde{x}'} (s , \tilde{X}_s) \, dN_s , \; t \geq 0 , \tilde{x}' \in \tilde{\mathfrak{X}}' , n \geq 1 ,$$

where

$$\varphi^n_{\tilde{x}'} (s , \tilde{X}_s) = \int_{\tilde{\mathfrak{X}} \setminus \{0\}} [e^{i < \tilde{x}' | \tilde{y} >} - 1 - i < \tilde{x}' | \tilde{y} >] \, \tilde{\pi}_n (s , d\tilde{y}) +$$

$$+ i < A_{n,1} (s , X_s) , h > - \frac{1}{2} \sum_{j=1}^{\infty} [(B^j_n (s , X_s) , h)_H +$$

$$+ 2 (B^j_n (s , X_s) , h)_H \, x_j + x_j^2] \, Q^j_s$$

and

$$\tilde{\pi}_n (s , \tilde{\Gamma}) = \int_{R^\infty \setminus \{0\}} \chi_{\tilde{\Gamma}} (\sum_{j=1}^{\infty} B^j_n (s , X_s) z_j , z) \, \pi (s , dz) , \; \tilde{\Gamma} \in \mathcal{B} (\tilde{X} \setminus \{0\}) .$$

According to theorem 1 in [6] and (2), it remains to check, using the assumptions of the theorem, that for each $\varepsilon > 0$, $t > 0$ and $\tilde{x}' \in \tilde{\mathfrak{X}}'$

$$\lim_{n \to \infty} \bar{P}^n \{ | M_t^{n,\tilde{x}'} - M_t^{\tilde{x}'} | > \varepsilon \} = \lim_{n \to \infty} \bar{P}^n \{ | \int_0^t e^{i < \tilde{x}' | \tilde{X}_s >} [\varphi^n_{\tilde{x}'} (s , \tilde{X}_s) -$$

$$- \varphi_{\tilde{x}'} (s , \tilde{X}_s)] \, dN_s | > \varepsilon \} = 0 .$$

We shall omit the technical details.

§ 3 - EXISTENCE AND UNIQUENESS OF WEAK SOLUTIONS.

Existence of weak solutions of the stochastic evolution equation (1) can be proved using the standard method of Galerkin's approximations, from existence results for the finite dimensional stochastic equations and applying the above stability results.

We shall use the following assumptions :

II' - For each $v \in V$, $t \geq 0$

$$2 < A(t,v), v > + |B(t,v)|_t^2 + \varepsilon |v|_V^2 \leq g_t (1 + |v|_H^2),$$

where $\varepsilon > 0$, $g \in \mathcal{P}^+(\mathbb{F})$ and P-a.e.

$$\int_0^t g_s \, dN_s < \infty, \quad t > 0;$$

III' - For each $v \in V$, $t \geqslant 0$

$$|A(t,v)|_{V'}^2 < g_t (1 + |v|_H^2) + R |v|_V^2, \quad R > 0;$$

VI - For each $v_1, v_2 \in V$, $t \geq 0$

$$2 < A(t,v_1) - A(t,v_2), v_1 - v_2 > + |B(t,v_1) - B(t,v_2)|_t^2 + \varepsilon_0 |v_1 - v_2|_V^2 \leq \ell_t |v_1 - v_2|_H^2$$

where $\varepsilon_0 > 0$, $\ell \in \mathcal{P}^+(\mathbb{F})$ and P-a.e.

$$\int_0^t \ell_s \, dN_s < \infty, \quad t > 0.$$

The following statements hold.

Theorem 3 - It the assumptions II', III' and IV are fulfiled, then there exists a weak solution of the stochastic evolution equation (1).

We shall omit the full details of the proof.

Theorem 4 - If, in addition to the assumptions of Theorem 3, the assumption VI is satisfied, then there exist a unique weak solution of the stochastic evolution equation (1).

Proof : the uniqueness of the weak solution $\overline{P}$ follows immediately from the fact, that under the conditions of Theorem 4 there exists a unique strong solution X of stochastic equation (1) on the filtered probability space $(\Omega, \mathcal{F}, \mathbb{F}, P)$ and the relation $\overline{P}(d\omega, dx) = P(d\omega)\, \varepsilon_X(dx)$ holds, where $\varepsilon_a(dx)$ is the Dirac measure (see [2], [5]).

Remark 1 - The $\mathcal{X}$-marginal $\overline{P}_{|\mathcal{X}}$ of a weak solution $\overline{P}$ of the stochastic evolution equation (1) is obviously a solution of the martingale problem, related to the semimartingale with values in a rigged Hilbert space and defined by the equation (1). Therefore from Theorems 3 and 4 we can obtain the corresponding conditions for existence and uniqueness of such semimartingales with the given triples of their predictable characteristics. These conditions can be extended in the standard way by means of the analogue of the Girsanov transformation and the known criteria of uniform integrability of positive local matingales (see, e.g., [13], [14]).

Remark 2 - It is obvious, that the convergence of the sequence $\{\overline{P}^n, n \geq 1\}$ in the space $M_{mc}(\overline{\Omega})$ implies the weak convergence of the $\mathcal{X}$- marginals $\{\overline{P}^n_{|\mathcal{X}}, n \geq 1\}$ in the space $M_c(\mathcal{X})$. Thus combining the results on stability, existence and uniqueness of the weak solutions of the stochastic evolution equation (1), we can obtain criterias of weak convergence of solutions of the corresponding martingale problems (see [6] and therein more complete references on the functionnal limit theorem).

Remark 3 - The case, when N is an increasing $\mathcal{P}(\mathbb{F})$-measurable process with discontinuities, can be considered in the analoguous way, applying the results from the papaer [9]. The case, when the driving local martingale M takes values in more general Frechet- spaces, can be reduced to the case of space R^∞ (see [5]).

REFERENCES

[1] J. Jacod, Calcul stochastique et problèmes de martingales.
Lecture Notes in Math. 714, Springer Verlag, Berlin, 1979.

[2] J. Jacod, J. Ménin, Weak and strong solutions of stochastic differential equations : existence and stability
Lecture Notes in Math., vol. 851, Stochastic integrals, Springer, 1981, p. 169-212.

[3] M. Viot, Solutions faibles d'équations aux dérivées partielles stochastiques non linéaires.
Thèse doct. sci. Univ. P. et M. Curie, Paris, 1976.

[4] Krylov N.V., B.L. Rozovskii, On stochastic evolution equation.
Modern Problems in Math., vol 14, VINITI, M., 1979, p. 71-146.

[5] B. Grigelionis, R. Mikulevičius, Stochastic evolution equations and densities of the conditional distributions.
Lecture Notes in Control and Inform. Theory, 49, Springer, 1983.

[6] B. Grigelionis, R. Mikulevičius, On the functional limit theorems.
Proc. Conf. Stochastic space-time models and limit theorems, Bremen, 1983 (to appear).

[7] B. Grigelionis, R. Mikulevičius, On weak convergence of semimartingales.
Lietuvos matem. rinkinys, 1981, vol. 21, N° 4 , p. 9-24 .

[8] J. Jacod, J. Mémin, Sur un type de convergence intermédiaire entre la convergence en loi et la convergence en probabilité.
Lecture Notes in Math., 850, Sem. de Probab. XV, Springer Verlag, Berlin, 1981, p. 529-546.

[9] R. Mikulevičius, On weak convergence of measures.
Lietuvos matem. rinkinys, 1985, t. XXV (to appear).

[10] Bourbaki N. Intégration, ch. IX, Hermann, Paris.

[11] Aldous D. Stopping times and tightness.
Ann. Probab., 1978, vol. 6, N° 2, p. 335-340.

[12] E. Lenglart, Relation de domination entre deux processus.
Ann. Inst. H. Poincaré, 1977, vol. 13, p. 171-179.

[13] S.M. Kozlov, Some questions on stochastic partial differential equations.
Proc. Petrovskii Sem., vol. 4, 1978, p. 147-172.

[14] D. Lepingle, J. Mémin, Sur l'intégrabilité uniforme des martingales exponentielles.
Z. Wahr. verw. Geb., 1978, B. 42, S. 175-204.

Stability of Parabolic Equations with Boundary and Pointwise Noise

Akira Ichikawa
Faculty of Engineering, Shizuoka University,
Hamamatsu 432, Japan

1. Introduction.

It is known that parabolic equations with boundary control can be described by a semigroup model [1], [12] and some important problems in systems theory such as quadratic control and stabilizability have been extensively studied by Lasiecka and Triggiani [13], [14]. Following their approach, we have proposed a semigroup model for parabolic equations with boundary and pointwise noise [11] and obtained existence, uniqueness and regularity results. Semigroup models for boundary noise can be also found in [2], [15]. In this paper we study the stability of our system. First we consider the linear case and establish the equivalence of (i) mean square stability, (ii) the existence of a solution to the Liapunov equation and (iii) the exponential stability of the second moment. This equivalence was originally established for a class of linear stochastic evolution equations with bounded noise operators [5], [7], [16]. This implies that noise for parabolic equations is of distributed nature. We then give a sufficient condition for stability which is of practical use. We shall also show that the stability of a certain class of nonlinear systems is equivalent to that of the linear system in the class. This is the so called absolute stability of a system and has been considered in [8] in the case of bounded noise operators. To illustrate our abstract results, we give three simple examples.

2. The mathematical model.

Let $S(t)$ be a strongly continuous analytic semigroup of negative type on a real separable Hilbert space Y and let $-A$ be its infinitesimal generator. Then fractional powers A^γ, $0<\gamma<1$, are well defined and that for each $0<T<\infty$ there exists a constant $K=K(T)>0$ such that

$$|A^\gamma S(t)| \leq K/t^\gamma \quad \text{for any} \quad 0<\gamma<1 \quad \text{and} \quad t \in (0,T]. \tag{2.1}$$

Consider the formal stochastic evolution equation

$$dy + Aydt = A^\theta bf(y)dw(t), \quad y(0) = y_0 \in Y, \tag{2.2}$$

where $0\leq\theta<1/2$, $b\in Y$, $f(y)$ is a real Lipschitz continuous function on $\mathcal{D}(A^\eta)$ for some $0\leq\eta<1$ and $w(t)$ is a real standard Wiener process defined on $(\Omega,\mathcal{F},\mathcal{F}_t,P)$. We say that $y(t)$ is a mild solution of (2,2) if it is adapted to $\mathcal{F}_t$ and satisfies

$$y(t) = S(t)y_0 + \int_0^t A^\theta S(t-r)bf(y(r))dw(r). \tag{2.3}$$

In [11] we have shown

Proposition 2.1.

If $0\le\beta=\theta+\eta<1/2$, then there exists a unique mild solution to (2.2) in $L_2((0,T)\times\Omega;\mathcal{D}(A^\eta))\cap C([0,T];L_2(\Omega;Y))$. Moreover, for $t\in(0,T]$

$$E||y(t)||_\eta^2 \le N|y_0|^2/t^{2\eta} \text{ for some } N=N(T)>0 \tag{2.4}$$

and $A^{-\theta}y(t)$ has continuous sample paths in Y.

Remark 2.1.

If $\theta=0$, we can take $\eta=1/2$ in some cases see [3], [4].

Now consider the operator differential equation

$$d/dt\langle P(t)y,y\rangle + 2\langle P(t)y,-Ay\rangle + \langle My,y\rangle + f^2(y)\langle (A^*)^\theta P(t)A^\theta b,b\rangle = 0,$$

$$P(T) = F, \quad y \in \mathcal{D}(A), \tag{2.5}$$

and its integrated version

$$\langle P(t)y,y\rangle = \langle FS(T-t)y,S(T-t)y\rangle$$
$$+ \int_t^T [\langle MS(r-t)y,S(r-t)y\rangle + f^2(S(r-t)y)\langle (A^*)^\theta P(r)A^\theta b,b\rangle]dr,$$
$$y \in Y, \tag{2.6}$$

where $0\le M$, $F\in\mathcal{L}(Y)$. We denote by $\mathcal{D}(A^{-\theta})$ the dual space of $\mathcal{D}((A^*)^\theta)$, $\theta\ge0$. Then we have

Proposition 2.2.

There exists a unique solution to (2.6) with properties

(i) $0\le P(t)\in\mathcal{L}(\mathcal{D}(A^{-\theta}),\mathcal{D}((A^*)^\theta))$ and is strongly continuous on $[0,T)$ with $||P(t)||_\theta \le C/(T-t)^{2\theta}$ for some $C=C(T)>0$.

(ii) $0\le P(t)\in\mathcal{L}(Y)$ and is strongly continuous on $[0,T]$.

If, in particular, $F=0$, then $P(t)\in\mathcal{L}(\mathcal{D}(A^{-\theta}),\mathcal{D}((A^*)^\theta))$ is strongly continuous on $[0,T]$.

To prove this we introduce

$$\langle Q(t)z,z\rangle = \langle FS(T-t)A^\theta z,S(T-t)A^\theta z\rangle$$
$$+ \int_t^T [\langle MS(r-t)A^\theta z,S(r-t)A^\theta z\rangle + \langle Q(r)b,b\rangle f^2(S(r-t)A^\theta z)]dr. \tag{2.7}$$

Proposition 2.3.

There exists a unique solution to (2.7) which is nonnegative and strongly continuous on $[0,T)$ in $\mathcal{L}(Y)$ with $|Q(t)|\le C/(T-t)^{2\theta}$. Moreover, if $F=0$, then $Q(t)$ is strongly continuous on $[0,T]$.

Proof.

To show uniqueness we consider

$$\langle R(t)z,z\rangle = \int_t^T \langle R(r)b,b\rangle f^2(S(r-t)A^\theta z)dr.$$

This yields

$$|R(t)| \le C\int_t^T |R(r)|dr/(r-t)^{2\beta} \quad \text{for some} \quad C > 0.$$

Thus by Gronwall's inequality [6,p.6] we obtain $|R(t)|=0$ a.e.t. To establish a solution we set

$$\langle Q_0(t)z,z\rangle = \langle FS(T-t)A^\theta z,S(T-t)A^\theta z\rangle + \int_t^T \langle MS(r-t)A^\theta z,S(r-t)A^\theta z\rangle dr$$

and iterate a sequence of nonnegative operators

$$\langle Q_n(t)z,z\rangle = \langle Q_0(t)z,z\rangle + \int_t^T \langle Q_{n-1}(r)b,b\rangle f^2(S(r-t)A^\theta z)dr.$$

Then $Q_n(t)$ is well defined and is continuous on $[0,T)$. Moreover for each $t\varepsilon[0,T)$, $Q_n(t)$ is monotone increasing in n. But we have an estimate

$$|Q_n(t)| \le C_0/(T-t)^{2\theta} + C_1\int_t^T |Q_{n-1}(r)|dr/(r-t)^{2\beta}$$

for some $C_0,C_1>0$. Thus we obtain [6]

$$|Q_n(t)| \le C/(T-t)^{2\theta} \quad \text{for some} \quad C=C(T)>0 \text{ independent of } n.$$

Hence for each t there exists a strong limit $Q(t)$ of $Q_n(t)$ and $Q(t)$ has required properties.

The relation between $P(t)$ and $Q(t)$ is given by

Lemma 2.1.

$P(t)=(A^*)^{-\theta}Q(t)A^{-\theta}$ has all properties of Proposition 2.2. and is a solution of (2.6). Conversely, if $P(t)$ is a solution of (2.6), then $Q(t)=(A^*)^{\theta}P(t)A^{\theta}$ is well defined and is a solution of (2.7) with required properties.

Next we shall show that a certain quadratic functional of the mild solution is given in terms of $P(t)$.

Lemma 2.2.

Let $b_n \varepsilon D(A^\theta)$ with $b_n \to b$ in Y. Then $y_n(t;y_0) \to y(t;y_0)$ in $C([0,T];L_2(\Omega;Y))$ where $y_n(t;y_0)$ is the mild solution of (2.1) with b replaced by b_n.

Lemma 2.3.
Let $P_n(t)$ be the solution of (2.6) with b replaced by b_n. Then $P_n(t) \to P(t)$ strongly in $\mathcal{L}(\mathcal{D}(A^{-\theta}), \mathcal{D}((A^*)^{\theta})) \cap \mathcal{L}(Y)$.

Proposition 2.4.
For each $0 \leq t \leq T$

$$E\langle P(t)y(t;y_0),y(t;y_0)\rangle = E\langle Fy(T;y_0),y(T;y_0)\rangle + \int_t^T E\langle My(r;y_0),y(r;y_0)\rangle dr. \tag{2.8}$$

Proof.
If $b \varepsilon \mathcal{D}(A^{\theta})$, then the proof is standard [9]. The general case can be established by approximating b by $b_n \varepsilon \mathcal{D}(A^{\theta})$ and by using Lemmas 2.2 and 2.3. To examine the stability of mild solutions, it is useful to take more general initial value y_0. As in [11] we can show

Proposition 2.5.
For each $y_0 \varepsilon \mathcal{D}(A^{-\nu})$, $0 \leq \nu+\eta < 1/2$ there exists a unique adapted solution to (2.3) in $L_2((0,T)\times\Omega;\mathcal{D}(A^{\eta})) \cap C((0,T];L_2(\Omega;Y))$. Moreover,

$$E||y(t;y_0)||_{\eta}^2 \leq C||y_0||_{-\nu}^2/t^{2(\nu+\eta)}.$$

We extend Proposition 2.4.

Proposition 2.6.
For each $0 < t \leq T$ and $y_0 \varepsilon \mathcal{D}(A^{-\nu})$, $0 \leq \nu+\eta < 1/2$, the equality (2.8) holds. Moreover,

$$(P(0)y_0,y_0) = E\langle Fy(T;y_0),y(T;y_0)\rangle + \int_0^T E\langle My(t;y_0),y(t;y_0)\rangle dt,$$

where (,) denotes the duality between $\mathcal{D}(A^{-\nu})$ and $\mathcal{D}((A^*)^{\nu})$.
This follows readily from

Lemma 2.4.
Let $y_n \varepsilon Y \to y_0$ in $\mathcal{D}(A^{-\nu})$. Then

$$y(t;y_n) \to y(t;y_0) \text{ in } L_2((0,T)\times\Omega;\mathcal{D}(A^{\eta})) \cap C((0,T];L_2(\Omega;Y))$$

where the convergence in $C((0,T];L_2(\Omega;Y))$ means convergence in $C([\varepsilon,T];L_2(\Omega;Y))$ for any $\varepsilon > 0$ small.

3. Stability of mild solutions.
In this section we study asymptotic behavior of mild solutions. Our main result is the following.

Proposition 3.1.

The three statements below are equivalent:

(i) The mild solutions of (2.2) are stable i.e.,

$$\int_0^\infty E|y(t;y_0)|^2 dt < \infty \quad \text{for any} \quad y_0 \in D(A^{-\theta}). \tag{3.1}$$

(ii) There exists a solution $0 \leq P \in L(D(A^{-\theta}), D((A^*)^\theta)) \cap L(Y)$ to the Liapunov equation

$$2\langle Py, -Ay\rangle + \langle (A^*)^\theta P A^\theta b, b\rangle f^2(y) = -|y|^2, \quad y \in D(A). \tag{3.2}$$

(iii) For each $0 < T < \infty$

$$E|y(t;y_0)|^2 \leq \begin{cases} N||y_0||_{-\theta}^2/t^{2\theta}, \ 0 < t \leq T, \\ \qquad\qquad \text{for some } N=N(T) > 0. \\ Ne^{-at}||y_0||_{-\theta}^2 \ , \ t > T, \ a > 0, \end{cases} \tag{3.3}$$

Corollary 3.1.

The condition (3.1) may be replaced by

$$\int_0^\infty E|y(t;y_0)|^2 dt \leq K||y_0||_{-\theta}^2 \quad \text{for some} \quad K > 0.$$

If $y_0 \in Y$, then (i) or (ii) implies

$$E|y(t;y_0)|^2 \leq Ne^{-at}|y_0|^2 \quad \text{for some} \quad N \geq 1 \quad \text{and} \quad a > 0.$$

Corollary 3.2.

If P is a solution assumed in (ii), then

$$(Py_0, y_0) = \int_0^\infty E|y(t;y_0)|^2 dt.$$

Hence there exists at most one solution to (3.2).

Proof.

We assume (i), then

$$\int_0^\infty E|y(t;A^\theta z)|^2 dt < \infty \quad \text{for any} \quad z \in Y.$$

Thus from Proposition 2.5 with F=0, M=I, we have

$$\langle (A^*)^\theta P_T(0) A^\theta z, z\rangle = \int_0^T E|y(t;A^\theta z)|^2 dt,$$

where $P_T(t)$ is the solution of (2.6).

Thus $(A^*)^\theta P_T(0) A^\theta \uparrow Q \geq 0$ in $L(Y)$. Now let $P=(A^*)^{-\theta} Q A^{-\theta}$, then $P_T(0) \to P$ strongly in $L(Y) \cap L(D(A^{-\theta}), D((A^*)^\theta))$. Since $P_T(t)=P_{T-t}(0)$, we conclude that P satisfies (3.2). Hence (i) implies (ii). Conversely, suppose P is a solution of (3.2). Then by Proposition 2.5 we have

$$(Py_0,y_0) = E\langle Py(T;y_0),y(T;y_0)\rangle + \int_0^T E|y(t;y_0)|^2dt.$$

Thus (i) holds. Now we assume (iii), then (i) follow immediately. Finally we assume (i) and show (iii). Using Theorem 2.1 or Theorem 2.2 in [10] we obtain

$$E|y(t;y_0)|^2 \le \tilde{N}e^{-at}|y_0|^2 \quad \text{for any} \quad y_0 \in Y.$$

Then for each $0 < T < \infty$, we have

$$E|y(t;y_0)|^2 \le \tilde{N}e^{-a(t-T)}E|y(T;y_0)|^2 \quad \text{for any} \quad t \ge T.$$

On the other hand we have from Proposition 2.5

$$E|y(t;y_0)|^2 \le \tilde{C}||y_0||_{-\theta}^2/t^{2\theta},\ t\varepsilon(0,T] \quad \text{for some} \quad \tilde{C}=\tilde{C}(T) > 0.$$

Combining these two estimates we obtain (3.3).

Remark 3.1.

If $\int_0^\infty E|y(t;A^\theta b)|^2dt < \infty$, then $(P_T(0)A^\theta b,A^\theta b)$ converges to some number $\rho \ge 0$ and $P_T(0) \uparrow P \ge 0$ in $L(Y)$. Moreover

$$2\langle Py,-Ay\rangle + \rho f^2(y) = -|y|^2 \quad \text{for} \quad y \in D(A).$$

But we cannot in general conclude $\rho = (PA^\theta b,A^\theta b)$.

Remark 3.2.

In section 2 we have assumed that fractional powers A^γ are well defined. If this is not the case, we need to replace A by A+kI for some k>0 and to modify the system (2.2). But using Proposition 3.1 we can show that if the modified system is stable -A generates an exponentially stable semigroup. Thus as far as stability is concerned our assumption is not restrictive.

Now we study more general stability problems. First consider

$$\langle P(t)y,y\rangle = \int_t^T [|A^\nu S(r-t)y|^2 + (P(r)A^\theta b,A^\theta b)f^2(S(r-t)y)]dr. \qquad (3.4)$$

Proposition 3.2.

Suppose $\theta+\nu<1/2$ and $\theta\le\mu<1/2-\eta$. Then there exists a unique strongly continuous solution $0\le P(t)\varepsilon L(Y) \cap L(D(A^{-\mu}),D((A^*)^\mu))$ to (3.4). Moreover, for each $y_0 \in \mathcal{D}(A^{-\mu})$

$$(P(0)y_0,y_0) = \int_0^T E|A^{\nu}y(t;y_0)|^2 dt. \qquad (3.5)$$

Proof.
Similar to Propositions 2.2 and 2.6.

Proposition 3.3.
The following statements are equivalent :

(i) $\int_0^\infty E|A^{\nu}y(t;y_0)|^2 dt < \infty$ for any $y_0 \in \mathcal{D}(A^{-\mu})$. (3.6)

(ii) There exists a unique solution $0 \le P \in \mathcal{L}(Y) \cap \mathcal{L}(\mathcal{D}(A^{-\mu}),\mathcal{D}((A^*)^{\mu}))$ to

$$2<Py,-Ay>+<(A^*)^{\theta}PA^{\theta}b,b>f^2(y) = -|A^{\nu}y|^2, \quad y \in \mathcal{D}(A). \qquad (3.7)$$

(iii) $$E||y(t;y_0)||_{\nu}^2 \le \begin{cases} N||y_0||_{-\mu}^2/t^{2\alpha}, \quad \alpha=\theta+\nu, & t \le T \\ \qquad\qquad \text{for some } N > 0. & \\ Ne^{-at}||y_0||_{-\mu}^2, & t > T \end{cases} \qquad (3.8)$$

Proof.
Similar to Proposition 3.1.

Next we give sufficient conditions for stability. Note that the Liapunov equation (3.2) is equivalent to

$$(Py_0,y_0) = \int_0^\infty [|S(t)y_0|^2 + <(A^*)^{\theta}PA^{\theta}b,b>f^2(S(t)y_0)]dt, \quad y_0 \in \mathcal{D}(A^{-\theta}). \qquad (3.9)$$

As in section 2 we may equally consider

$$<Qy,y> = \int_0^\infty [|S(t)A^{\theta}y|^2 + <Qb,b>f^2(S(t)A^{\theta}y)]dt, \quad y \in Y. \qquad (3.10)$$

Proposition 3.4.
If the inequality

$$|b|^2 \int_0^\infty |f(S(t)A^{\theta}y)|^2 dt \le \rho|y|^2, \quad y \in Y \quad \text{for some} \quad 0 \le \rho < 1. \qquad (3.11)$$

holds, then there exists a unique nonnegative solution to (3.10) and hence to (3.9).

Corollary 3.3.
Suppose $S(t)y = \sum_{n=1}^{\infty} e^{-\lambda_n t}\psi_n<\psi_n,y>$ for some orthonormal basis $\{\psi_n\}$ and

$\lambda_n > 0$. If $f(y)=<f,y>$ for some $f \in Y$, then (3.11) is written

$$|b|^2 \sum_{n=1}^{\infty} f_n^2/\ 2\lambda_n^{1-2\theta} \leq \rho < 1,\ f_n = <f,\psi_n> . \qquad (3.12)$$

We may obtain sufficient conditions in terms of Liapunov functions. Let v(y) be twice Fréchet differentiable and define the formal differential generator of (2.2)

$$Lv(y) = <v_y(y),-Ay> + <(A^*)^\theta v_{yy}(y)A^\theta b,b>f^2(y),\ y \in D(A),$$

whenever the right hand side makes sense. We assume that A is self-adjoint and that $f(y)=<f,y>$, $f \in Y$. Then

$$\begin{aligned}|A^{-\theta}y|^2 &= -2<A^{1-2\theta}y,y> + |b|^2<f,y>^2, \quad y \in D(A)\\ &= -2|A^{1/2-\theta}y|^2 + |b|^2<A^{-(1/2-\theta)}f,A^{1/2-\theta}y>^2\\ &\leq -(2-|b|^2|A^{-(1/2-\theta)}f|^2)|A^{1/2-\theta}y|^2\\ &\leq -\alpha_\theta(2-|b|^2A^{-(1/2-\theta)}f|^2)|y|^2 \quad \text{for some} \quad \alpha_\theta > 0. \qquad (3.13)\end{aligned}$$

Hence if $|b|^2|A^{-(1/2-\theta)}f|^2 < 2$, then the system (2.3) is stable [9], [10]. Under the assumption of Corollary 3.3 this is equivalent to (3.12).

Remark 3.3.

Note that $v(y)=|y|^2$ cannot be used as a Liapunov function since $L|y|^2 = 2<y,Ay> + <A^\theta b,A^\theta b>f^2(y)$ does not make sense for $b \notin D(A^\theta)$.

Finally let $f \in D((A^*)^{-\eta})$ and let L_k be the class of real Lipschitz continuous functions g with $|g(x)|\leq k|x|$, $k>0$. Consider

$$dy + Aydt = kA^\theta b(f,y)dw(t), \quad y(0) = y_0 , \qquad (3.14)$$

$$dy + Aydt = A^\theta bg((f,y))dw(t), \quad y(0) = y_0 . \qquad (3.15)$$

Proposition 3.5.

The system (3.15) is absolutely stable in the class L_k i.e., stable for each $g \in L_k$, if and only if the system (3.14) is stable.

Proof.

Using Proposition 3.1 and the Liapunov function $<Py,y>$ we can show this as in [8].

Remark 3.4.

We may replace (2.2) by a more general system as in [11]. We may also consider control problems as in [4].

4. Examples.

We shall give a few simple examples.

Example 4.1.

Consider the stochastic parabolic equation of mixed type :

$$\begin{aligned} &\partial y(x,t)/\partial t = -A(x,\partial)y(x,t), \quad x \in O, \\ &y(x,0) = y_0(x), \\ &\partial y(x,t)/\partial n + a(x)y(x,t) = g(x)f(y(x,t))\dot{w}(t), \quad x \in \Gamma, \end{aligned} \tag{4.1}$$

where O is a bounded open domain in R^d with smooth boundary Γ, $-A(x,\partial)$ is a uniformly strongly elliptic operator of order two with smooth coefficient and $\dot{w}(t)$ is a white noise. We take $Y = L_2(O)$ and $D(A)$ as the closure in $H^2(O)$ of the subspace of $C^2(\overline{O})$ which consists of functions ψ satisfying the boundary condition $\partial\psi/\partial n + a\psi = 0$. Let $-A$ be the restriction of $-A(x,\partial)$ on $D(A)$, then it generates an analytic semigroup $S(t)$. Let M be the map : $L_2(\Gamma) \to L_2(O)$ defined by $y = Mg$ where y is the solution of $A(x,\partial)y = 0$, $\partial y/\partial n + ay = g$. The semigroup model in [11] for (4.1) is

$$\begin{aligned} y(t) &= S(t)y_0 + \int_0^t AS(t-r)Mgf(y(r))dw(r) \\ &= S(t)y_0 + \int_0^t A^\theta S(t-r)bf(y(r))dw(r), \end{aligned} \tag{4.2}$$

where $\theta = 1/4+\varepsilon$, $\varepsilon > 0$ and $b = A^{1-\theta}Mg \in Y$.

If we assume $f(y) = \langle f,y\rangle$, $f \in Y$, then there exists a unique adapted solution of (4.2) in

$$L_2((0,T)\times\Omega;D(A^{1/4-})) \cap C((0,T];L_2(\Omega;D(A^{1/4-}))) \cap C([0,T];L_{4-}(\Omega;Y)).$$

If we take $O=(0,1)$ and $A=-d^2/dx^2+1$, $D(A)=\{y\in H^2(0,1) : y'(0)=y'(1)=0\}$, then (4.1) may represent

$$\begin{aligned} &\partial y(x,t)/\partial t = \partial^2 y(x,t)/\partial x^2 - y(x,t) \\ &y(x,0) = y_0(x) \\ &\partial y(0,t)/\partial x = -k\langle f,y(x,t)\rangle\dot{w}(t), \quad \partial y(1,t)/\partial x = 0. \end{aligned} \tag{4.3}$$

Recall that the solution of

$$y'' - y = 0, \quad -y'(0) = a, \quad y'(1) = 0$$

is given by

$m(x) = a\cosh(1-x)/\sinh 1.$

Note that $S(t)y = e^{-t}<y,1> + \sum_{n=1}^{\infty} 2e^{-(1+n^2\pi^2)t}\cos n\pi x<\cos n\pi x,y>$

and $m_0 = \int_0^1 m(x)dx = a$

$$m_n = \int_0^1 \sqrt{2}\cos n\pi x\, m(x)dx = \sqrt{2}\, a/(1+n^2\pi^2).$$

If we assume $|f| = 1$, then (3.12) with $\theta = 1/2$ yields

$$k^2[1/2 + \sum_{n=1}^{\infty} \{n\pi/(1+n^2\pi^2)\}^2] < 1.$$

This is satisfied if $k^2[1/2 + (1/\pi^2)\sum_{n=1}^{\infty} 1/n^2] < 1.$

Thus if $k^2 < 3/2$, then the mild solution of (4.3) is stable.

Example 4.2.

Consider the stochastic parabolic equation

$$dy(x,t) = [\partial^2 y(x,t)/\partial x^2]dt + k\delta(x-\xi)<f,y>dw(t), \quad 0 < x,\ \xi < 1$$

$$y(0,t) = y(1,t) = 0, \quad y(x,0) = y_0(x). \tag{4.4}$$

In this case we take $Y=L_2(0,1)$ and $A=-d^2/dx^2$, $D(A)=H_0^1(0,1)\cap H^2(0,1)$. The semigroup model for (4.4) is

$$y(t) = S(t)y_0 + \int_0^t AS(t-r)gk<f,y(r)>dw(r), \tag{4.5}$$

where g is the Green's function

$$g(x,\xi) = \begin{cases} (1-\xi)x, & 0 \le x < \xi \\ (1-x)\xi, & \xi \le x \le 1. \end{cases}$$

Since $\int_0^1 \sqrt{2}\sin n\pi x\, g(x,\xi)dx=\sqrt{2}\sin n\pi\xi/(n\pi)^2$,

we have $g \in D(A^{3/4-})$. Hence (4.5) is written

$$y(t) = S(t)y_0 + \int_0^t A^{\theta}S(t-r)bk<f,y(r)>dw(r),$$

where $\theta=1/4+\varepsilon$, $\varepsilon>0$ and $b=A^{1-\theta}g$. Hence all the conditions of Proposition 2.1 are satisfied. To obtain a sufficient condition for stability we assume $|f|=1$ and apply (3.12) with $\theta=1/2$.

$$k^2 \sum_{n=1}^{\infty} (1/n^2)/\pi^2 < 1.$$

Hence if $k^2 < 6$, then the system (4.4) is stable.

Example 4.3.
Consider the stochastic parabolic equation

$$dy(x,t) = [\partial^2 y(x,t)/\partial x^2]dt + k\,b(x)y(x_0,t)dw(t), \quad 0 < x, x_0 < 1$$
$$y(x,0) = y_0(x), \; y(0,t) = y(1,t) = 0. \tag{4.6}$$

We take Y and A as in Example 4.2. Since $f(y) = y(x_0)$ is continuous on $D(A^\eta)$, $\eta>1/4$, we can apply Proposition 2.1 with $\theta = 0$. To obtain a sufficient condition for stability we assume $|b| = 1$ and apply Proposition 3.4. Then

$$k^2|b|^2\int_0^\infty |f(S(t)y)|^2 dt = k^2\int_0^\infty \Big(\sum_{n=1}^{\infty} \sqrt{2}\sin n\pi x_0 e^{-n^2\pi^2 t} \langle \sqrt{2}\sin n\pi x, y\rangle\Big)^2 dt$$

$$\le k^2|y|^2\int_0^\infty \sum_{n=1}^{\infty} 2(\sin n\pi x_0)^2 e^{-2n^2\pi^2 t} dt$$

$$\le k^2|y|^2/6 .$$

Hence if $k^2 < 6$, then the system (4.6) is stable.

References.

1. A.V. Balakrishnan, Applied Functional Analysis, Springer-Verlag, New York, 1976.
2. R.F. Curtain, Stochastic distributed systems with point observation and boundary control : An abstract theory, Stochastics, 3(1979), 85-104.
3. G. Da Prato, Some results on linear stochastic evolution equations in Hilbert spaces by semigroup method, Stochastic Analysis and Applications, 1(1983), 57-88.
4. G. Da Prato and A. Ichikawa, Stability and quadratic control for linear stochastic equations with unbounded coefficients, Boll. UMI (submitted).
5. U.G. Haussmann, Asymptotic stability of the linear Ito equation in infinite dimensions, J.Math.Anal.Appl., 65(1978), 219-235.
6. D. Henry, Geometric Theory of Semilinear Parabolic Equations, Lecture Notes in Math., 840, Springer-Verlag, Berlin, 1981.
7. A. Ichikawa, Dynamic programming approach to stochastic evolution equations, SIAM J.Control Optimiz., 17(1979), 152-174.
8. __________, Absolute stability of a stochastic evolution equation, Stochastics, 11(1983), 143-158.
9. __________, Semilinear stochastic evolution equations : Boundedness, stability and invariant measures, Stochastics, 12(1984), 1-39.

10. A. Ichikawa, Equivalence of L_p stability and exponential stability for a class of nonlinear semigroups, Nonlinear Analysis TMA, 8(1984) to appear.
11. __________, A semigroup model for parabolic equations with boundary and pointwise noise, Workshop "Stochastic Space-Time Models and Limit Theorems", University of Bremen, November, 1983.
12. I. Lasiecka, Unified theory for abstract parabolic boundary problems: A semigroup approach, Appl.Math.Optimiz., 6(1980), 281-333.
13. I. Lasiecka and R. Triggiani, Feedback semigroups and cosine operators for boundary feedback parabolic and hyperbolic equations, J. Diff.Eqns., 47(1983), 246-272.
14. __________________________, Dirichlet boundary control problem for parabolic equations with quadratic cost: Analyticity and Riccati feedback synthesis, SIAM J.Control Optimiz., 21(1983), 41-67.
15. J. Zabczyk, On decomposition of generators, SIAM J.Control Optimiz., 16(1978), 523-534.
16. _________, On stability of infinite dimensional stochastic systems, Probability Theory, Banach Center Publications, Vol. 5(1979), Z. Ciesielski (ed.), Warsaw, 273-281.

STOCHASTIC PARTIAL DIFFERENTIAL EQUATIONS AND RENORMALIZATION THEORY (STOCHASTIC QUANTIZATION)

G. JONA-LASINIO(*)

P.K. MITTER

Laboratoire de Physique Théorique et Hautes Energies
4, Place Jussieu, Tour 16, 1er Etage
F-75230 PARIS CEDEX 05 (France)

In recent years one has seen a growing interest in the theory of stochastic partial differential equations stimulated by problems in theoretical and mathematical physics.

The equations considered are typically of the form

$$\frac{\partial \varphi}{\partial t} = -\frac{1}{2}(-\Delta+1)\varphi - \frac{1}{2}V'(\varphi) + \frac{dw}{dt} \tag{1}$$

where φ is a field variable depending on time and a d-dimensional space variable $\underline{x}$, Δ is the Laplacian, $V(\varphi)$ is a non linear function of φ bounded below, e.g. an even polynomial in φ, $\frac{dw}{dt}$ is a white noise formally defined by the correlation function

$$E\left(\frac{dw(\underline{x},t)}{dt}\frac{dw(\underline{x}',t')}{dt'}\right) = \delta(t-t')\delta(\underline{x}-\underline{x}') \tag{2}$$

For d=1 it is possible to develop a rigorous theory of eq.(1) along the lines of the classical theory of Ito stochastic ordinary differential equations and for this case we refer the reader to the paper /1/ where one can find also motivations and a bibliography.

The real difficulties are encountered for $d \geq 2$ and are closely connected with those met by physicists over several decades in the attempt of constructing a quantum theory of fields. In fact equations of the form (1) have been introduced /2/ as a possible approach to quantum field theory (QFT) where one is interested in the properties of a measure formally defined by

$$d\mu = e^{-S(\varphi)}\prod_{\underline{x}} d\varphi(\underline{x}) \tag{3}$$

where

$$S(\varphi) = \int d\underline{x}\left[\frac{1}{2}\langle\varphi,(-\Delta+1)\varphi\rangle + V(\varphi)\right] \tag{4}$$

(*)Permanent Address : Dipartimento di Fisica, Università "La Sapienza", Piazza A. Moro 2, 00185 ROMA (Italy)

The connection between (3), (4) and (1) is given by the fact that (3) is a formal stationary measure for the process described by (1).

It is therefore natural to try in the study of eq.(1) or related equations the methods developed by rigorous QFT, the so-called constructive approach /3/.

It is convenient to formulate our problem in a slightly more general form. Everything we say for the moment is formal. We consider the equation

$$\frac{\partial\varphi}{\partial t} = -\frac{1}{2}A\varphi - \frac{1}{2}B(\varphi) + \sigma\frac{dw}{dt} \tag{5}$$

where A and σ are linear operators independent of φ, while $B(\varphi)$ is non linear. Consider now the solution φ_0 of the linear problem

$$\frac{\partial\varphi_0}{\partial t} = -\frac{1}{2}A\varphi_0 + \sigma\frac{dw}{dt} \tag{6}$$

Then the solution of (5) can be described by a measure μ_φ which is connected to μ_{φ_0} by the Girsanov-Cameron-Martin formula

$$\frac{d\mu_\varphi}{d\mu_{\varphi_0}} = \exp\left\{-\frac{1}{2}\int_0^t <\sigma^{-1}B(\varphi_0), dw> - \frac{1}{8}\int_0^t <\sigma^{-1}B(\varphi_0), \sigma^{-1}B(\varphi_0)>dt'\right\} \tag{7}$$

We say (following a well established tradition) that eq.(5) possesses a weak solution if the right hand side of (7) makes sense for some initial condition $\varphi(\underline{x},0)$ and

$$E\left(\frac{d\mu_\varphi}{d\mu_{\varphi_0}}\right) = 1 \tag{8}$$

In the case of eq.(1), where $A = -\Delta+1$, $\sigma=1$ and $B(\varphi) = V'(\varphi)$ it is easily seen that (7) has no meaning for $d \geqslant 2$.

It is at this point that one tries the renormalization procedures of QFT.

However instead of considering the special difficulties of eq.(1) we want to observe that if our aim is to study equations which admit the measure (3) as stationary distribution, the general class described by (5) and (7) contains other possibilities which may be easier than eq.(1). In fact it is easily seen that if $\sigma^{-2}A$ is a symmetric operator and B is of the form $\sigma^2\frac{\delta V}{\delta\varphi}$, the formal equilibrium measure of (5) is

$$d\mu = \exp\left\{-\int d\underline{x}\left[\frac{1}{2}<\varphi, \sigma^{-2}A\varphi> + V(\varphi)\right]\right\}\prod_{\underline{x}} d\varphi(\underline{x}) \tag{9}$$

Therefore for $\sigma^{-2}A = -\Delta + 1$ we obtain (3).

We can then exploit the freedom in the choice of σ and A to minimize the difficulties.

We now discuss an example in d = 2.

Our choice will be, A = +1 and $\sigma = (-\Delta+1)^{-1/2}$. Furthermore, to be definite, we take

$$V(\varphi) = \frac{\lambda}{4} :\varphi^4: \qquad \lambda > 0 \tag{10}$$

The dots indicate the Wick product /3/ with respect to the Gaussian measure of covariance $(-\Delta + 1)^{-1}$, that is

$$V(\varphi) = \frac{\lambda}{4}\left[\varphi^4 - 6\,\sigma^2(0)\,\varphi^2 + 3(\sigma^2(0))^2\right] \tag{11}$$

Of course (11) is a formal expression as $\sigma^2(0)$ is infinite. The Girsanov formula (7) now becomes (*)

$$\frac{d\mu_\varphi}{d\mu_{\varphi_0}} = \exp\Big\{-\frac{\lambda}{2}\int_0^t \langle(-\Delta+1)^{-1/2} :\varphi_0^3:, dw\rangle_\Omega - \frac{\lambda}{8}\int_0^t dt' \,\|(-\Delta+1)^{-1/2} :\varphi_0^3:\|_\Omega^2\Big\} \tag{12}$$

where Ω is a domain in two-dimensional space.

Using the methods of constructive field theory it is now possible to give a precise meaning to (12). In order to apply such methods it is necessary to make (12) more explicit by performing the stochastic integral. This can be done and we obtain

$$\int_0^t \langle(-\Delta+1)^{-1/2} :\varphi_0^3:, dw\rangle_\Omega = \frac{1}{4}\int_\Omega d\underline{x} :\varphi_0^4(\underline{x},t): - \frac{1}{4}\int_\Omega d\underline{x} :\varphi_0^4(\underline{x},0): + \frac{1}{2}\int_0^t dt' \int_\Omega d\underline{x} :\varphi_0^4(\underline{x},t'): \tag{13}$$

$\varphi_0(\underline{x},0)$ is the initial configuration.

From the well known Nelson estimates /3/ it now follows that all the terms in the exponential (12) are well defined random variables with respect to μ_{φ_0}, and that the exponential itself is in $L^p(d\mu_{\varphi_0})$ for $\forall p$. It can also be verified that $E_{\mu_{\varphi_0}}(d\mu_\varphi/d\mu_{\varphi_0}) = 1$. This establishes the existence of a weak solution for the equation

$$d\varphi = -\frac{1}{2}\left(\varphi + \lambda(-\Delta+1)^{-1} :\varphi^3:\right)dt + (-\Delta+1)^{-1/2}dw \tag{14}$$

The analysis can be pushed much further. For additional results and a rigorous mathematical treatment we refer the reader to /4/.

(*) The Wick product is now with respect to the covariance of φ_0 .

REFERENCES

/1/ W.G. FARIS, G. JONA-LASINIO, "Large Fluctuations for a Nonlinear Heat Equation with Noise", J. Phys. A 15, 3025 (1982);

/2/ G. PARISI, WU YONG-SHI, Scientia Sinica 24, 483 (1981);

/3/ B. SIMON, "The $P(\phi)_2$ Euclidean (Quantum) Field Theory", Princeton, 1974 ;

J. GLIMM, A. JAFFE, "Quantum Physics", Springer, New York 1981.

/4/ G. JONA-LASINIO, P.K. MITTER, "Stochastic Quantization in Field Theory : a Rigorous Study", Preprint 1984.

ON THE REGULARITY OF THE SOLUTIONS OF STOCHASTIC PARTIAL DIFFERENTIAL EQUATIONS

by

A.S.Ustunel

Introduction

In many branches of mathematics,one meets often the so-called stochastic partial differential equations:quantum physics,nonlinear filtering, transport theory,etc.In this work we expose some regularity results for such equations of even order governed by the finite dimensional,standard Wiener processes.In order to study these equations in a satisfactory way,we treat them as the stochastic differential equations in the space of the distributions and use some techniques of the stochastic calculus of the semimartingales with values in nuclear spaces elaborated in[8]-[11] .

I.Preliminaries

Let p and q_i be random partial differential operators of constant order 2m and m_i,i=1,...,N, with

$$p=p(t,x,\omega,\partial_x),\quad q_i=q_i(t,x,\omega,\partial_x)$$

having C^∞-coefficients with uniformly bounded derivatives on the compact sets of $\mathbb{R}^d$. Suppose that u is a $\mathcal{D}'(\mathbb{R}^d)$-valued semimartingale satisfying the following equation:

(I.1) $du_t=-pu_t\,dt + q_iu_t\,dW_t^i + dh_t$

where h is any given semimartingale with values in the space of the C^∞-functions on $\mathbb{R}^d$, denoted by $\mathcal{E}(\mathbb{R}^d)$.We say that $(p,q_i;i=1,\ldots,N)$ is hypoelliptic, if any solution of (I.1) is undistinguishable from a semimartingale with values in $\mathcal{E}(\mathbb{R}^d)$.Since we are dealing with a very special type of semimartingales,we shall adapt our notations consequently: Let u be a semimartingale with values in a separable Hilbert space H having the following decomposition

$$u_t=u_0+\int_0^t a_s\,ds + \int_0^t b_s^i\,dW_s^i \quad ,\ t\in[0,1] \quad ,$$

where $W_t=(W_t^1,\ldots,W_t^N)$ is a standard Wiener process in $\mathbb{R}^N$,$N<\infty$,a and b^i are adapted,measurable processes in H satisfying

$$\int_0^1\|a_s\|_H^2\,ds + \sum_i\int_0^1\|b_s^i\|^2\,ds < +\infty \quad \text{a.s.}$$

We shall denote by $S_2(H)$ the space of such semimartingales for which the following norm is finite

$$|u|_H^2= E(\|u_0\|_H^2+\int_0^1(\|a_s\|_H^2+\sum_i\|b_s^i\|_H^2)ds\ ) \ .$$

A weaker topology on $S_2(H)$ is defined by

$$\|u\|_H^2 = E\int_0^1\|u_s\|_H^2\,ds$$

and the corresponding scalar product is denoted by $(.,.)_H$. $S_2^0(H)$ is the subspace of $S_2(H)$ such that $u\in S_2^0(H)$ if $u_0=0$ a.s. If H is a Sobolev space then instead of putting the whole notation as above we will just write the index of the space at the place of H. For $u\in S_2(H)$, a and b^i are denoted respectively by D_+u and $\partial_{W^i}u, i=1,\dots,N$, and for such semimartingales, D_+u and $\partial_{W^i}u$ are well defined (cf. [4]). If $H=\mathbb{R}$ we note $S_2(\mathbb{R})$ by S_2 and if F is any nuclear space, $S_2(F)$ and $S_2\tilde{\otimes}F$ will be the equivalent notations, where $S_2\tilde{\otimes}F$ denotes the completed projective tensor product topology when S_2 is equipped with $|.|$-topology. The nuclearity of F implies that $S_2\tilde{\otimes}F$ is topologically isomorphic to the completed tensor product of S_2 and F under the topology of bi-equicontinuous convergence (cf. [6]) and any u $S_2\tilde{\otimes}F$ can be represented as a Hilbert space-valued (say H) semimartingale such that H can be continuously imbedded in F and $u\in S_2(H)$ (cf. [9]).

II.A condition for hypoellipticity

Let us give first the precise hypothesis on p and q_i: let U be an open domain of $\mathbb{R}^d$ and $p(t,x,\omega,\partial_x)$ be a (random) partial differential operator of constant degree 2m, m>0. We suppose that p has measurable, adapted, C^∞-coefficients $a_{\alpha,\beta}(t,x,\omega)$ such that, for any $\mu\in\mathbb{N}^d$, $K\subset\subset U$ (i.e., K is a compact subset of U)

$$\sup_{x\in K}\sup_{t\in[0,1]}|\partial_x^\mu a_{\alpha,\beta}(t,x,\omega)|\leqslant c \quad \text{a.s.},$$

where c is a constant depending only on K and μ. We suppose that $q_i = q_i(t,x,\omega,\partial_x)$ satisfy the same kind of hypothesis as p with degree $p_i=m_i\leqslant m$. We make also the following, essential hypothesis about p :

(H) There exists s>0 such that, for any $K\subset\subset U$, there exists constants $c,\bar{c}=c,\bar{c}(K,s)>0$ with

$$\|u\|^2_{m+s-1}\leqslant c(B(u,u)+\bar{c}\|u\|_o^2)$$

for any u

for any u $S_2^0\tilde{\otimes}\mathcal{D}_K(U)$ ($\mathcal{D}_K(U)=\{\varphi\in\mathcal{D}(U);\operatorname{supp}\varphi\subset K\}$) where B(u,u) is defined by the following bilinear form:

$$B(u,v)=E\int_0^1((D_++p_s)u_s,v_s)_o\,ds+1/2\sum_i E\int_0^1(\partial_{W^i}u_s,\partial_{W^i}v_s)_o\,ds\ .$$

In the following, for $\alpha\in\mathbb{R}$, we shall denote by T^α the properly supported part of the pseudodifferential operator $(I-\Delta)^{\alpha/2}$ who defines the Sobolev norm of order α (cf. [7]).

Proposition II.1

Let $\alpha\in\mathbb{R}$ and suppose that (H) is satisfied. Then, for any $u\in S_2^0\tilde{\otimes}\mathcal{D}_K(U)$, one has $\|u\|^2_{\alpha+m+s-1}\leqslant c(B(u,T^{2\alpha}u)+c_1\|u\|^2_{\alpha+m-1})$

where c and c_1 are some positive constants depending only on α, s and K.

The proof is an application of the calculus of the pseudodifferential operators. Similarly one can also prove the following

Proposition II.2

Suppose that u $S_2^0 \tilde{\otimes} \mathcal{D}'(U)$ is a solution of the following stochastic partial differential equation:

(II.1) $du_t = (-p + 1/2(-1)^{m_i+1} q_i^2) u_t \, dt + q_i u_t \, dW_t^i + dh_t$

where $h \in S_2 \tilde{\otimes} \mathcal{E}(U)$. Let φ, $\varphi_1 \in \mathcal{D}(U)$ with $\varphi_1 = 1$ on the support of φ. If, for some $\alpha \in \mathbb{R}$, $\|\varphi_1 u\|_{\alpha+m-1} < +\infty$

then we have

$$\|J_\varepsilon \varphi u\|^2_{\alpha+m+s-1} \leq c_{\alpha,m} (\|\varphi_1 u\|^2_{\alpha+m-1} + \|\varphi_1 u\|_\alpha \|\varphi_1 D_+ h\|_\alpha + \sum_i \|\varphi_1 \partial_{W^i} h\|^2_{\alpha+m})$$

where $c_{\alpha,m} > 0$ is independent of $\varepsilon > 0$ and $u^\varepsilon = J_\varepsilon \varphi u$, J_ε being a Friedrich's mollifier (cf. [7]).

This result is tedious to prove and its proof will appear in [12] . If we iterate this result and apply the closed graph theorem, we obtain:

Theorem II.1

For any solution of the equation II.1 in $S_2 \tilde{\otimes} \mathcal{D}'(U)$, there exists a semi-martingale $\hat{u}$ in $S_2 \tilde{\otimes} \mathcal{E}(U)$ which is undistinguishable from u (once injected in $\mathcal{D}'(U)$).

Remark

Note that there is a supplementary term at the drift of the equation II.1 which is analogous to Stratonovitch's correction in spite of the fact that q_i's are not semimartingales. In particular, if we want to study the equation without this term, we have to replace p by $\tilde{p}$ in hypothesis (H), where

$$\tilde{p} = p + (-1)^{m_i+1} \, 1/2 \, q_i^2 .$$

This means that the stochastic integral part "diminishes" the hypo-ellipticity of the equation.

Upto now we have supposed that the solutions of the equation II.1 were in $S_2 \tilde{\otimes} \mathcal{D}'(U)$, i.e., that $D_+ u$ and $\partial_{W^i} u$ were square integrable with respect to $dt \times dP$ on $[0,1] \times \Omega$ for i=1,...,N. In order to treat the general case we will work with the processes indexed by $\mathbb{R}_+$ and modify the hypothesis (H) in the following manner:

(H') For any bounded stopping time T, for any $K \subset\subset U$, one has

$$\|u\|^2_{m+s-1,T} \leq c(B_T(u,u) + \bar{c}\|u\|^2_{0,T})$$

for any $u \in S_2^0 \tilde{\otimes} \mathcal{D}_K(U)$, where

$$\|u\|^2_{r,T} = E\int_0^T \|u_s\|^2_r \, ds , \quad r \in \mathbb{R} ,$$

$$B_T(u,v)=E\int_0^T((D_+ + p_s)u_s,v_s)_0\,ds+(1/2)E\int_0^T\sum_i(\partial_{Wi}u_s,\partial_{Wi}v_s)_0\,ds$$

and c and $\bar{c}$ are positive constants depending only on K,s and T .
Now we have the following

Theorem II.2

Suppose that u is a semimartingale with values in $\mathcal{D}'(U)$ satisfying the following equation:

$$du_t=(-p+(-1)^{m_i+1}\ (1/2)\ q_i^2)u_t dt+q_i u_t\ dW_t^i+dh_t$$

where h is an $\mathcal{E}(U)$-valued semimartingale having the following decomposition:

$$dh_t= D_+h_t dt\ + \partial_{Wi}h_t\ dW_t^i\ .$$

If the hypothesis (H') is satisfied,then there exists a semimartingale $\hat{u}$ with values in $\mathcal{E}(U)$ which is undistinguishable from u.

An example:

The simplest example of the operator p satisfying the hypothesis (H') is any dterministic operator $p(x,\partial_x)$ which is elliptic of order 2m.In fact,using Garding's inequality,we have

$$\|u\|_{m,T}^2 \leqslant c((pu,u)_{0,T}+\|u\|_{0,T}^2\ )$$

for any bounded stopping time T and $u\in S_2^0\tilde{\otimes}\mathcal{D}_K(U)$ .Using Itô's formula:

$$\|u_t\|_0^2 =2\int_0^t(u_s,du_s)_0+\int_0^t\sum_i\|\partial_{Wi}u_s\|_0^2\ ds\ ,$$

hence

$$E\int_0^T(D_+u_s,u_s)_0 ds\ =(1/2)E(\|u_T\|_0^2\ -\int_0^T\sum_i\|\partial_{Wi}u_s\|_0^2\ \ ds\ )$$

and finally

$$\|u\|_{m,T}^2\leqslant c(((p+D_+)u,u)_{0,T}\ +\|u\|_{0,T}^2\ +(1/2)\ \sum_i\|\partial_{Wi}u\|_{0,T}^2\ )$$

$$= c(B_T(u,u)+\ \|u\|_{0,T}^2\ ).$$

II.Q.E.D.

References:

[1] Chaleyat-Maurel,M.and Michel,D.(1984).Hypoellipticity theorems and conditional laws.Zeitschrift f.Wahrsch.und verw.Gebiete,Vol.65, p.573-599.

[2] Martias,C.(1984).Sur les supports des processus à valeurs dans des espaces nucléaires.Preprint.

[3] Métivier,M.and Pellaumail,J.(1980).Stochastic Integration.Academic Press,New York.

[4] Nelson,E.(1967).Dynamical Theories of Brownian Motion.Princeton University Press,Princeton,New Jersey.

[5] Pardoux,E.(1982).Equation du filtrage nonlinéaire de la prédiction et du lissage.Stochastics,Vol.6,p.192-231.

[6] Schaefer,H.H.(1971).Topological Vector Spaces.Graduate Texts in Math.Springer.Berlin-Heidelberg-New York.

[7] Treves,F.(1980).Introduction to Pseudodifferential and Fourier Integral Operators,Vol.1.Plenum Press.New York and London.

[8] Ustunel,A.S.(1982).Some applications of stochastic integration in infinite dimensions.Stochastics,7,p.255-288.

[9] Ustunel,A.S.(1982).A characterization of semimartingales on nuclear spaces.Z.f.W.,60,p.21-39.

[10] Ustunel,A.S.(1982).Stochastic integration on nuclear spaces and its applications.Annales de l'Inst.Henri Poincaré,Vol.18,p.165-200.

[11] Ustunel,A.S.(1984).Distributions-valued semimartingales and applications to control and filtering.Lecture Notes in Control and Information Sciences,Vol.61,p.314-325.

[12] Ustunel,A.S.(1984).Hypoellipticity of the stochastic partial differential operators.Preprint.

Author's adress: Laboratoire de Probabilités. Université Paris VI
4. place Jussieu - 75230 PARIS CEDEX 05

2. FLUCTUATIONS AND ASYMPTOTIC ANALYSIS OF FINITE AND INFINITE DIMENSIONAL SYSTEMS

ASYMPTOTIC ANALYSIS OF MULTILEVEL STOCHASTIC SYSTEMS

Donald A. Dawson

Department of Mathematics and Statistics,
Carleton University,
Ottawa, Canada K1S 5B6.

1. INTRODUCTION.

Ensembles of exchangeably interacting subsystems arise in many fields such as statistical physics, chemical kinetics, population biology, neurobiology, economics and sociology. Most of the pioneering work on such systems has taken place in the context of statistical physics; however in recent years the resulting ideas and methodology have begun to permeate other fields including those listed above.

In statistical physics there are at least three separate sources for interest in these models. The first, which was pioneered by Kac, McKean and Tanaka, arose from stochastic models associated with the classical Boltzmann equation (refer to Sznitman (1983) for an up-to-date account of this development). A second but related source is the study of kinetic equations such as the Vlasov equation; the mathematical foundations of this development have been studied in recent years by Dobrushin (1979) and Braun and Hepp (1977). The third source of these models arises in the study of phase transitions and critical phenomena such as those arising in models of ferromagnetism. The study of critical point behavior and fluctuations lead to difficult and deep problems in even the simplest lattice models such as the Ising model. For this reason a family of models known as "mean-field" or Curie-Weiss models have been used to explore these phenomena in a simpler and more tractable setting. On the other hand in chemical kinetics, Prigogine and others (c.f. Nicolis and Prigogine (1977)) have introduced mean-field stochastic models to investigate transitions in dissipative systems. Closely related models have been proposed in economics by Aoki (1980) and in neurobiology by Amari (1972).

The objective of this paper is to introduce a natural extension of the mean-field-like family of models. In particular we consider a system which is organized into a multilevel hierarchy. Each level of this hierarchy consists of a number of interacting subsystems at the next lower level. For example, P. Auger (1983) has proposed a hierarchical model in ecology with three levels of organization cor-

---

(1) This research was supported by the Natural Sciences and Engineering Research Council of Canada.

responding to individuals, populations and ecosystems. In this paper we consider a general formulation of multilevel systems with emphasis on the case in which at each level there are a large number of interacting subsystems at the next lower level. This emphasis differs from that of Dyson's hierarchical model which consists of infinitely many hierarchical levels each of which consists of two subsystems at the next lower level (c.f. Collet and Eckmann (1978)). As an illustration we consider in detail a two-level ferromagnetic model and investigate the nature of the fluctuations. The ultimate goal of this research is to describe the nature of the flow of organization and control between levels in the hierarchy.

The approach of this paper is to formulate the relevant stochastic models in the context of measure-valued processes and then to establish a number of probabilistic limit theorems. In real hierarchical systems there are a finite number of individual subsystems at each level of the hierarchy. Nevertheless insight into the qualitative behavior of such systems can be obtained by studying the idealized limit as the number of subsystems per level approaches infinity. In this setting the finite population effects are captured in the study of the random perturbations, fluctuations and large deviations about the infinite limit.

In this paper we present statements of a number of results and indicate some of the basic ideas involved in the proofs. Detailed proofs and a systematic exposition of this material will appear in Dawson (1984).

2. THE MULTILEVEL SYSTEM AND LAW OF LARGE NUMBERS.

2.1. FORMULATION OF A TWO-LEVEL SYSTEM.

To model a two-level system consider a system of $M.N$ individuals which are organized into M subsystems or ensembles each containing N individuals. Let D denote the state space for individuals at the lower level.

Let $x_{ij}(t) \in D$ denote the state of the jth individual in the ith cell at time t. Assume that D is a complete separable metric space. Let M_1 denote the family of probability measures on D furnished with the topology of weak convergence. Then M_1 can again be metrized to be a complete separable metric space. Let M_2 denote the family of probability measures on M_1 again with the topology of weak convergence.

Let

$$X_i := N^{-1} \sum_{j=1}^{N} \delta_{x_{ij}(t)} \in M_1, \quad \text{and} \tag{2.1}$$

(2.2) $X(t) := M^{-1} \sum_{j=1}^{M} \delta_{X_i(t)} \in M_2.$

Let $\Omega := C([0,\infty),M_2)$. The canonical process $X:[0,\infty) \to M_2$ is defined by $X(\omega,t) = \omega(t)$. The distribution of an M_2-valued diffusion process is prescribed by a mapping $\mu \to P_\mu$ from M_2 into $P(\Omega)$, the space of probability measures on Ω with initial condition $P_\mu(X(0) = \mu) = 1$.

Let $\{x_{ij}:i = 1,\dots,M;\ j = 1,\dots,N\}$ form a collection of biexchangeable D-valued random variables, that is,

$$\{x_{ij}:\ i = 1,\dots,M;\ j = 1,\dots,N\} \overset{\text{law}}{\sim} \{x_{\Pi(i,j)}:\ i = 1,\dots,M;\ j = 1,\dots,N\}$$

where $\Pi(i,j) = \pi_0(i)\pi_i(j)$, π_0 is a permutation of $\{1,\dots,M\}$ and for each $i = 1,\dots,M$, $\pi_i(.)$ is a permutation of $\{1,\dots,N\}$.

Let

(2.3) $X(0) = M^{-1} \sum_{i=1}^{M} \delta_{X_i(0)} \in M_2,$ where

$$X_i(0) = N^{-1} \sum_{j=1}^{N} \delta_{x_{ij}(0)} \in M_1.$$

$M_2^{M,N} \subset M_2$ denotes the collection of measures on M_1 of the form (2.3). Note that there is a natural equivalence between $M_2^{M,N}$ and collections of M,N bi-exchangeable random variables.

For $i = 1,\dots,M$ and $j = 0,\dots,M$ let $\{w_{ij}(t):t \geq 0\}$ denote a family of independent d'-dimensional Wiener processes. The system $\{x_{ij}:t \geq 0\}$ is assumed to satisfy the system of stochastic differential equations:

(2.4)
$$dx_{ij}(t) = \beta_{M,N}(x_{ij}(t),X_i(t),X(t))dt + \sigma_0(x_{ij}(t),X_i(t),X(t))dw_{ij}(t) + \sigma_1(X_i(t),X(t))dw_{i0}(t),$$

for $i = 1,\dots,M$, $j = 1,\dots,N$, where $D = R^d$ and

$\beta_{M,N}:R^d \times M_1 \times M_2 \to R^d$ and for $k = 0,1$,

$\sigma_k:R^d \times M_1 \times M_2 \to d'\times d$ matrices.

To keep the exposition brief, in the remainder of the paper we restrict our attention to the case in which σ_0,σ_1 are constant and

$$\beta_{M,N}(x,\mu,\nu) = \int b_1^{(N)}(x,y)\mu(dy) + \iint b_2^{(M,N)}(x,y)\mu(dy)\nu(d\mu)$$

where $b_1^{(N)}$ and $b_2^{(M,N)}$ are Lipschitz continuous functions on R^{2d}. The term simple ensemble refers to the case in which $M = 1$ and $b_2^{(M,N)} = 0$, $b_1^{(N)} = b_1$.

PROPOSITION 2.1. Let $D \in R^d$. Then the solution of the system (2.4) with M_2-valued initial condition can be represented as an M_2-valued Markov diffusion process.

Proof. The existence of a solution to the system (2.4) follows from the Lipschitz assumption. The solution $\{X(t):t \geq 0\}$ is an $M_2^{M,N}$-valued stochastic process with continuous sample paths. In view of the form of the coefficients and the biexchangeability of the system of Wiener processes $\{w_{ij}: i = 1,\ldots,M,\ j = 0,\ldots,N\}$, the conditional distribution $P(X(t+s) \in .|\{x_{ij}(t)\}) = P(X(t+s) \in .|X(t))$. Therefore $\{X(t):t \geq 0\}$ is an $M_2^{M,N}$-valued Markov process and the proof is complete. □

Let $f,h_1 \in C_b^2(R^1)$, the space of functions on R^1 having bounded second derivatives and $h_2 \in C_K^2(R^d)$, the space of functions having continuous second derivatives and compact support. For $\nu \in M_2$, define:

$$(2.5)\quad F_{f,h_1,h_2}(\nu) := f[\int_{M_1} h_1(\int_{R^d} h_2(x)\mu(dx))\,\nu(d\mu)] = f(<< h_1(< h_2,.>),\nu>>).$$

Let D_2 denote the collection:

$$D_2 = \{F_{f,h_1,h_2}: f,h_1 \in C_b^2(R^1),\ h_2 \in C_K^2(R^d)\}.$$

For $F_{f,h_1,h_2} \in D_2$, $\nu \in M_2^{M,N}$, define

$$(2.6)\quad \underline{\underline{G}}_{M,N} F_{f,h_1,h_2}(\nu) :=$$

$$f'(\nu)<<h_1'(\mu)<\beta_{M,N}(.,\mu,\nu).\nabla h_2,\mu>,\nu>> + \tfrac{1}{2}\sigma_0^2 f'(\nu)<<h_1'(\mu)<\Delta h_2,\mu>,\nu>>$$

$$+ (\sigma_0^2/2N)f'(\nu)<< h_1''(\mu)<|\nabla h_2|^2,\mu>,\nu>> + (\sigma_0^2/2MN)f''(\nu)<<h_1'(\mu)<|\nabla h_2|^2,\mu>,\nu>>$$

$$+ (\sigma_1^2/M)f''(\nu)<<h_1'(\mu)<h_2,\mu>^2,\nu>> + \tfrac{1}{2}\sigma_1^2 f'(\nu)<<h_1''(\mu)|<\nabla h_2,\mu>|^2,\nu>>$$

$$+ \tfrac{1}{2}\sigma_1^2 f'(\nu)<<h_1'(\mu)<\Delta h_2,\mu>,\nu>> .$$

PROPOSITION 2.2.

Assume that the coefficients $b_1^{(N)}$ and $b_2^{(M,N)}$ satisfy a Lipschitz condition. Then the probability law $\{P_\mu:\mu \in M_2^{M,N}\}$ of $\{X(t):t \geq 0\}$ defined by (2.2), (2.4) is characterized as the unique solution to the martingale problem on $C([0,\infty),M_2^{M,N})$ associated with the pair $(D_2,\underline{\underline{G}}_{M,N})$, that is, for $F_{f,h_1,h_2} \in D_2$,

$$F_{f,h_1,h_2}(X(t)) - \int_0^t \underline{\underline{G}}_{M,N} F_{f,h_1,h_2}(X(s))ds \quad \text{is a } P_\mu\text{-martingale.}$$

Proof. From the Lipschitz condition the system of stochastic differential equations (2.4) has a unique strong solution. The uniqueness to the martingale problem follows by a Yamada-Watanabe theorem. To verify that $X(.)$ is a solution to the martingale problem let:

$$\Psi(\{x_{ij}(t)\}) := f(M^{-1}\sum_{i=1}^{M} h_1[N^{-1}\sum_{j=1}^{N} h_2(x_{ij}(t))]) .$$

The by Itô' lemma,

$$
\begin{aligned}
&d\Psi(t) = \\
&(MN)^{-1}\sum_{i,j} f'h_1'\,(\nabla h_2.dx_{ij}(t)) + \tfrac{1}{2}\sigma_0^2\{(MN)^{-2}\sum_{i,j} f''h_1^2|\nabla h_2|^2(x_{ij}(t)) \\
&+ (MN^2)^{-1}\sum_{i,j} f'h_1''|\nabla h_2|^2(x_{ij}(t)) + (MN)^{-1}\sum_{i,j} f'h_1'\,\Delta h_2(x_{ij}(t))\}dt \\
&+ \tfrac{1}{2}\sigma_1^2\{(MN)^{-2}\sum_{i,j,k} f''(h_1')^2(\nabla h_2(x_{ij}).\nabla h_2(x_{ik})) \\
&+ (MN^2)^{-1}\sum_{i,j,k} f'h_1''\,(\nabla h_2(x_{ij}).\nabla h_2(x_{ik})) + (MN)^{-1}\sum_{i.j} f'h_1'\,\Delta h_2(x_{ij})\}\,dt \\
&+ (MN)^{-1}[\sigma_0\sum_{i,j} f'h_1'(\nabla h_2(x_{ij}).dw_{ij}(t)) + \sigma_1\sum_{i,j} f'h_1'(\nabla h_2(x_{ij}).dw_{i0}(t))] \\
&= \underset{=}{G}_{M,N}\, F_{f,h_1,h_2}(X(t))dt.
\end{aligned}
$$

Therefore

$$
\begin{aligned}
&F_{f,h_1,h_2}(X(t)) - \int_0^t \underset{=}{G}_{M,N}\, F_{f,h_1,h_2}(X(s))ds \\
&= (MN)^{-1}\sum_{i=1}^{M}\sum_{j=0}^{N}\int_0^t f'h_1'(\,h_2(x_{ij}(t)).dw_{ij}(t))
\end{aligned}
$$

and the latter is a martingale. This completes the proof. □

2.2 EXAMPLES.

EXAMPLE 2.1. THE CONTINUOUS MODEL OF FERROMAGNETISM.

This one dimensional example is an extension of the model studied by Dawson (1983) and is given by:

$$\text{(2.7)}\qquad \beta_{M,N}(x,\mu,\nu) := x - x^3 + \theta_1(\langle x,\mu\rangle - x) + (\theta_2/N^{\alpha})(\langle\langle\langle x,\mu\rangle,\nu\rangle\rangle - \langle x,\mu\rangle)$$

where $\theta_1,\theta_2 \geq 0$ and $\alpha \geq 0$. The term $(x - x^3)$ corresponds to the motion of particles in a two-well potential. The term with coefficient θ_1 describes the mean-field interaction and when $\theta_1 > 0$ there is a force acting on each particle towards the centre of gravity of the ensemble. Finally, the term with coefficient $\theta_2 > 0$ corresponds to an attractive force between the ensembles forming the higher level. Although the term (θ_2/N^α) drops out in the law of large numbers limit below when $\alpha > 0$, it does play a key role in the study of the critical fluctuations.

EXAMPLE 2.2. A GIBB'S MODEL.

This model which has been studied by Tamura (1983) generalizes the form of the previous example but imposes fairly stringent regularity conditions on the coefficient. It is given by:

$$(2.8)\quad \beta_{M,N}(x,\mu,\nu) := -\nabla\Phi_1(x) - \theta_2\int\nabla\Phi_2(x,y)\mu(dy) - (\theta_3/N^\alpha)\int\nabla\Phi_3(\mu,\mu_1)\nu(d\mu_1)$$

where Φ_1 and Φ_2 are given by

$$\Phi_1(x) = (\alpha/2)|x|^2 + \phi_1(x),\ \alpha > 0,\ \phi_1 \in \mathcal{S}(R^d), \text{ and}$$

$$\Phi_2(x,y) = \phi_2(x-y)\ ,\ \theta_3 \geq 0 \text{ and } \phi_2 \in \mathcal{S}(R^d) \text{ and}$$

$$\Phi_3(\mu_1,\mu_2) = \int\phi_3(x_1-x_2)\mu_1(dx_1)\mu_2(dx_2)\ ,\ \phi_3 \in \mathcal{S}(R^d).$$

2.3 THE MULTILEVEL LAW OF LARGE NUMBERS (THE $M,N \to \infty$ LIMIT).

STATEMENT: Let $\{X_{M,N}(t):t \geq 0\}$ denote the $M_2^{M,N}$-valued process with generator $\underline{\underline{G}}_{M,N}$ defined by (2.6). Assume that as $M,N \to \infty$,

$$(2.9)\quad X_{M,N}(0) \to \delta_\mu \text{ with } \mu \in M_1(R^d).$$

Then (under conditions described below),

$$(2.10)\quad X_{M,N} \to Y \text{ as } M,N \to \infty,$$

in the sense of weak convergence of probability measures on $C([0,\infty),M_2)$ where $Y(.)$ is a deterministic M_2-valued process given by the probability law of the colution of the M_1-valued martingale problem associated with the stochastic evolution equation:

$$(2.11)\quad \langle\mu(t),\phi\rangle - \langle\mu(0),\phi\rangle - \int_0^t \langle\mu(s),L(\mu(s))\phi\rangle ds = \sigma_1\int_0^t (\nabla\mu(s).dw(s))$$

where $\{w(s):s \geq 0\}$ is a d-dimensional Wiener process and

$$(2.12)\quad L(\mu)\phi = \beta(.,\mu,P(\mu)).\nabla\phi + \tfrac{1}{2}(\sigma_0^2 + \sigma_1^2)\Delta\phi\ , \text{ where}$$

$$\beta(.,.,,) = \lim_{M,N\to\infty}\beta_{M,N}(.,.,.) \text{ is assumed to exist,}$$

and where $P(\mu)$ denotes the probability law of the random measure μ.

Outline of Proof. The first step involves using a standard criteria to verify that the laws $P^{M,N}$ of $X_{M,N}$ are relatively compact.

The second step is to prove that any limit point P_μ satisfies the $(D_2,\underline{\underline{G}}_\infty)$-martingale problem where:

$$(2.13) \quad \underline{\underline{G}}_\infty F_{f,h_1,h_2}(\nu) := f'(\nu)<<h_1'(\mu)<\beta(x,\mu,\nu).\nabla h_2(x),\mu>,\nu>>$$
$$+ \tfrac{1}{2}(\sigma_0^2 + \sigma_1^2)f'(\nu)<<h_1'(\mu)<\Delta h_2,\mu>,\nu>>$$
$$+ \tfrac{1}{2}\sigma_1^2 f'(\nu)<<h_1''(\mu)|<\nabla h_2,\mu>|^2,\nu>> .$$

To verify this first note that

$$(2.14) \quad F_{f,h_1,h_2}(X(t)) - \int_0^t \underline{\underline{G}}_{M,N} F_{f,h_1,h_2}(X(s))ds$$

is a P_μ-martingale. Let P_μ be any limit point of the $\{P_\mu^{M,N}:M,N = 1,2,3,\ldots\}$. In virtue of the Lipschitz hypothesis on β we can verify a uniform integrability condition on the functional (2.14). In addition we know that $\underline{\underline{G}}_{M,N} F_{f,h_1,h_2}(\nu) = \underline{\underline{G}}_\infty F_{f,h_1,h_2}(\nu) + o(M+N)$. Thes facts together with the continuity of β on the sets $\{|x| < k\}\times M_1\times M_2$ can be used to show that

$$F_{f,h_1,h_2}(X(t)) - \int_0^t \underline{\underline{G}}_\infty F_{f,h_1,h_2}(X(s))ds \quad \text{is a } P_\mu\text{-martingale.}$$

The final step involves proving that the $\underline{\underline{G}}_\infty$-martingale problem has a unique solution. This can be done under two sets of conditions. In the first case we assume that β does not depend on ν. The the uniqueness of the $\underline{\underline{G}}_\infty$-martingale problem is established by first using results from the theory of stochastic evolution equations to show that the solution of (2.11) has a unique strong solution (c.f. Krylov and Rozovskii (1979)) and then using an infinite dimensional version of the Yamada-Watanabe theorem. The second case is under the assumption that $\sigma_1 = 0$. In the latter case the limit $Y(.)$ is given by $Y(t) = \delta_{\mu(t)}$ where $\mu(t)$ is the solution of the weak form equation:

$$(2.15) \quad <\mu(t),\phi> - <\mu(0),\phi> = \int_0^t <\mu(s),L(\mu(s))\phi> ds$$

where $\phi \in C_K^2(R^d)$ and

$$L(\mu)\phi = \beta(.,\mu,\delta_\mu).\nabla\phi + \tfrac{1}{2}\sigma_0^2 \Delta\phi .$$

In the latter case uniqueness is proved as for the McKean Vlasov limit (see Funaki (1983), Gärtner (1983)).

2.4. THE $M \to \infty$ LIMIT.

In this section we consider the case $\sigma_1 = 0$ and in which the number of lower level ensembles is effectively infinite, that is, $M \to \infty$.

PROPOSITION 2.3.

As $M \to \infty$, the processes $\{X_{M,N}(t):t \geq 0\}$ converge in the sense of weak convergence of M_2-valued processes to a process of the form:

$$(2.17) \quad Z(t) := \delta_{X_N(t)} ,$$

where $\{X_N(t):t\geq 0\}$ is the M_1-valued Markov diffusion process with nonlinear generator as follows. For $f \in C_b^2(R^1)$, $g \in C_K^2(R^d)$,

$$F_{f,g}(\mu) := f(\langle\mu,g\rangle),$$

$$(2.18) \quad \underline{\underline{G}}_{\infty N}F_{f,g}(\mu) = f'(\langle\mu,g\rangle)[\langle\mu,\beta(.,\mu,\delta_\mu).\nabla g\rangle + (\sigma_0^2/2)\langle\mu,\Delta g\rangle]$$
$$+ (\sigma_0^2/2N)f''(\langle\mu,g\rangle)\langle\mu,\nabla g.\nabla g\rangle .$$

Then $\{X_N(t):t \geq 0\}$ is characterized as the unique solution to the $\underline{\underline{G}}_{\infty N}$-martingale problem.

Proof. Refer to Dawson (1984).

In the case of the ferromagnetic model the limit $X_N(.)$ corresponds to the solution of the system of nonlinear stochastic differential equations:

$$(2.19) \quad dx_j(t) = (-x_j^3 + x_j)dt + \sigma_0 dw_j(t) + \theta_1(\bar{x}(t) - x_j)dt$$
$$+ (\theta_2/N^\alpha)\, E(\bar{x}(t) - x_j)dt$$

where $\bar{x}(t) := N^{-1} \sum_{j=1}^{N} x_j(t)$.

The term involving (θ_2/N^α) can be interpreted as a "macroscopic input signal".

3. RANDOM PERTURBATION, LINEARIZED EVOLUTION AND GAUSSIAN FLUCTUATIONS.

In Section 2 we considered the $M_2^{M,N}$-valued process $\{X^{M,N}(t):t \geq 0\}$ whose probability law can be characterized as the unique solution to the martingale problem associated with $(D_2,\underline{\underline{G}}_{M,N})$. From (2.6), $\underline{\underline{G}}_{M,N}$ has the form:

$$(3.1) \quad \underline{\underline{G}}_{M,N} = \underline{\underline{G}}_\infty + (1/N)\underline{\underline{G}}_1 + (1/M)\underline{\underline{G}}_2 + (1/MN)\underline{\underline{G}}_3 ,$$

where $\underline{\underline{G}}_\infty$ is a "first order" operator and corresponds to a deterministic M_2-valued evolution $Y(.)$. In this section we assume that $\sigma_1 = 0$ so that $Y(.)$ satisfies (2.15).

For large M and N we can view $\{X^{M,N}(t):t \geq 0\}$ as a "small random perturbation" of the deterministic evolution associated with (2.15), that is, $\underline{\underline{G}}_\infty$. The finite system effects can then be studied by investigating the fluctuations and large deviations around the limiting evolution. In this section we present a brief introduction to this viewpoint and describe some preliminary results.

It is assumed in this section that there is an invariant probability measure $p_0(dx) = p_0(x)dx$ and we investigate the perturbations about this fixed point for the limit dynamics.

3.1. LINEARIZED EVOLUTION.

Let $p_0(x)$ denote an invariant probability density and assume that

$$(3.2)\qquad \int_{R^d} \Phi_2(x,y)p_0(y)dy = 0.$$

The limit dynamics for the Gibb's model is prescribed by: (assuming $\sigma_0^2 = 1$)

$$(3.3)\qquad \partial u/\partial t = \nabla.[\tfrac{1}{2}\nabla u(t,x) + u(t,x)\nabla\Phi_1(x) - \theta_2 u(t,x)\nabla\int_{R^d} \Phi_2(x,y)u(t,y)dy].$$

Under assumption (3.2) the linearized evolution around $p_0(x)dx$ is given by:

$$(3.4)\qquad \partial u/\partial t = L^*_{p_0} u, \quad \text{where}$$

$$L^*_{p_0} u := \nabla.[\tfrac{1}{2}\nabla u + u\nabla\Phi_1 - \theta_2 p_0 \nabla(\int_{R^d} \Phi_2(x,y)u(t,y)dy)].$$

3.2. THE INVARIANCE PRINCIPLE IN REAL TIME FOR THE MULTILEVEL SYSTEM.

In this section we consider the multilevel system defined by (2.6) under the assumption that $\sigma_1 = 0$ and investigate the fluctuations around the invariant probability $p_0(x)dx$. Let $h(\mu) := h_1(\langle h_2,\mu\rangle)$. Then the linearized evolution $\nu(t)$ around $p_0(x)dx$ can be described by:

$$(3.5)\qquad \langle\langle h,\nu(t)\rangle\rangle - \langle\langle h,\nu(0)\rangle\rangle$$

$$= \tfrac{1}{2}\sigma_0^2 \int_0^t \langle\langle h_1'(\langle h_2,\mu\rangle)\langle\Delta h_2,\mu\rangle,\nu(s)\rangle\rangle ds$$

$$+ \int_0^t \langle\langle h_1'(\langle h_2,\mu\rangle)\langle\beta(x,p_0,\delta_{p_0}).\nabla h_2(x),\mu\rangle,\nu(s)\rangle\rangle ds$$

$$+ \int_0^t \langle\langle h_1'(\langle h_2,\mu\rangle)\langle(\delta\beta(x,\mu,\nu)/\delta\nu),\nu(s)).\nabla h_2(x),\mu\rangle,\delta_{p_0}\rangle\rangle\, ds$$

$$:= \int_0^t \langle\langle L_{p_0} h,\nu(s)\rangle\rangle\, ds.$$

In order to investigate the "centred fluctuations" around δ_{p_0} it suffices to consider the functions in D_2 of the form F_{f,h_1,h_2} satisfying $F_{f,h_1,h_2}(\delta_{p_0}) = 0$. In particular we assume that f,h_1,h_2 are chosen to satisfy $\langle h_2,p_0\rangle = 0$, $h_1(0) = 0$ and $f(0) = 0$. Note that for such a function

$$F_{f,h_1,h_2}(X_{M,N}(t) - \delta_{p_0}) = F_{f,h_1,h_2}(X_{M,N}(t)).$$

PROPOSITION 3.1. (THE SIMPLE ENSEMBLE INVARIANCE PRINCIPLE)

Let $\{X_N(t):t \geq 0\}$ denote the simple ensemble process (i.e. $M = 1$) and consider the centred process:

$$(3.6)\quad Y_N(t) := N^{\frac{1}{2}}[X_N(t) - p_0(.)dx] .$$

Then as $N \to \infty$, Y_N converges to a limit process $Y(.)$ in the sense of weak convergence of probability measures on $C([0,\infty),\mathcal{S}')$. The generalized process Y is Gaussian and is given as the solution of the linear stochastic evolution equation

$$(3.7)\quad dY(t) = L_{p_0}^{*}Y(t)dt + dW(t) ,$$

where $\{W(t):t \geq 0\}$ is an $\mathcal{S}'$-valued Wiener process with

$$(3.8)\quad \mathrm{Cov}(\langle W(t),\phi\rangle,\langle W(t),\psi\rangle) = \sigma_0^2\, t\int(\nabla\phi(x).\nabla\psi(x))p_0(x)dx.$$

Proof. Refer to Dawson (1983), Kusuoka and Tamura (1983), Sznitman (1983) and Tanaka (1983) for several methods of proof.

Tanaka (1983) and Sznitman (1983) have studied an extension of this result. They considered the fluctuations around the McKean-Vlasov limit of the resulting empirical measures on $C([0,T],R^d)$; that is, they proved that for a class of functionals F on $C([0,T],R^d)$,

$$(3.9)\quad Y_N^F := N^{-\frac{1}{2}} \sum_{j=1}^{N} [F(x_j(.)) - \int F(x(.))\mu_{p_0}(dx(.))]$$

is asymptotically normal.

PROPOSITION 3.2 (THE WITHIN CELL FLUCTUATION LIMIT FOR THE MULTILEVEL SYSTEM)

Let

$$(3.10)\quad Z_{M,N}(t) := X_{M,N}(t,N^{\frac{1}{2}}d\mu) - \delta_{p_0}(N^{\frac{1}{2}}d\mu).$$

Then as $M,N \to \infty$,

$$Z_{M,N}(t) \to \delta_{z(t)} ,\quad z(t) \in M_1,$$

where $z(t)$ denotes the law of $Z(t)$ where $\{Z(t):t \geq 0\}$ is a Gaussian $\mathcal{S}'$-valued process. The process $Z(.)$ can be represented as the solution of the stochastic evolution equation (3.7), where

$$L_{p_0}^{*}u := - \nabla.(\beta(.,p_0,\delta_{p_0})u) + \tfrac{1}{2}\sigma_0^2\,\Delta u - \nabla.\langle(\delta\beta(.,p_0,\delta_{p_0})/\delta\mu(y)),u(y)\rangle .$$

Proof. Details will appear in Dawson (1984). Is is based on implementing the scaling by setting

$$F_{f,h_1,h_2}(Z_{M,N}(t)) = F^N_{f,h_1,h_2}(X_{M,N}(t)) \quad\text{where}\quad F^N_{f,h_1,h_2} := F_{f,h_1,N^{\frac{1}{2}}h_2} .$$

There are two other scaling which lead to Gaussian fluctuations about $p_0(x)dx$.

These are:

(3.11) $Z^{B}_{M,N}(t) := (MN)^{\frac{1}{2}} [X_{M,N}(t,d\mu) - \delta_{p_0}(d\mu)]$ (BETWEEN CELL FLUCTUATIONS)

$$F_{f,h_1,h_2}(Z^{B}_{M,N}(t)) = f((MN)^{\frac{1}{2}} \langle\langle h_1(\langle h_2,\mu\rangle, X_{M,N}(t)\rangle\rangle), \text{ and}$$

(3.12) $Z^{T}_{M,N}(t) := M^{\frac{1}{2}}[X_{M,N}(t,N^{\frac{1}{2}}d\mu) - \delta_{p_0}(N^{\frac{1}{2}}d\mu)]$ (TWO LEVEL FLUCTUATIONS)

$$F_{f,h_1,h_2}(Z^{T}_{M,N}(t)) = F_{f,M^{\frac{1}{2}}h_1,N^{\frac{1}{2}}h_2}(X_{M,N}(t)) .$$

<u>REMARK 3.1.</u>

When the linearized system $L^{*}_{p_0}$ is stable, that is, except for the simple eigenvalue 0 the spectrum of $L^{*}_{p_0}$ is contained in $(-\infty,-\varepsilon]$ for some $\varepsilon > 0$, then the fluctuation process $\{Y(t):t \geq 0\}$ defined by (3.7) evolves towards a Gaussian generalized random field which is invariant under the evolution as $t \to \infty$. This implies that the fluctuations are approximately Gaussian at <u>all</u> time scales. In Dawson (1984) it is demonstrated that this is not always the case. In fact it is shown that at a critical point where the set of invariant probabilities for the McKean-Vlasov limit has a bifurcation, the linearized system is not stable and has a non-trivial null space (complementary to p_0).

<u>REFERENCES.</u>

1. S.I. Amari. Characteristics of random nets of neuron-like elements, IEEE Trans. Systems Man Cybernetics, SMC-2 (1972), 643-657.

2. M. Aoki. Dynamics and control of systems composed of a large number of similar subsystems, in Dynamic Optimization and Mathematical Economics, P.T. Liu, ed., (1980), Plenum Press, New York.

3. P. Auger. Hierarchically organized populations: interactions between individual, population and ecosystem levels, Math. Biosciences 65 (1983), 269-289.

4. W. Braun and K. Hepp. The Vlasov dynamics and its fluctuation in the 1/N limit of interacting particles, Comm. Math. Phys. 56 (1977), 101-113.

5. P. Collet and J.-P. Eckmann. A renormalization group analysis of the hierarchical model in statistical mechanics, Lecture Notes in Physics 74 (1978), Springer-Verlag.

6. D.A. Dawson. Critical dynamics and fluctuations for a mean-field model of cooperative behavior, J. Stat. Phys. 31 (1983), 29-85.

7. D.A. Dawson. Stochastic ensembles and hierarchies, (1984), in preparation.

8. R.L. Dobrushin. Vlasov equations, Funct. Anal. and Appl. 13 (1979), 115.

9. T. Funaki. A certain class of diffusion processes associated with non-linear parabolic equations, Tech. Rep. 43, Center for Stochastic Processes, Univ. of North Carolina, (1983), Chapel Hill.

10. N.V. Krylov and B.L. Rozovskii. Stochastic evolution equations, Itogi Nauki i Techniki, Seriya Sovremennye Problemy Matematiki 14 (1979), 71-146. (Translated in J. Soviet Math. (1981), 1233-1277).

11. S. Kusuoka and Y. Tamura. The convergence of Gibbs measures associated with mean field potentials, preprint.

12. H.P. McKean, Jr. Propagation of chaos for a class of nonlinear parabolic equations, in Lecture Series in Differential Equations, Vol. 2 (1969), 41-57, Van Nostrand Reinhold.

13. G. Nicolis and I. Prigogine. Self-organization in non-equilibrium systems, Wiley-Interscience.

14. K. Oëlschlager. A martingale approach to the law of large numbers for weakly interacting stochastic processes, Preprint 181, SFB123, Universitat Heidelberg, 1982.

15. A.S. Sznitman. Equations de type Boltzmann spatialment homogenes, (1983), to appear.

16. A.S. Sznitman. An example of nonlinear diffusion process with normal reflecting boundary conditions and some related limit theorems, Laboratoire de Probabilités CNRS 224 (1983), Paris.

17. A.S. Sznitman. A fluctuation result for non linear diffusions,(1983), preprint.

18. Y. Tamura. On asymptotic behavior of the solution of a non-linear parabolic equation associated with a system of diffusing particles with interactions, (1983), preprint.

19. H. Tanaka. Limit theorems for certain diffusion processes with interaction, (1982), to appear.

20. H. Tanaka and M. Hitsuda (1981). Central limit theorem for a simple diffusion model of interacting particles, Hiroshima Math. J. 11, 415-423.

SPACE SCALING LIMIT THEOREMS FOR INFINITE PARTICLE BRANCHING BROWNIAN MOTIONS WITH IMMIGRATION*

Luis G. Gorostiza
Centro de Investigación y de Estudios Avanzados, I.P.N.
México 07000, D.F., México

INTRODUCTION

The limit behavior of infinite particle branching diffusions in $\mathbb{R}^d$, with and without immigration, has been studied extensively by Dawson and Ivanoff [2,3,5,10,11], and others. Dawson and Ivanoff have proved in particular central limit theorems under the scaling $x \to Kx$ of $\mathbb{R}^d$ as $K \to \infty$, the time scale and all parameters of the system being held fixed; these limit theorems concern the random field at a fixed time t, including $t = \infty$ when a steady-state random field exists.

In this paper we study the limit behavior of essentially the same models considered by Dawson and Ivanoff, under the same space scaling, but our emphasis is put on the time evolution of the system; thus our framework is weak convergence of probability measures on a space of trajectories. This approach yields additional information on the limit behavior of the system. Our methods are different: they apply for fixed $t < \infty$ Brillinger's central limit theorem for spatially homogeneous random fields subject to a mixing condition [1], which is based on cumulant calculations (see also Ivanoff [13]), while we prove convergence of finite-dimensional distributions by Lévy's continuity theorem. Tightness of the processes is obtained by results of Holley and Stroock [7,8], and Mitoma [17].

We restrict our attention to the case where the particle migration process is Brownian motion, which allows us to make use of calculations contained in [6]. Dawson and Ivanoff consider a more general migration: a time-homogeneous Markov process with transition density function $p(t,x,y)$, symmetric in x,y, and depending only on $x - y$; since these are the only properties we use in the case of Brownian motion, it should be possible to extend our results to the more general case. Nevertheless, as we shall see, under the present scaling the limit fluctuation process of the system is independent of the particular migration process; moreover, in the critical case it makes no difference whether the particles migrate or not.

A special feature of the central limit theorems under the space scaling considered by the abovementioned authors and here is that the covariance functional of the limit is concentrated on the diagonal. This, which reflects the asymptotic independence of fluctuations at large distances, is in sharp contrast to the space-time scaling limit of the critical case (see Dawson [3], and Holley and Stroock [7]), and to the high density limit without any scalings (see Gorostiza [6]), where the covariance functional is spread throughout the whole $\mathbb{R}^d$, representing long range dependence of fluctua-

*Research supported in part by CONACyT grant PCCBBNA 002042.

tions.

Ivanoff has extended her results to the multitype case [12]. It would be desirable to study also the asymptotic time evolution of such a model.

Here we give some of our results and a sketch of the methods of proof. A detailed version, including other scalings, will appear elsewhere.

THE BRANCHING RANDOM FIELD WITH IMMIGRATION

The model under consideration consists of an infinite system of particles in d-dimensional Euclidean space $\mathbb{R}^d$ evolving in the following way. At time $t=0$ the particles are distributed according to a homogeneous Poisson random field with intensity $\gamma>0$. As time elapses, each particle independently migrates according to a time-homogeneous Markov process which preserves the spatial homogeneity of the system and which here is taken to be Brownian motion, and independently undergoes branching after an exponentially distributed lifetime with parameter V, with branching law $\{p_n\}_{n=0,1,\dots}$, which is assumed to have a finite third moment. The offspring particles appear at the same location where their parent branched, and evolve obeying the previous rules. In addition, particles from an external source immigrate into $\mathbb{R}^d$ according to a homogeneous space-time Poisson field with intensity $\beta>0$, i.e., the probability that a particle appears in an element of volume Δx in the time interval $(t,t+\Delta t)$ is $\beta\Delta x\Delta t+o(\Delta x\Delta t)$. Each immigrant particle generates an independent branching Brownian motion as described above. The object of study is the measure-valued process $N\equiv\{N_t,\ t\geq 0\}$, where $N_t(A)$ is the number of particles contained in the bounded Borel set $A\subset\mathbb{R}^d$ at time t. The process N is called a branching random field with immigration. The intensities β and γ can be made 0 in order to isolate the effects of the initial field and the immigration field, respectively.

Under the scaling $x\to Kx$ of $\mathbb{R}^d$, $K\geq 1$, we denote $\phi^K(x)=\phi(x/K)$ for a real function ϕ on $\mathbb{R}^d$, and N^K stands for N on the scaled space; thus

$$\int\phi\, dN_t^K=\int\phi^K dN_t. \tag{1}$$

The mean EN_t^K being a Borel measure on $\mathbb{R}^d$, we may consider the centered normalized random field $M^K\equiv\{M_t^K,\ t\geq 0\}$ defined by

$$M_t^K=K^{-d/2}(N_t^K-EN_t^K). \tag{2}$$

Both processes N^K and M^K have versions in the Skorohod-type space $D([0,\infty),\mathcal{S}'(\mathbb{R}^d))$, where $\mathcal{S}'(\mathbb{R}^d)$ is the Schwartz space of tempered distributions on $\mathbb{R}^d$, i.e. the topological dual of the space $\mathcal{S}(\mathbb{R}^d)$ of infinitely differentiable functions on $\mathbb{R}^d$ which are rapidly decreasing at infinity. The reason for working with $\mathcal{S}'(\mathbb{R}^d)$ is that although M^K takes values in the space $\mathcal{M}(\mathbb{R}^d)$ of signed tempered Radon measures on $\mathbb{R}^d$, the limit as $K\to\infty$ takes values in $\mathcal{S}'(\mathbb{R}^d)\setminus\mathcal{M}(\mathbb{R}^d)$.

THE LIMIT THEOREMS

We use the following notations.

$\langle\ \rangle$: the canonical bilinear form on $\mathcal{S}'(\mathbb{R}^d)\times\mathcal{S}(\mathbb{R}^d)$.

W: the standard Gaussian white noise on $\mathbb{R}^d$, i.e. W is the random element of $\mathcal{S}'(\mathbb{R}^d)$ whose characteristic functional is

$$Ee^{i\langle W,\phi\rangle} = \exp\{-\frac{1}{2}\int\phi^2(x)dx\}, \quad \phi\in\mathcal{S}(\mathbb{R}^d).$$

λ: the Lebesgue measure on $\mathbb{R}^d$; thus $\langle\lambda,\phi\rangle = \int\phi(x)dx$ for integrable ϕ.

m_1 and m_2: the mean and the second factorial moment, respectively, of the branching law $\{p_n\}$.

$\alpha = V(m_1 - 1)$: the Malthusian parameter of the branching process; the cases $\alpha = 0, >0$ and <0 are termed critical, supercritical and subcritical, respectively.

$\Longrightarrow$: convergence in distribution.

$$p(t,x,y) = e^{-\|x-y\|^2/2t}/(2\pi t)^{d/2}, \ t>0; \quad p(0,x,y) = \delta(x-y), \quad x,y\in\mathbb{R}^d.$$

The first result is the law of large numbers for the system.

THEOREM 1. *For each* $\phi\in\mathcal{S}(\mathbb{R}^d)$,

$$K^{-d}\langle N_t^K,\phi\rangle \to \mathcal{J}(\alpha,\beta,\gamma;t)\langle\lambda,\phi\rangle, \quad 0\le t<\infty, \quad \text{as} \quad K\to\infty$$

in the mean square, where

$$\mathcal{J}(\alpha,\beta,\gamma;t) = \begin{cases} e^{\alpha t}[\gamma+\beta(1-e^{-\alpha t})/\alpha] & \text{if } \alpha\neq 0 \\ \gamma+\beta t & \text{if } \alpha = 0 \end{cases}$$

The next result gives the spatial functional central limit theorem for the system and properties of the limit process.

THEOREM 2. $M^K \Longrightarrow M$ *as* $K\to\infty$, *where* $M\equiv\{M_t, t\ge 0\}$ *is an* $\mathcal{S}'(\mathbb{R}^d)$*-valued centered Gaussian process with covariance functional*

$$\mathrm{Cov}(\langle M_s,\phi\rangle,\langle M_t,\psi\rangle) = \mathcal{K}(\alpha,\beta,\gamma;s,t)\langle\lambda,\phi\psi\rangle, \quad 0\le s\le t<\infty, \quad \phi,\psi\in\mathcal{S}(\mathbb{R}^d),$$

where

$$\mathcal{K}(\alpha,\beta,\gamma;s,t) = \begin{cases} e^{\alpha t}\{\gamma[1+m_2V(e^{\alpha s}-1)/\alpha]+\beta[(1-e^{-\alpha s})/\alpha+m_2V(e^{\alpha s}+e^{-\alpha s}-2)/2\alpha^2]\} & \text{if } \alpha\neq 0 \\ \gamma(1+m_2Vs)+\beta(s+m_2Vs^2/2) & \text{if } \alpha = 0 \end{cases}$$

In particular, $M_t = \mathcal{K}(\alpha,\beta,\gamma;t,t)^{1/2}W_t$, $0\le t<\infty$, *where* $W_t = W$ *in distribution.*

Moreover, the process M *is Markovian, has a norm-continuous version, and satisfies the generalized Langevin equation*

$$\partial M/\partial t = \alpha M + \mathfrak{w}, \quad t\ge 0$$
$$M_0 = \gamma^{1/2}W,$$

where $\mathfrak{w}\equiv\{\mathfrak{w}_t, t\ge 0\}$ *is an* $\mathcal{S}'(\mathbb{R}^d)$*-valued centered Gaussian noise process with covariance functional*

$$\mathrm{Cov}(\langle\mathfrak{w}_s,\phi\rangle,\langle\mathfrak{w}_t,\psi\rangle) = \delta(s-t)\mathcal{N}(\alpha,\beta,\gamma;t)\langle\lambda,\phi\psi\rangle, \quad 0\le s,t<\infty, \quad \phi,\psi\in\mathcal{S}(\mathbb{R}^d),$$

where

$$\eta(\alpha,\beta,\gamma;t) = \begin{cases} \gamma e^{\alpha t}(m_2V-\alpha) + \beta[2-e^{\alpha t}+m_2V(e^{\alpha t}-1)/\alpha] & \text{if} \quad \alpha \neq 0 \\ \gamma m_2V+\beta(1+m_2Vt) \quad \text{if} \quad \alpha = 0 \end{cases}$$

The norm-continuity referred to above means that for each $T \in (0,\infty)$ there is an integer $p = p_T > 0$ such that M_t is $\| \|_{-p}$- continuous on $[0,T]$ almost-surely, where $\| \|_{-p}$ is the dual norm on $\mathcal{S}_p'(\mathbb{R}^d)$ (see Ito [9]). In the present case p is in fact independent of T.

The generalized Langevin equation is understood in a weak space-time sense, as in [6].

We note that the space scaling $x \to Kx$ annahilates the migration process in the limit $K \to \infty$ while the branching is preserved, and that this is reflected in the Langevin equation, which contains information coming from the underlying branching process but no information on the particle migration process. In fact this happens for any particle migration process which is time-homogeneous Markov and preserves the spatial homogeneity of the system. Moreover, the migration plays no rôle at all in the limit; this can be seen by considering the model which is the same as above except that the particles remain at the locations of their births. The limit results for such a model are exactly the same as above.

Dawson and Ivanoff [5,10,11] have proved the existence of a steady-state random field N_∞^K for the process N^K in the cases 1) $\alpha = 0$, $\gamma > 0$, $\beta = 0$, with $d \geq 3$, and 2) $\alpha < 0$, $\gamma \geq 0, \beta > 0$; and they have obtained spatial central limit theorems for N_∞^K as $K \to \infty$. In case 1) a larger normalization is needed $(K^{-d/2-1}(N_\infty^K - EN_\infty^K))$ because the fluctuations remain important at macroscopic scales (Brillinger's theorem does not apply in this case), and therefore the limit cannot be obtained by taking the limit of M_t as $t \to \infty$ in Theorem 2, which clearly does not exist. In case 2) we see from Theorem 2 that

$$M_t \Rightarrow (-\beta/\alpha + m_2V\beta/2\alpha^2)^{1/2}W \quad \text{as} \quad t \to \infty.$$

We remark that this effect is due entirely to the immigration; all the progenie of the initial particles dies out by subcriticality. This limit coincides with that obtained by Ivanoff [11] for $K^{-d/2}(N_\infty^K - EN_\infty^K)$ as $K \to \infty$. It is not a steady state random field of the process M.

METHODS OF PROOF

The branching random field with immigration $N \equiv \{N_t, t \geq 0\}$ is a point measure-valued Markov process with infinitesimal generator $\mathcal{L}$ such that

$$\mathcal{L}f(\langle\mu,\phi\rangle) = f'(\langle\mu,\phi\rangle)\langle\mu,\tfrac{1}{2}\Delta\phi\rangle + \tfrac{1}{2}f''(\langle\mu,\phi\rangle)\langle\mu,|\nabla\phi|^2\rangle \tag{3}$$
$$+ V\int\sum_{n=0}^{\infty} p_n[f(\langle\mu,\phi\rangle + (n-1)\phi(x)) - f(\langle\mu,\phi\rangle)]\mu(dx)$$
$$+ \beta\int[f(\langle\mu,\phi\rangle + \phi(x)) - f(\langle\mu,\phi\rangle)]dx,$$

where μ is a point measure, $f: \mathbb{R} \to \mathbb{R}$ has bounded continuous derivatives of up to second order, and $\phi \in \mathcal{S}(\mathbb{R}^d)$; hence

$$f(\langle N_t,\phi\rangle) - \int_0^t \mathcal{L}f(\langle N_s,\phi\rangle)ds, \quad t \geq 0 \tag{4}$$

is a martingale, and in particular

$$\langle N_t,\phi\rangle - \int_0^t \langle N_s,(\tfrac{1}{2}\Delta + \alpha)\phi\rangle ds - \beta\langle\lambda,\phi\rangle t, \quad t \geq 0 \tag{5}$$

and

$$(\langle N_t,\phi\rangle - \int_0^t \langle N_s,(\tfrac{1}{2}\Delta+\alpha)\phi\rangle ds - \beta\langle\lambda,\phi\rangle t)^2 \tag{6}$$
$$- \int_0^t \langle N_s, |\nabla\phi|^2 + V(m_2-m_1+1)\phi^2\rangle ds - \beta\langle\lambda,\phi^2\rangle t, \quad t \geq 0$$

are martingales (the last two terms in (6) constitute the increasing process of the martingale (5)).

By results of Mitoma [15,16] one sees from (5) that N has a version in $D([0,\infty), \mathcal{S}'(\mathbb{R}^d))$.

Proceeding as in [5], one can show that the joint characteristic function of $\langle N_{t_1},\phi_1\rangle,\ldots,\langle N_{t_m},\phi_m\rangle$, $t_1 < \ldots < t_m$, $\phi_1,\ldots,\phi_m \in \mathcal{S}(\mathbb{R}^d)$ is given by

$$E\exp\{i\sum_{j=1}^m u_j\langle N_{t_j},\phi_j\rangle\} \tag{7}$$
$$= \exp\{\gamma\int E[\exp(i\sum_{j=1}^m u_j\langle N^x_{t_j},\phi_j\rangle) - 1]dx$$
$$+\beta\int_0^{t_m}\int E[\exp(i\sum_{j=1}^m u_j\langle N^x_{t_j-s},\phi_j\rangle)-1]dxds\}, \qquad u_1,\ldots,u_m \in \mathbb{R},$$

where $\{N^x_t, t \geq 0\}$ denotes the branching random field generated by a single particle located initially at $x \in \mathbb{R}^d$, and in the second integral N^x_t is taken as 0 for $t < 0$.

It follows from (7) and the Lemma in [6] that

$$E\langle N_t,\phi\rangle = \gamma\int E\langle N^x_t,\phi\rangle dx + \beta\int_0^t\int E\langle N^x_{t-s},\phi\rangle dxds \tag{8}$$
$$= (\gamma e^{\alpha t} + \beta(e^{\alpha t} - 1)/\alpha)\langle\lambda,\phi\rangle, \quad t \geq 0$$

and

$$\mathrm{Cov}(\langle N_s,\phi\rangle, \langle N_t,\psi\rangle) \tag{9}$$

$$= \gamma\int E\langle N^x_s,\phi\rangle\langle N^x_t,\psi\rangle dx + \beta\int_0^s\int E\langle N^x_{s-r},\phi\rangle\langle N^x_{t-r},\psi\rangle dxdr$$

$$= e^{\alpha t}(\gamma + \beta\alpha^{-1}(1-e^{-\alpha s}))\iint\phi(x)\psi(y)p(t-s,x,y)dxdy$$

$$+ e^{\alpha t}\gamma m_2 V\int_0^s e^{\alpha(s-r)}\iint\phi(x)\psi(y)p(t+s-2r,x,y)dxdydr$$

$$+ e^{\alpha t}\beta m_2 V\int_0^s \alpha^{-1}(1-e^{-\alpha r})e^{\alpha(s-r)}\iint\phi(x)\psi(y)p(t+s-2r,x,y)dxdydr, \quad s\leq t.$$

Consider the centered process $M \equiv \{M_t, \ t\geq 0\}$:

$$M_t = N_t - EN_t = N_t - (\gamma e^{\alpha t} + \beta(e^{\alpha t}-1)/\alpha)\lambda$$

(not to be confused with the limit process M in Theorem 2; notice that M^K in (2) is the M above scaled and normalized). Clearly M also has a version in $D([0,\infty), \mathcal{S}'(\mathbb{R}^d))$. From (3), (4), (5) and (6) one finds that

$$\langle M_t,\phi\rangle - \int_0^t\langle M_s,(\tfrac{1}{2}\Delta+\alpha)\phi\rangle ds, \quad t\geq 0, \tag{10}$$

and

$$(\langle M_t,\phi\rangle - \int_0^t \langle M_s,(\tfrac{1}{2}\Delta+\alpha)\phi\rangle ds)^2 \tag{11}$$

$$- \int_0^t \langle N_s, |\nabla\phi|^2 + V(m_2-m_1+1)\phi^2\rangle ds - \beta\langle\lambda,\phi^2\rangle t, \quad t\geq 0$$

are martingales.

Now we introduce the scaling and normalization (1) - (2) and we observe that

$$\iint\phi(x)\psi(x)K^d p(t,Kx,Ky)dxdy \to \langle\lambda,\phi\psi\rangle, \quad t>0, \quad \text{as} \quad K\to\infty. \tag{12}$$

Theorem 1 follows easily from (8), (9) and (12).

To prove Theorem 2 it suffices by Mitoma [17] to show that

$$(\langle M^K_{t_1},\phi_1\rangle,\ldots,\langle M^K_{t_m},\phi_m\rangle) \Longrightarrow (\langle M_{t_1},\phi_1\rangle,\ldots,\langle M_{t_m},\phi_m\rangle) \quad \text{as} \quad K\to\infty \tag{13}$$

for any $t_1<\ldots<t_m$, and $\phi_1,\ldots,\phi_m\in\mathcal{S}(\mathbb{R}^d)$, and that

$$\{\langle M^K,\phi\rangle\}_{K\geq 1} \text{ is tight in } D([0,\infty),\mathbb{R}) \text{ for any } \phi\in\mathcal{S}(\mathbb{R}^d). \tag{14}$$

From (7), (8) and (9) we obtain

$$E\exp\{i\sum_{j=1}^m u_j\langle M^K_{t_j},\phi_j\rangle\} = \exp\{A^K + B^K\}, \tag{15}$$

where

$$A^K = -\frac{1}{2}\{\sum_{j=1}^m u_j^2 e^{\alpha t_j}(\gamma+\beta\alpha^{-1}(1-e^{-\alpha t_j}))\int\phi_j(x)\psi_j(x)dx + \tag{16}$$

$$+ \sum_{\substack{j,k=1 \\ j\neq k}}^{m} u_j u_k e^{\alpha t_j \vee t_k}(\gamma+\beta\alpha^{-1}(1-e^{-\alpha t_j\wedge t_k}))\iint \phi_j(x)\psi_k(y)K^d p(|t_j-t_k|,Kx,Ky))dxdy$$

$$+ \sum_{j,k=1}^{m} u_j u_k e^{\alpha t_j \vee t_k} m_2 V[\gamma\int_0^{t_j\wedge t_k} e^{\alpha r}\iint \phi_j(x)\psi_k(y)K^d p(|t_j-t_k|+2r,Kx,Ky)dxdydr$$

$$+ \beta\int_0^{t_j\wedge t_k}\alpha^{-1}(1-e^{-\alpha r})e^{\alpha(t_j\wedge t_k-r)}\iint \phi_j(x)\psi_k(y)K^d p(t_j\wedge t_k+t_j\vee t_k-2r,Kx,Ky)dxdydr]\},$$

and B^K is the error term.

Now, from (12), (15) and (16) we see that the finite-dimensional distributions of M^K tend to those of the generalized process M described in Theorem 2 as $K\to\infty$, provided that the error term B^K converge to 0.

We note that it is due to (12) that the covariance functional of M is concentrated on the diagonal and that M is independent of the particular migration process; this phenomenon does not take place in the high density limit [6], nor in the space-time scaling limit in [3] and [7].

In the expressions above we have assumed $\alpha\neq 0$. Part of the result in the critical case $\alpha=0$ is a special case of the general one, but a separate computation is necessary in order to obtain the last term of $\mathcal{K}(0,\beta,\gamma;s,t)$.

In order to prove that the error term B^K in (15) tends to 0 it does not seem sufficient that the branching law have a finite second moment (in contrast to the high-density limit [6]). Having a finite third moment and denoting Z_t^x the total progenie at time t of a particle initially located at $x\in\mathbb{R}^d$, it is easy to see that B^K is bounded by a finite sum of terms of the form

$$K^{-3d/2}L\int E Z_r^x Z_s^x \langle N_t^x,\psi^K\rangle\, dx,$$

where $r\leq s\leq t$, L is a finite constant, and ψ is a nonnegative function such that $\langle\lambda,\psi\rangle<\infty$, and it is not hard to show, by a renewal argument as in the proof of the Lemma in [6], that

$$\int E Z_r^x Z_s^x \langle N_t^x,\psi^K\rangle dx \leq L_1\int\psi^K(x)dx = K^d L_1\langle\lambda,\psi\rangle$$

where L_1 is a finite constant. Hence $B^K\to 0$ as $K\to\infty$.

Therefore (13) is proved.

To prove (14), by Holley and Stroock [7,8] it suffices to show that $\{\langle M_0^K,\phi\rangle\}_{K\geq 1}$ is tight (which it is, since $\langle M_0^K,\phi\rangle \Longrightarrow \langle M_0,\phi\rangle$ as $K\to\infty$), and that for each $T\in(0,\infty)$,

$$\sup_{K\geq 1} E \sup_{0\leq t\leq T}(\theta_{\phi,i}^K(t))^2<\infty,\quad i=1,2, \tag{17}$$

where

$$\theta_{\phi,1}^K(t) = \langle M_t^K,(\tfrac{1}{2}\Delta+\alpha)\phi\rangle,\quad t\geq 0$$

and

$$\theta^K_{\phi,2}(t) = K^{-d}\langle N^K_t, |\nabla\phi|^2 + V(m_2-m_1+1)\phi^2\rangle + K^{-d}\beta\langle\lambda,(\phi^K)^2\rangle, \quad t \geq 0,$$

i.e. $\theta^K_{\phi,i}$, $i=1,2$, are the random functions appearing in the martingales (10) and (11) after scaling and normalizing. The proof of (17) is analogous to those in [4] and [6].

To prove the Markov property of the limit process M we must show that given $t_0 < t$ and $\phi \in \mathcal{S}(\mathbb{R}^d)$, there is a $\hat{\phi} \in \mathcal{S}(\mathbb{R}^d)$ such that

$$E((\langle M_t,\phi\rangle - \langle M_{t_0},\hat{\phi}\rangle)\langle M_s,\psi\rangle = 0 \tag{18}$$

for all $s \leq t_0$ and $\psi \in \mathcal{S}(\mathbb{R}^d)$. Inspection of the covariance in Theorem 2 shows that (18) is satisfied with $\hat{\phi} = e^{\alpha(t-t_0)}\phi$.

From the covariance in Theorem 2 it is easy to see that given $T \in (0,\infty)$ and $\phi \in \mathcal{S}(\mathbb{R}^d)$ there is a finite constant $L_{\phi,T}$ such that

$$E(\langle M_t,\phi\rangle - \langle M_s,\phi\rangle)^2 \leq L_{\phi,T}(t-s)$$

for $0 \leq s \leq t \leq T$. Then by Mitoma's extension of the Dudley-Fernique theorem [14] M has a norm-continuous version. Moreover, it is possible to find a positive locally bounded function f on $[0,\infty)$ such that

$$\sup_{T\geq 0} E(\sup_{0\leq t\leq T} \langle M_t,\phi\rangle^2)/f(T) < \infty \tag{19}$$

for every $\phi \in \mathcal{S}(\mathbb{R}^d)$; therefore by a criterion of Mitoma [14] M_t is $\|\ \|_{-p}$-continuous for some p, for all $t \geq 0$, almost-surely. (19) can be proved using the fact that

$$\langle M_t,\phi\rangle - \alpha\int_0^t \langle M_s,\phi\rangle ds, \quad t \geq 0 \tag{20}$$

is a martingale (it is the limit of the scaled and normalized martingale (10) as $K \to \infty$). That (20) is a martingale also follows from the fact that

$$e^{-\alpha t}\langle M_t,\phi\rangle, \quad t \geq 0 \tag{21}$$

is a martingale, which in turn can be shown using the Gaussian distribution of $\langle M,\phi\rangle$. It is interesting to note the analogy between the martingale (21) and the martingale $e^{-\alpha t}Z_t$ of the underlying branching process of the system.

The proof that M satisfies the generalized Langevin equation is analogous to that in [6].

REFERENCES

1. Brillinger, D.R. (1975). Statistical inference for stationary point processes. In Stochastic Processes and Related Topics (M.L. Puri, Ed.), 55-99, Academic Press, New York.
2. Dawson, D.A. (1977). The critical measure diffusion process, Z. Wahrsch. Verw. Gebiete, Vol. 40, 125-145.

3. Dawson, D.A. (1980). Limit theorems for interaction free geostochastic systems. Seria Coll. Math. Soc. Janos Bolyai, Vol. 24, 27-47.

4. Dawson, D.A. and Gorostiza, L.G. (1984). Limit theorems for supercritical branching random fields, Math. Nachr., Vol. 119, 19-46.

5. Dawson, D.A. and Ivanoff, B.G. (1978). Branching diffusions and random measures. In Advances in Probability (A. Joffe and P. Ney, Eds.), Vol. 5, 61-104, Dekker, New York.

6. Gorostiza, L.G. (1983). High density limit theorems for infinite systems of unscaled branching Brownian motions, Ann. Probab., Vol. 11 (2), 374-392.

7. Holley, R. and Stroock, D. (1978). Generalized Ornstein-Uhlenbeck processes and infinite particle branching Brownian motions, Publ. RIMS, Kyoto Univ., Vol. 14, 741-788.

8. Holley, R. and Stroock, D. (1981). Generalized Ornstein-Uhlenbeck processes as limits of interacting systems. In Stochastic Integrals, Lecture Notes in Math., Vol. 851. Springer-Verlag, Berlin.

9. Ito, K. (1978). Stochastic analysis in infinite dimensions. In Stochastic Analysis (A. Friedman and M. Pinsky, Eds.), 187-197. Academic, New York.

10. Ivanoff, B.G. (1980). The branching random field, Adv. Appl. Probab., Vol. 12, 825-847.

11. Ivanoff, B.G. (1980). The branching diffusion with immigration, J. Appl. Probab., Vol. 17, 1-15.

12. Ivanoff, B.G. (1981). The multitype branching diffusion, J. Multiv. Anal., Vol. 11(3), 289-318.

13. Ivanoff, B.G. (1982). Central limit theorems for point processes, Stoch. Proc. Appl., Vol. 12, 171-186.

14. Mitoma, I. (1981). On the norm-continuity of $\mathfrak{g}'$-valued Gaussian processes, Nagoya Math. J., Vol. 82, 209-220.

15. Mitoma, I. (1981). Martingales of random distributions, Mem. Fac. Sci., Kyushu Univ., Ser. A, Vol. 35(1), 185-197.

16. Mitoma, I. (1983). On the sample continuity of $\mathfrak{g}'$-processes, J. Math. Soc. Japan Vol. 35(4), 629-636.

17. Mitoma, I. (1983). Tightness of probabilities in $C([0,1], \mathfrak{g}')$ and $D([0,1], \mathfrak{g}')$, Ann. Probab., Vol. 11(4), 989-999.

AN INVARIANCE PRINCIPLE FOR MARTINGALES WITH VALUES IN SOBOLEV SPACES

Michel METIVIER*

Ecole Polytechnique
F-91128 Palaiseau Cedex

Abstract

Starting with an example occuring in the modeling of a chemical reaction we introduce a wide class of situations where one has to study the convergence in law of an "accompanying martingale" taking its values in a Sobolev space.

We present a sufficient condition for tightness and an "invariance principle", stating only the results, the proofs of which will be exposed in detail in a subsequent paper [7].

I. Introduction : example and problem

I.1 - Example.

We start with presenting the following situation studied by L. Arnold and M. Theodosopulu in [3] and P. Kotelenez in [4].

We denote by $\mathbb{H}_o$ the Hilbert space $L^2[0,1]$. For each $N \in \mathbb{N}$, X^N is a pure jump Markov process with values in a finite dimensional subspace $\mathbb{H}_o^N$ of $\mathbb{H}_o$. More precisely, if h_i^N denotes the indicator function of $[i/N, \frac{i+1}{N}[$,

(1.1) $$X_t^N := \sum X_t^{N,i} h_i^N$$

where the real processes $X^{N,i}$, $i=0,\dots,N-1$, are the coordinates of a pure jump Markov process in $\mathbb{R}^N$.

For each $p \in [i/N, \frac{i+1}{N}[$, $X_t^N(p)$ is intended to represent the "density of presence" of "ions" of a given type present at time t in $[i/N, \frac{i+1}{N}[$ during a specific chemical reaction occuring in a domain which is for simplification here assumed of dimension 1. If ε_N is the mass of each ion and $\xi_t^{N,i}$ is the number

---

(*) Research partially supported by a grant F49620-82-C-009 of the Air Force Office of Scientific Research, during the summer 1983, at the Center for Stochastic Processes. University of North Carolina. Chapel Hill, U.S.A.

of those present in $[i/N, \frac{i+1}{N}[$ one has $X_t^{n,i} = N \varepsilon_N \xi_t^{N,i}$. The ions appear or disappear in the ith cell as an effect of the chemical reaction. They can also migrate by jump from one cell to a neighbouring one. No jump outside or coming from outside is permitted. The jump of the process X^N can thus easily be described as being of the form $N\varepsilon_N(h_{i+1}^N - h_i^N)$, $N\varepsilon_N(h_{i-1}^N - h_i^N)$, $N\varepsilon_N h_i^N$ and $-N\varepsilon_N h_i^N$. The intensity $\gamma^N(g,m)$ of the jump m for a state $g \in \mathbb{H}_o^N$ of the process is defined by the chemical potential as to birth and death and by the speed of migration as to jumps of the form $N\varepsilon_N(h_j^N - h_i^N)$, $j = i \pm 1$. The hypotheses are so that

$$\gamma^N(g,m) = \frac{1}{N\varepsilon_N} \lambda(g^{N,i}) \quad \text{if} \quad m = \varepsilon_N N h_i^N$$

$$\gamma^N(g,m) = \frac{1}{N\varepsilon_N} \mu(g^{N,i}) \quad \text{if} \quad m = -\varepsilon_N N h_i^N$$

$$\gamma^N(g,m) = \frac{DN}{2\varepsilon_N} (g^{N,i}) \quad \text{if} \quad m = \varepsilon_N N(h_j^N - h_i^N),\ j = i \pm 1$$

λ and μ being two positive functions on $\mathbb{R}^+$ with a global Lipschitz property (and moreover coinciding with polynomials in a proper bounded set of $\mathbb{R}^+$) and with $\mu(0) = 0$. D is a constant.

The infinitesimal generator L^N of X^N is given by

$$L^N\varphi(g) = \sum_m \gamma^N(g,m)(\varphi(g+m) - \varphi(g)).$$

Applying the Dynkin Formula to $\varphi(X_t^N)$ with $\varphi(g) = g$ gives

$$X_t^N = X_o^N + \int_o^t b^N(X_s^N)ds + M_t^N \tag{1.2}$$

where M_t^N is a local martingale and $b^N(g)$ is for every g an element of $\mathbb{H}_o$ defined in the following way :

$$b^N(g) = C(g) + D\Delta^N g \tag{1.3}$$

where we have set

$$C(u) := \lambda(u) - \mu(u) \tag{1.4}$$

and

$$(1.5)\qquad \Delta^N g(x) := \begin{cases} N^2(g(x+\frac{1}{N}) + g(x-\frac{1}{N}) - 2g(x) & \text{if } \frac{1}{N} \leqslant x < 1-\frac{1}{N} \\ N^2(g(x+\frac{1}{N}) - g(x) & \text{if } 0 \leqslant x < \frac{1}{N} \\ N^2(g(x-\frac{1}{N}) - g(x) & \text{if } 1-\frac{1}{N} \leqslant x \leqslant 1 \end{cases}$$

I.2 - Problem.

One is interested in the "accompanying martingale" $(M_t^N)_{t\geqslant 0}$ of formula (1.2), which more or less represent the fluctuation of the process around its averaged behaviour. More precisely, the processes M^N being considered as processes with cad-lag paths in a suitable vector space $\mathbf{V}'$ containing $\mathbb{H}_o$, their laws $\tilde{P}^N$ are probability laws on the corresponding skorokhod space $\mathbb{D}(\mathbb{R}^+ ; \mathbf{V}')$. Do the sequence of laws $\tilde{P}^N$ converge weakly to a law $\tilde{P}$ which can moreover be characterized ?

We want to emphazise that this type of result can be easily obtained from the properties of the processes $\langle\!\langle M^N \rangle\!\rangle$, whose definition we recall now.

$\mathbb{E}$ being a Hilbert space, we denote by $(.;.)_{\mathbb{E}}$ the scalar product in $\mathbb{E}$.

If M is a martingale taking its values in a Hilbert space $\mathbb{E}$, we call $\langle\!\langle M \rangle\!\rangle$ the process with values in $\mathcal{L}(\mathbb{E} ; \mathbb{E})$, the space of bounded operators in $\mathbb{E}$, such that for every u and v in $\mathbb{E}$

$$(1.6)\qquad (u ; \langle\!\langle M \rangle\!\rangle_t v)_{\mathbb{E}} = \langle\!\langle <M ; u>, <M ; v> \rangle\!\rangle_t$$

where the symbol $\langle\!\langle , \rangle\!\rangle$ in the second member of (1.6) denotes the Meyer process of the two real valued martingales $(M;u)$ and $(M;v)$.

Let us remark (see [6]) that the process $\langle\!\langle M \rangle\!\rangle$ actually *takes its values in the space* $\mathcal{L}(\mathbb{E} ; \mathbb{E})$ *of positive nuclear operators*.

We may thus define

$$(1.7)\qquad \{\!\!\{ M \}\!\!\}_t := \text{trace } \langle\!\langle M \rangle\!\rangle_t \ .$$

Let us remark also that the process $\langle\!\langle M \rangle\!\rangle_t^{\frac{1}{2}}$ is well defined because of the positivity of $\langle\!\langle M \rangle\!\rangle_t$ and is a Hilbert-Schmidt operator-valued process. We call $\mathcal{L}_2(\mathbb{E} ; \mathbb{E})$ the space of Hilbert-Schmidt operators and $\|.\|_2$ the Hilbert-Schmidt norm.

In the example (I.1) the process $\langle\!\langle M \rangle\!\rangle$ is easily calculated from the Dynkin formula applied to $(X^N;u)\,(X^N;v)$ which gives

$$(M^N;u)(M_t^N;v) = \int_0^t (u;a^N(X_s^N)v)ds + \text{Local Martingale}$$

and therefore

$$(1.8)\qquad (u; \langle\!\langle M^N \rangle\!\rangle_t v) = \int_0^t (u;a^N(X_s^N)v)ds \ .$$

II. Tightness of sequences of Hilbert-valued martingales

Definition 1 : A sequence $(X^N)_{n\geq 0}$ of processes with values in (E,d) (a polish space) is said to satisfy [A] if [A] $\forall \varepsilon, \eta > 0 \ \exists \delta > 0$ such that for every sequence (τ^n), where τ^n is a stopping time of X^n,

$$\lim_{n\to\infty} \sup \sup_{0<\theta<\delta} P^n\{d(X^n_{\tau^n+\theta}, X^n_{\tau^n}) > \eta\} \leq \varepsilon .$$

We may then state the following theorem for a sequence $(M^n)_{n\geq 0}$ of martingales with values in the Hilbert space G. This theorem extends to this situation results of Rebolledo for real values martingales (see [7]).

Theorem 1

Assume : a) $(M^n_0)_n$ tight in G

b) $(\langle\langle M^n_T\rangle\rangle^{\frac{1}{2}})_{n\geq 0}$ tight in $\mathcal{L}_2(G,G)$

c) [A] holds for the increasing processes (trace $\langle\langle M^n\rangle\rangle)_{n\geq 0}$

Then : $(M^n)_{n\geq 0}$ is tight in $D([0,T] ; G)$.

Proof

The new difficulty which arises in the infinite dimensional case concerns the weak compactness of the sequence $(M^n_t)_{n\geq 0}$ of G-valued random variables for each t. The fundamental step of the proof is to show that the tightness of the sequences $(\langle\langle M^n\rangle\rangle^{\frac{1}{2}}_T)_{n\geq 0}$ implies the tightness of the sequence $(M^n_t)_{n\geq 0}$ for each $t \leq T$. The reader is referred to [7] .

The interest of such a theorem is that the tightness of the sequence $(\langle\langle M^n\rangle\rangle^{\frac{1}{2}}_t)_{n\geq 0}$ is sometime easily obtained from a representation of the type (1.8). The following proposition gives a sufficient condition for the tightness of $(\langle\langle M^n\rangle\rangle^{\frac{1}{2}}_t)_{n\geq 0}$.

Proposition 1 - Let us assume :

i) $\lim_{\rho\to\infty} \sup_n P^n\{\text{trace} \langle\langle M^n\rangle\rangle_t > \rho\} = 0$

ii) there exists a complete orthonormal system $(h_k)_{k\subset \mathbb{N}}$ in $\mathbb{G}$ such that

$$\lim_{m\to\infty} \sup_n P^n\{\sum_{k=m}^{\infty} (h_k ; \langle\langle M^n\rangle\rangle_t h_k)_{\mathbb{G}} > \eta\} = 0 .$$

Then the sequence $(\ll M^n \gg_t^{\frac{1}{2}})_{n \geq 0}$ is tight in $\mathcal{L}_2(\mathbb{E} ; \mathbb{E})$.

The hypotheses of the proposition allow to show easily for each $\varepsilon > 0$ and $\eta > 0$ the existence of a finite dimensional subspace L_ε of $\mathcal{L}_2(\mathbb{E} ; \mathbb{E})$ and a bounded set B_ε in $\mathcal{L}_2(\mathbb{E} ; \mathbb{E})$ such that

$$\sup_n P^n\{\ll M^n \gg_t^{\frac{1}{2}} \notin B_\varepsilon\} \leq \varepsilon$$

and

$$\sup_n P^n\{\| \ll M^n \gg_t^{\frac{1}{2}} - \Pi(\ll M^n \gg_t^{\frac{1}{2}})\|_\varepsilon > \eta\} \leq \varepsilon$$

where Π denotes the projection operator from $\mathcal{L}_2(\mathbb{E} ; \mathbb{E})$ onto L_ε .

III. An invariance principle

We consider the following situation, which generalizes the one of the example I.1.

$(\mathbf{V},\mathrm{H},\mathbf{V}')$ is a Gelfand triple of Hilbert spaces where $\mathbf{V}'$ is the dual of $\mathbf{V}$, $\mathbb{H}$ is identified with its dual and we have continuous injection $\mathbf{V} \hookrightarrow \mathbb{H} \hookrightarrow \mathbf{V}'$.

We also assume the existence of a Banach space W with continuous injections $\mathbf{V} \hookrightarrow W \hookrightarrow \mathbf{V}'$.

We assume that X^N is a process such that the paths of X^N belong to $L^\infty([0,T] ; W) \cap \mathbb{D}([0,T]; \mathbb{H})$ and

$$X_t^N = X_o^N + \int_o^t b^N(X_s^N)ds + M_t^N \tag{3.1}$$

M^N being a local martingale,

$$<u , \ll M^N \gg_t v> = \int_o^t <u , a^N(X_s^N)v > ds \tag{3.2}$$

with

$$\forall\, g \in W \quad b^N(g) \in \mathbf{V}' \ , \ a^N(g) \in \mathcal{L}_1(\mathbf{V} ; \mathbf{V}') . \tag{3.3}$$

Hypotheses

(H_1) (i) $\forall\ T > 0\quad \varepsilon > 0 \qquad \limsup_N E^N\{ \sup_{s\leq T} \| X^N_s\|_W\} < \infty$.

(ii) $\forall\ T > 0 \quad \lim_N P^N \{ \sup_{t\leq T} \| \frac{\Delta X^N_t}{\sqrt{\varepsilon_N}} \|_{\mathbf{V}'} > \rho\} = 0$, for every $\rho > 0$.

(iii) $\sup_N \sup_{t\leq T} \| \frac{1}{\sqrt{\varepsilon_N}} \Delta X^N_t \|_{\mathbf{V}'} < \infty$.

(H_2) $\exists\ \{v_k\}$ orth. basis in $\mathbf{V}$ for which

$$<v_k , a^N(g)v_k> \leq D\ \varepsilon_N \| g\|_W \alpha_k$$

with $\varepsilon_N \downarrow 0 \quad \sum_k \alpha_k < \infty$.

(H_3) $\lim_N <u , [\frac{1}{\varepsilon_N} a^N(g) - a(g)]u> = 0 \quad \forall\ g \in W \quad u \in \mathbf{V}$

$a(g) \in \mathcal{L}_1(\mathbf{V} , \mathbf{V}')$

(H_4) $\exists\ f \in C(\mathbb{R}^+ ; \mathbb{H}) \cap L^\infty([0,T] , W)$, $\forall\ u$, $u \in \mathbf{V}$

$$\lim_N E<u , [\frac{1}{\varepsilon_N} a^N(X^N_s) - \frac{1}{\varepsilon_N} a^N(f(s))]u> = 0 \quad \text{a.s.}$$

Theorem 2

$(\frac{1}{\sqrt{\varepsilon_N}} M^N)_{N\in\mathbb{N}}$ converges weakly in $\mathbb{D}(\mathbb{R}^+ ; \mathbf{V}')$ to the continuous process with independent gaussian increments and covariance

$$C(t) \in \mathcal{L}_1(\mathbf{V} , \mathbf{V}') : <u ; C(t)v> = \int_0^t <u ; a(f(s))v>ds \ .$$

Sketch of the proof (for details see [7]).

The condition (H_2) essentially gives the sufficient condition of tightness of proposition 1.

To identify the limit, we derive, using (H_3) and (H_4) that any limit $\tilde{P}$ of the sequence $\tilde{P}^N$ is such that the canonical process ξ in $\mathbb{D}(\mathbb{R}^+ ; \mathbf{V}')$ is a continuous martingale such that for every $u \in \mathbf{V}$ and $v \in \mathbf{V}$ the process $(<u , \xi_t><v , \xi_t> - \int_0^t <u ; a(f(s))v>ds)_{t\geq 0}$ is a local martingale.

Application

Let us call H_q the subspace of $L^2[0,1]$ consisting of functions with distributional derivatives which are functions in $L^2[0,1]$ up to order q, with the additional property :

$$\left.\frac{\partial^k g}{\partial x^k}\right|_{x=0} = \left.\frac{\partial^k g}{\partial x^k}\right|_{x=1} = 0 \quad \text{for} \quad 1 \leqslant k \leqslant q-1$$

and, with the scalar product :

$$(u\,;\,v)_q = \sum_{0 \leqslant k \leqslant q} \left(\frac{\partial^k u}{\partial x^k}\,;\,\frac{\partial^k v}{\partial x^k}\right)_{\mathbb{H}_o}.$$

f being the unique solution of

$$\frac{\partial f}{\partial t}(t,x) = D\,\frac{\partial^2 f}{\partial x^2}(t,x) + C(f(t,x))$$

$$\frac{\partial f}{\partial x}(t,0) = \frac{\partial f}{\partial x}(t,1) = 0$$

$$f(0,x) \in \mathbb{H}_2$$

Theorem 2 applies immediately to the situation considered in example I.1 with, for every $u,v \in \mathbb{H}_q$

$$< u\,,\,a(g)v > = (|C|(g)\;u\,;\,v)_{\mathbb{H}_o} + 2D\left(g\,\frac{\partial u}{\partial x}\,;\,\frac{\partial v}{\partial x}\right)_{\mathbb{H}_o}$$

where $|C|(u) := \lambda(u) + \mu(u)$.

If the processes X^N are stopped in such a way they stay in $L^\infty[0,1]$ as in [3] one takes $W = L^\infty$ and obtains the convergence in $\mathbb{D}(\mathbb{R}^+\,;\,\mathbb{H}_{-2})$. If such a stopping procedure is not used one takes $W = \mathbb{H}_{-2}$ and the convergence is in $\mathbb{D}(\mathbb{R}^+\,;\,\mathbb{H}_{-4})$. For details see [7].

References

[1] D. ALDOUS.- Stopping times and tightness.- Ann. of Prob. 6, n° 2, 1978, 335-40.

[2] L. ARNOLD.- Mathematical models of chemical reactions. in: Stochastic systems. (ed.) M. Hazewinkel, J. Willems. Dordrecht, 1981.

[3] L. ARNOLD, M. THEODOSOPULU.- Deterministic limit of the stochastic model of chemical reactions with diffusion. Adv. Appl. Prob. 12, 1980,

[4] P. KOTELENEZ.- Law of large numbers and central limit theorem for chemical reactions with diffusions. Universität Bremen, 1982.

[5] E. LENGLART.-Relations de domination entre deux processus. Ann. Inst. H. Poincaré, B XIII, 1977, 171-179.

[6] M. METIVIER.- Semimartingales. De Gruyter ed. Berlin, New York, 1982.

[7] M. METIVIER.- Convergence faible et principe d'invariance pour des martingales à valeurs dans des espaces de Sobolev. To appear in Ann. Inst. H. Poincaré.

[8] R. REBOLLEDO.- La méthode des martingales appliquée à la convergence en loi des processus. Mémoires de la S.M.F. n° 62, 1979.

LARGE DEVIATIONS FOR STATIONARY GAUSSIAN PROCESSES*

M. D. Donsker and S.R.S. Varadhan

Courant Institute of Mathematical Sciences
New York University
251 Mercer Street - NEW YORK, N.Y. 10012

Let $\{X_k\}$, $-\infty < k < \infty$, be a stationary Gaussian process with $E\{X_k\} = 0$ and $E\{X_0 X_j\} = \rho_j = \frac{1}{2\pi} \int_0^{2\pi} e^{ij\theta} f(\theta)\, d\theta$. We assume that the spectral density function $f(\theta)$ is continuous on $[0,2\pi]$, $f(0) = f(2\pi)$, and

$$\text{(1)} \qquad \int_0^{2\pi} \log f(\theta)\, d\theta > \infty \; .$$

Let $\Omega = \prod_{j=-\infty}^{\infty} \mathbb{R}_j$ where, for each j, $\mathbb{R}_j$ is the real line, i.e., Ω is the space of doubly infinite sequences of real numbers. We specify a point $\omega \in \Omega$ by $\omega = \{x_k\}$, $-\infty < k < \infty$, and let $\omega(j) = x_j$ for $-\infty < j < \infty$. The process $\{X_k\}$ induces a probability measure P on Ω . We will denote integration over Ω with respect to P measure by $E^P\{\ \}$.

For each positive integer n and each $\omega \in \Omega$, let $\omega^{(n)}$ be the point in Ω obtained by the periodic extension in both directions of the elements $x_1, x_2 \ldots, x_n$ of ω , i.e., if $\omega = \{x_k\}$, $-\infty < k < \infty$, then $\omega^{(n)}$ is the point

$$\ldots, x_1, \ldots, x_{n-1}, x_n, x_1, x_2, \ldots, x_n, x_1, x_2, \ldots x_n, \ldots$$

Let T be the shift operator acting on the points of Ω . For each $\omega \in \Omega$ and each positive integer n , we define a probability measure on Ω , call it $\pi_n(\omega)$, ad follows:

$$\text{(2)} \qquad \pi_n(\omega) = \frac{1}{n} (\delta_{\omega^{(n)}} + \delta_{\omega^{(n)}} + \ldots + \delta_{T^{n-1}\omega^{(n)}}) \; .$$

* The research in this paper was supported by the National Science Foundation, Grant No. MCS-80-02568.

In other words, given $\omega \in \Omega$ and $n \geq 1$, $\pi_n(\pi_n(\omega)$ is the probability measure on Ω which assigns mass $1/n$ to $\omega^{(n)}$ and to each of its $(n-1)$ translates. We note that $\pi_n(\omega)$ is a translation invariant measure on Ω, i.e., a stationary measure on Ω. Let $\underline{M}_S$ be the space of all stationary measures on Ω and impose on $\underline{M}_S$ the topology of weak convergence. We can use the mapping $\pi_n: \Omega \to \underline{M}_S$ and the measure P on Ω to construct a probability measure on $\underline{M}_S$ by defining, for each n, $Q_n = \pi_n P^{-1}$, i.e., if A is a set of stationary measures in $\underline{M}_S$, then

(3) $$Q_n(A) = P\{\omega \in \Omega: \pi_n(\omega) \in A\} .$$

The assumptions we made on the Gaussian process $\{X_k\}$ imply that it is an ergodic process, and it then follows from the ergodic theorem that, for almost all ω (P-measure), the measure $\pi_n(\omega)$ converges weakly to the measure P as $n \to \infty$, i.e.,

(4) $$P\{\omega \in \Omega: \pi_n(\omega) \Rightarrow P\} = 1 .$$

Thus, with reference to the Q_n measure on $\underline{M}_S$ given by (1.3), we expect that if the set A contains the measure P, then $Q_n(A) \to 1$ as $n \to \infty$, whereas if P is not in the closure of A, we expect $Q_n(A) \to 0$ as $n \to \infty$. In this paper we show that in this latter case $Q_n(A)$ approaches zero expenentially fast as $n \to \infty$ and we determine the constant in the exponential rate. To be more specific, let R denote a general element of $\underline{M}_S$, i.e., a stationary on Ω, and recall that f is the spectral density of our basic stationary Gaussian process $\{X_k\}$. We define a functional $H_f(R)$ which is, in fact, the entropy of the stationary process R with respect to the stationary Gaussian process $\{X_k\}$, and $H_f(R): \underline{M}_S \to [0,\infty]$ is such that:

If $C \subset \underline{M}_S$ is closed,

(5) $$\overline{\lim_{n\to\infty}} \frac{1}{n} \log Q_n(C) \leq - \inf_{R\varepsilon C} H_f(R) ,$$

and if $G \subset \underline{M}_S$ is open,

(6) $$\underline{\lim_{n\to\infty}} \frac{1}{n} \log Q_n(G) \geq - \inf_{R\varepsilon G} H_f(R) .$$

Consistent with remarks made above it will indeed be true that $H_f(P) = 0$. We give an explicit formula for $H_f(R)$ in (9) below.

Let $\Phi: \underline{M}_S \to \mathbb{R}$ be bounded and continuous, and let $E^{Q_n}\{\ \}$ denote integration over $\underline{M}_S$ with respect to Q_n measure. From (5) and (6) it follows easily that

(7) $$\lim_{n\to\infty} \frac{1}{n} \log E^{Q_n}\{e^{n\Phi(R)}\} = \sup_{R\varepsilon \underline{M}_S} [\Phi(R) - H_f(R)] .$$

Since $Q_n = \pi_n P^{-1}$, equivalent to (1.7) is

(8) $$\lim_{n\to\infty} \frac{1}{n} \log E^P\{e^{n\Phi(\pi_n(\omega))}\} = \sup_{R\varepsilon \underline{M}_S} [\Phi(R) - H_f(R)] .$$

The main results of this paper then are (5), (6) their implication (8), and the formula (9) for $H_f(R)$ which we discuss now.

Let $R \varepsilon \underline{M}_S$ and let, for A $\mathbb{R}$, $R(A|\omega) = R(X_0 \varepsilon A | X_{-1}, X_{-2}, \ldots)$ be the regular conditional probability distribution of X_0 given the entire past. Denote by $r(y|\omega)$ the corresponding density. If $G(\theta)$ is the spectral measure of the stationary process R, then the formula for $H_f(R)$ is

(9) $$H_f(R) = E^R\{\int_{-\infty}^{\infty} r(y|\omega) \log r(y|\omega) dy\}$$
$$+ \frac{1}{2} \log 2\pi + \frac{1}{4\pi} \int_0^{2\pi} \frac{dG(\theta)}{f(\theta)} + \frac{1}{4\pi} \int_0^{2\pi} \log f(\theta)\, d\theta ,$$

where it is understood that $H_f(R)$ is defined to be $+\infty$ if for any reason we cannot define any of the ingredients in (9).

The authors have developed a theory of large deviations for Markov processes (see [1], [2], for theoretical results and e.g. [3], [4], [5] for some applications thereof). In theory, our methods should apply to rather general stationary processes, but it is difficult to see in any great generality what the natural hypotheses to impose on a stationary process are in order to obtain the analogues of (5), (6) and (8) much less to obtain the analogue of (9) in anything like explicit form. For recent work in this direction see Orey [6].

The detailed proofs will appear in [7].

Bibliography

[1] Donsker, M. D., and Varadhan, S. R. S., Asymptotic Evaluation of Certain Wiener Integrals for Large Time, Functional Integration and its Applications, Proceedings of the International Conference held at the Cumberland Lodge, Windsor Great Park, London, in April 1974. A. M. Arthurs, Ed., Clarendon, Oxford, 1975.

[2] Donsker, M. D., and Varadhan, S. R. S., Asymptotic Evaluation of Certain Markov Process Expectations for Large Time, I, Comm. Pure Appl. Math. 28, 1976, pp. 1-47; II, Comm. Pure Appl. Mth. 28, 1975, pp. 279-301; III Comm. Pure Appl. Math. 29, 1976, pp. 389-461; IV, Comm. Pure Appl. Math. 36, 1983, pp. 183-212.

[3] Donsker, M. D., and Varadhan, S. R. S., Asymptotics for the Wiener Sausage, Comm. Pure Appl. Math. 28, 1975, pp. 525-565.

[4] Donsker, M. D., and Varadhan, S. R. S., On Laws of the Iterated Logarithm for Local Times, Comm. Pure Appl. Math. 30, 1977, pp. 707-753.

[5] Donsker, M. D., and Varadhan, S. R. S., Asymptotics for the Polaron, Comm. Pure Appl. Math. 36, 1983, pp. 505-528.

[6] Orey, Steven, Large Deviations and Shanon-McMillan Theorems, preprint.

[7] Donsker, M.D., and Varadhan, S.R.S., Large Deviations for Stationary Gaussian Processes, Comm. on Math. Physics (to appear).

ASYMPTOTIC EXPANSION OF THE LYAPUNOV EXPONENT AND THE ROTATION NUMBER FOR THE SCHRÖDINGER OPERATOR WITH RANDOM POTENTIAL

Volker Wihstutz
Universität Bremen
2800 Bremen 33, FRG

1. Introduction

This paper is an outcome of a joint work with L. Arnold, Universität Bremen, and G. Papanicolaou, Courant Institute of Mathematical Sciences, New York.

Let H with $Hy = -\ddot{y} + \sigma F(\xi)y$ be the one-dimensional Schrödinger operator with the random potential $\sigma F(\xi)$, where $\sigma \geq 0$, ξ a stationary and ergodic nice diffusion on the state space $Y = S^1$ (the unit circle) or $Y = [\alpha,\beta] \subset R$ with invariant measure $\nu(\xi)d\xi$. Let $F : \varphi \to \mathbb{R}$ be a smooth function s.t. $EF(\xi) = 0$.

We assume for the generator Q of ξ

(A) $Q = Q^*$, with pure point spectrum, $\mu_0 = 0 > \mu_1 > \dots$,
$Qe_n(\xi) = \mu_n e_n(\xi)$;

or

(B) $Q = a(\xi)\frac{\partial}{\partial\xi} + \frac{1}{2} b^2(\xi)\frac{\partial^2}{\partial\xi^2}$ with natural boundaries $(b(\alpha)=b(\beta)=0)$
and the zero-flux property $-a\nu + \frac{1}{2}\frac{d}{d\xi}(b^2\nu) = 0$.

Let $\gamma \in \mathbb{R}$ be an energy level. We consider the system

(1) $$-\ddot{y} + \sigma F(\xi)y = \gamma y$$

or, with $x = (y,\dot{y})^T$, $\dot{x} = \begin{pmatrix} 0 & 1 \\ \sigma F(\xi)-\gamma & 0 \end{pmatrix} x$.

The Lyapunov exponent of (1) then, in general, is defined pathwise as a random variable, depending also on the initial condition x_0,

$$\lambda(\omega,x_0) = \limsup_{t \to \infty} \frac{1}{t} \log|x(t,\omega,x_0)|$$

($x(t,\omega,x_0)$ a trajectory of the solution starting at x_0). Since $F(\xi)$ is a stationary and ergodic process with existing mean $E|F(S\xi)|$, by Oseledec' multiplicative ergodic theorem [10], for all $x_0 \neq 0$ and almost all ω $\lambda(\omega,x_0)$ takes one of the real values λ or $-\lambda$

(λ suitable). And, since ξ is a diffusion (thus (x,ξ) is, if we only permit non-anticipating solutions), by a theorem of L. Arnold and Kliemann, for all $x_o \neq 0$, almost surely

$$\lambda(\omega,x_o) = \max\{\lambda,-\lambda\} =: \lambda .$$

Moreover, the Lyapunov exponent can be represented as

$$(2) \qquad \lambda = (q,p) = \int_{\gamma}\int_{o}^{2\pi} q(\varphi,\xi)p(\varphi,\xi)\,d\varphi d\xi,$$

where φ is the angle corresponding to the projection $x/|x|$ onto the unit sphere S^1, obeying

$$(3) \qquad \dot{\varphi} = h(\varphi,\xi) = -[\sin^2\varphi + \gamma\cos^2\varphi - \sigma F(\xi)\cos^2\varphi],$$

$$(4) \qquad q(\varphi,\xi) = \frac{1}{\alpha}[1 + (\sigma\xi-\gamma)]\sin\alpha\varphi,$$

and $p(\varphi,\xi)$ the unique density (unique up to π-periodicity in φ) of the invariant measure associated to the pair (φ_t,ξ_t). See Kliemann and Arnold [6] for general dimension d and ξ living on a manifold without boundaries, or Arnold and Kliemann [2] for the 2x2-case. If $L = h\frac{\partial}{\partial\varphi} + Q$ denotes the generator of (φ_t,ξ_t), then $L^*p = 0$ or, in integrated form $0 = (Lf,p) = (h\frac{\partial}{\partial\varphi}f,p) + (Qf,p)$ for $f \in$ domain L. For $h\frac{\partial}{\partial\varphi}f = q$, (2) reads $\lambda = -(Qf,p)$.
Loparo and Blankenship [9] use a similar approach to represent λ.

We define the rotation number analogously to (2) as

$$(5) \qquad \alpha = (h,p).$$

2. Recalling Facts on the Lyapunov Exponent and Rotation Number

In the deterministic case $(\sigma=0)$, $\lambda = \lambda(\gamma) = 0$ for $\gamma \geq 0$, $\lambda = \sqrt{-\gamma}$ for $\gamma < 0$, $\alpha = \alpha(\gamma) = -\sqrt{\gamma}$ for $\gamma \geq 0$, $= 0$ for $\gamma < 0$, and $\lambda \pm i\alpha$ is the eigenvalue of $A = \begin{pmatrix} 0 & 1 \\ -\gamma & 0 \end{pmatrix}$ with the largest real part. This can be generalised in the following sense. There is a complex function w, which is holomorphic on the upper complex plane $\mathbb{C}^+$, such that $\lambda(\gamma) = -\operatorname{Re} w(\gamma+i0)$, $\alpha = \operatorname{Im} w(\gamma+i0)$. See Johnson and Moser [4] for almost periodic potential and Kotani [7] for a random potential with a stationary and ergodic process.
If (1) is viewed in a mechanical way as harmonic oscillator, α is the average of the angle velocity $\dot{\varphi}$, while λ describes the growth of $|x(t)| \sim \exp(\lambda t)$ in comparison with the exponential functions, thus being crucial for stability considerations.
In the context of the Schrödinger operator H those numbers are

characteristic for its spectral properties. A theorem of Pastur [11] says that $\alpha(\gamma) = \pi N(\gamma)$, where N is the integrated density of states of H, i.e.

$$N(\gamma) = \lim_{\ell\to\infty} N_\ell(\gamma),\ N_\ell(\gamma) = \frac{1}{2\ell}\{\lambda_i(\ell);\ \lambda_i \le \gamma\}$$

$\{\lambda_i(\ell)\}$ the set of of all eigenvalues of H w.r.t. functions on the intervall $[-\ell,\ell]$. The positivity of λ entails (Molčanov [8]) and is even equivalent to (Kotani [7]) the absence of an absolutely continuous spectrum of H. Moreover, λ is positive (for $\sigma > o$), if the noise is a stationary ergodic diffusion (Molčanov [8]) or, very general, a bounded non-deterministic stationary ergodic process (Kotani [7]).

3. Our Interest

Here we are interested in how λ and α behave as functions of the noise intensity σ, given γ, if σ is small $(\sigma\to o)$ or large $(\sigma\to\infty)$. We will expand λ and α in a suitable power of σ. For white noise and small σ this was done by Auslender and Mil'shtein [3]. They expanded λ for all possible 2x2-matrices.

4. Results

Theorem 1 (small noise intensity)
Let the generator Q of the stationary ergodic diffusion ξ satisfy condition (A). Then for small σ and each fixed $\gamma \neq 0$

(i) the invariant probability density

$p(\sigma) = p_o + \sigma p_1 + \sigma^2 p_2 + \ldots$ of (φ,ξ) satisfies

$$(f, p(\sigma) - [p_o + \sigma p_1 + \sigma^2 p_2]) = O(\sigma^3)$$

for all smooth functions f on $S^1 \times Y$;

(ii) the Lyapunov exponent of (1) has the expansion

$$\lambda(\sigma,\gamma) = \lambda_o(\gamma) + \sigma^2\lambda_2(\gamma) + O(\sigma^3)$$

where $\lambda_o(\gamma) = 0,\quad \lambda_2(\gamma) > 0$ for positive γ

and $\lambda_o(\gamma) = \sqrt{-\gamma},\ \lambda_2(\gamma) < 0$ for negative γ; and

(iii) the rotation number is

$\alpha(\sigma,\gamma) = -\sqrt{\gamma} + O(\sigma^3)$ for $\gamma > 0$, while vanishing for $\gamma < 0$.

(Note that the average rotational speed α hardly differs from the speed $-\sqrt{\gamma}$ of the deterministic system, if the noise is small; there is neither a first order nor a second order term.)

More precisely, for $\gamma > 0$

$$p_0(\varphi,\xi) = \sqrt{\gamma}\,\frac{1}{\pi}\,\frac{1}{h_0(\varphi)}\,\nu(\xi) ,$$

$h_0(\varphi) = \sin^2\varphi + \gamma \cos^2\varphi$.

If $\nu(\xi) = \text{const} = \nu$, the formular for λ_2 is

$$\lambda_2(\gamma) = \sum_{n=1}^{\infty} \sqrt{\gamma}\, c_n^2 \frac{\nu}{\pi} \int_0^{2\pi} \sin 2\varphi \cdot H_0(\varphi)^{-1} e^{M_n(\varphi)} [K_n - g_n(\varphi)]d\varphi$$

where c_n is defined by the Fourier expansion of $\frac{1}{2}F(\xi) = \Sigma c_n e_n(\xi)$, $Qe_n = \mu_n e_n$,

$$M_n(\varphi) = |\mu_n| \int_0^{\varphi} h_0(\alpha)^{-1} d\alpha = \frac{|\mu_n|}{\sqrt{\gamma}} \text{ arc tg}\left(\frac{\text{tg } \varphi}{\sqrt{\gamma}}\right)$$

$$g_n(\varphi) = \int_0^{\varphi} e^{-M_n(\alpha)} \sin 2\alpha \, h_0(\alpha)^{-2} \, d\alpha ,$$

$$K_n = (1-e^{-M_n(2\pi)})^{-1} g_n(2\pi) .$$

For $\gamma = 1$ and $\xi \in [\alpha,\beta]$, e.g.

$$\lambda_2 = (\beta-\alpha)^{-1} \sum_{n=1}^{\infty} c_n^2 |\mu_n| \; (\mu_n^2+4)^{-1} .$$

If $\gamma \to 0$, $\lambda_2(\gamma) \to 0$ at least with order $1/\sqrt{\gamma}$ (or faster) (see Wihstutz [12]) .

If $\gamma > 0$ and σ that small that $h = \sin^2 + \gamma \cos^2 f - \sigma F(\xi)\cos^2\varphi$ is bounded away from zero, the family of operators $L(\sigma) = Q + h(\sigma)\frac{\partial}{\partial\varphi}$, $\sigma \geq 0$, associated with (φ,ξ), is holomorphic in σ in such a sense that Kato's [5] analytic perturbation theory is applicable. $p(\sigma)$, $\lambda(\alpha)$ and $\alpha(\sigma)$ then even have analytic expansions (see Wihstutz [12]).

If $\gamma < 0$, $h_0(\varphi) = \sin^2\varphi + \gamma\cos^2\varphi$ vanishes at $\varphi_0 = \text{arc tg}\sqrt{-\gamma}$. (There is no ambignity; if σ is small, the support of the invariant measure is contained in $[0,\frac{\pi}{2}] \times Y$.) Here ($\delta$ denoting the Dirac function),

$$p_0 = \nu(\xi)\;\delta(\varphi-\varphi_0) ;$$

$$p_1 = r_1(\xi)\;\delta'(\varphi-\varphi_0)$$

with $r_1(\xi) = (1-\gamma)^{-1} R_o(F\nu)$, $R_o = (Q^*-2\sqrt{-\gamma})^{-1}$;

$$p_2 = r_{21}(\xi)\delta'(\varphi-\varphi_o) + r_{22}(\xi)\delta''(\varphi-\varphi_o)$$

with

$$\langle r_{21},1\rangle = c_1(\gamma)\ \langle F,-R_o(F\nu)\rangle,\quad \langle r_{22},1\rangle = c_2(\gamma)\ \langle F,-R_o(F\nu)\rangle,$$

$$c_1(\gamma) = (4\sqrt{-\gamma}(1-\gamma)^2)^{-1},\qquad c_2(\gamma) = (-4\gamma(1-\gamma)^2)^{-1}(1-3\gamma)\ ,$$

$\langle\cdot,\cdot\rangle$ denoting the inner product w.r.t. $d\xi$;

$$\lambda_2(\gamma) = \frac{1}{4\gamma}\ \langle F,\ -R_o(F\nu)\rangle < 0\ .$$

(since, $\sqrt{\nu}^{-1}\ R_o\ \sqrt{\nu}$ is negativ definit and $\gamma < 0$)

Remark 1

Although $\lambda(\gamma) > 0$ for $\sigma \geq 0$, $\lambda(\gamma,\sigma)$ is smaller than $\lambda_o(\gamma) = \sqrt{-\gamma}$. This stabilizing effect is due to a "mixture of the eigendirections of A". How to stabilize a linear system by noise, see Arnold et alii [1]. A criterian when the Lyapunov exponent is not effected by noise is given in [6].

Remark 2

We cannot treat the case $\gamma = 0$, σ small; but approaching (0,0) in the γ-σ-plane via $\gamma = \gamma_o\sigma$, $\gamma_o < 0$ we obtain

$$\lambda = \lambda_1(\gamma_o)\sqrt{\sigma} + \lambda_2(\gamma_o)\sigma + O(\sqrt{\sigma}^3),$$

where

$$\lambda_1(\gamma_o) = -\sqrt{-\gamma_o} + 2\sqrt{-\gamma_o}^{-3}\ \langle F,\ P(F\nu)\rangle\ ,$$

$$\lambda_2(\gamma_o) = \qquad - 2\sqrt{-\gamma_o}^{-2}\ \langle F,\ S(F\nu)\rangle$$

(P denoting the Projection of $L^2(Y)$ onto the eigenspace of Q^* w.r.t. 0 and S the reduced resolvent of Q^* w.r.t. 0 , i.e. $Q^*S = SQ^* = id - P$).

Theorem 2 (large noise intensity)

Let $\gamma = \gamma_o + \sigma\gamma_1$, γ_o, γ_1 real. Let F be the identical map on $\mathbb{R}$. If the generator Q of the stationary ergodic diffusion $\xi \in [\alpha,\beta]$ $\alpha < 0 < \beta$, satisfies condition (B), then

(i) for $\gamma_1 \geq \beta > 0$, γ_o arbitrary,

$$\lambda(\sigma,\gamma_1) = \qquad \lambda_o(\gamma_1) + O(\tfrac{1}{\sqrt{\sigma}})\ ,$$

$$\alpha(\sigma,\gamma_1) = \sqrt{\sigma}\,\alpha_{-1}(\gamma_1) + O(\tfrac{1}{\sqrt{\sigma}}) ,$$

where

$$\lambda_0(\gamma_1) = \int_\alpha^\beta [\tfrac{1}{4}a(\xi-\gamma_1)^{-1} - \tfrac{3}{16}b^2(\xi-\gamma_1)^{-2}]\nu d\xi$$

$$= \tfrac{1}{32} E_\nu\{b^2(\xi)(\xi-\gamma_1)^{-2}\} > 0 ,$$

$$\alpha_{-1}(\gamma_1) = E_\gamma\{\sqrt{-(\xi-\gamma_1)}\} > 0 ;$$

(ii) for $\gamma_1 \le \alpha < 0$, γ_0 arbitrary,

$$\lambda(\sigma,\gamma_1) = \sqrt{\sigma}\,\lambda_{-1}(\gamma_1) + \lambda_0(\gamma_1) + O(\tfrac{1}{\sqrt{\sigma}}) ,$$

$$\alpha(\sigma,\gamma_1) = 0 ,$$

where

$$\lambda_0(\gamma_1) = E_\nu\{\sqrt{(\xi-\gamma_1)}\} > 0 ,$$

$$\lambda_0(\gamma_1) = \int_\alpha^\beta [\tfrac{1}{4}a(\xi-\gamma_1)^{-1} - \tfrac{3}{16}b^2(\xi-\gamma_1)^{-2}]\nu d\xi$$

$$= \tfrac{1}{32} E_\nu\{b^2(\xi)(\xi-\gamma_1)^{-2}\} > 0 .$$

5. Ideas of proofs

If we are satisfied with weak convergence of the expansion for the density $p(\sigma)$ (and this is enough in order to obtain a convergent expansion for λ or α) , the proofs are given by very elementary perturbation theoretical arguments.

For small σ we put $L(\sigma) = L_0 + \sigma L_1$, where $L_0 = Q - h_0\frac{\partial}{\partial\varphi}$, $L_1 = -\sigma F(\xi)\cos^2\varphi\frac{\partial}{\partial\varphi}$ and solve the Fokker-Planck $0 = L^*p(\sigma)$ equation with the Ansatz

$$p(\sigma) = p_0 + \sigma p_1 + \sigma^2 p_2 + \ldots :$$

(6) $$L_0^*p_0 = 0 , \quad L_0^*p_k = -L_1^*p_{k-1} , \quad k = 1,2,\ldots .$$

Since both, q and h are of the form $f_0 + \sigma f_1$,

$$\lambda(\sigma) = (q_0,p_0), \ \lambda_1 = (q_1,p_0) + (q_0,p_1), \ \lambda_2 = (q_1,p_1) + (q_0,p_2)$$

and analogously for the α_k .

In order to justify this expansion, for given smooth f we solve the problem, adjoint to (6),

$$(7) \qquad L_0F_0 = f - f_0(\xi),\ L_0F_k = -L_1F_{k-1} - f_k(\xi)\ ,\ k = 1,2\ ,$$

where $\tilde{f}(\xi) = f_0(\xi) + \sigma f(\xi) + \sigma^2 f_2(\xi)$ is a function only depending on ξ and chosen such that the equations in (7) are solvable.

Then for $F = F_0 + \sigma F_1 + \sigma^2 F_2$, $LF = f - \tilde{f}(\xi) + \sigma^2 L_1 F_2$ and due to (6), $L^*(p_\sigma - [p_0 + \sigma p_1 + \sigma^2 p_2]) = -\sigma^2 L_1{}^* p_2$.

Therefore with $\tilde{p}_2 = p_0 + \sigma p_1 + \sigma^2 p_2$

$$(f, p_\sigma - [p_0 + \sigma p_1 + \sigma^2 p_2])$$
$$= (LF, p_\sigma - \tilde{p}_2) + (\tilde{f}, p_\sigma - \tilde{p}_2) - \sigma^2 (L_1 F_2, p_\sigma - \tilde{p}_2)$$
$$= -\sigma^2 (L_1 F, p_2) + (\tilde{f}, p_\sigma - \tilde{p}_2) - \sigma^2 (L_1 F_2, p_\sigma - \tilde{p}_2)$$

where the middle term vanishes. The order of convergence is σ^2, if one can show that $|(L_1 F, p_\sigma)|$, $|(L_1 F_2, \tilde{p}_2)|$, $\max |L_1 F_2| \le C$, C independent of σ.

<u>For large σ</u> we decompose $L(\sigma)$ into $\sqrt{\sigma}A + Q$, $A = \frac{h}{\sqrt{\sigma}} \frac{\partial}{\partial \varphi}$, and solve $A^* p_0 = 0$.

Then for f smooth

$$(f,p) = (f,p_0) - \frac{1}{\sqrt{\sigma}}(QF_0, p_0) - \frac{1}{\sigma}(QF_1, p_\sigma)$$

where $AF_0 = f - f_0(\xi)$, $AF_1 = -AF_0 - f_1(\xi)$.
It turns out that for $f = q$ or h, (f, q_0) vanishes or equals (f_0, p_0).

Now, (f_0, p_0) yields the $\sqrt{\sigma}$-term $+ O(\frac{1}{\sqrt{\sigma}})$ (if appearing), while $\frac{1}{\sqrt{\sigma}}(QF_0, p_0)$ is a constant $+ O(\frac{1}{\sqrt{\sigma}})$ and $\max |\frac{1}{-}QF_1| \le O(\frac{1}{\sqrt{\sigma}})$.

For both, σ small or large, the solution p_0 is not unique and need not to be independent of σ. The choice has to be made such that supp p_0 is close to supp p_σ, otherwise the expansion will not converge. Using support theorems and control the retical arguments we get hints for the support of the unknown p_σ fromthe vectorfield on $S^1 \times Y$.

For detailed proofs see the joint paper of L. Arnold, G. Papanicolaou and V. Wihstutz (in preparation).

REFERENCES

[1] L. Arnold, H. Crauel and V. Wihstutz,
Stabilization of Linear Systems by Noise,
SIAM J. Control and Optimization 21 (1983), 451-461

[2] L. Arnold and W. Kliemann
Qualitative Theory of Stochastic Systems in:
A. T. Bharucha-Reid(ed.): Probabilistic Analysis and
Related Topics, vol. 3, Academic Press, New York, 1981

[3] E. I. Auslender and G. N. Mil'shtein,
Asymptotic Expansion of the Lyapunov Index for
Linear Stochastic Systems with Small Noise,
Prikl. Matem. Mekhan 46 (1982), 358-365
engl.: PMM U.S.S.R. 46 (1983), 277-283

[4] R. Johnson and J. Moser,
The Rotation Number for Almost Periodic Potentials,
Comm. Math. Phys. 84 (1982), 403-438

[5] T. Kato,
Perturbation Theory for Linear Operators
Grundlehren der Mathematischen Wissenschaften
vol. 132, Springer, second ed. 1980

[6] W. Kliemann and L. Arnold,
Lyapunov Exponents and Linear Stochastic Systems
Report Nr. 93 (1983) des Forschungsschwerpunkts
Dynamische Systeme, Universität Bremen

[7] Sh. Kotani,
Lyapunov Indices Determine Absolutely
Continuous Spectra of Stationary Random
One-dimensional Schrödinger Operators
Proc. Stoch. Analys. Kyoto 1982

[8] S. A. Molčanov,
The Structure of Eigenfunctions of One-dimensional
Unordered Structures
engl.: Math. U.S.S.R. Izvestija 12 (1978), 69-101

[9] K. A. Loparo and G. L. Blankenship,
Almost Sure Instability of a Class of
Linear Stochastic Systems with Jump Process Coefficients
Preprint 1983

[10] V. I. Oseledec,
A Multiplicative Ergodic Theorem.
Lyapunov Characteristic Numbers for Dynamical Systems
engl.: Trans. Moscow Math. Soc. 19 (1968), 197-231

[11] L. A. Pastur,
Spectral Properties of Disordered Systems in the One-body
Approximation, Comm. Math. Phys. 75 (1980), 179-196

[12] V. Wihstutz,
Quantitative Results on Lyapunov Exponents
Report Nr. 99 (1983) des Foschungsschwerpunkts
Dynamische Systeme, Universität Bremen

HOMOGENEIZATION FOR EQUATIONS WITH RANDOM COEFFICIENTS.

V.V. YURINSKII
USSR, Novosibirsk, 90
Institute of Mathematics
Siberian Division of the
USSR Academy of Sciences.

1 - STATEMENT OF THE PROBLEM.

Let the coefficient $a_{ij} = a_{ji}$ of the operator

$$L_\varepsilon u \equiv \frac{1}{2} \Sigma a_{ij}(x/\varepsilon, \omega) D_i D_j u , \quad D_i = \partial/\partial x^{(i)} \tag{1}$$

where $y = \{ y^{(i)} \} \in R^d$, $i, j = 1, \dots, d$, $\omega \in \Omega$, be homogeneous random fields on a probability space $(\Omega, \mathfrak{M}, Pr)$. There are natural restrictions upon the coefficient field, providing G-convergence of L_ε for $\varepsilon \to 0$ to a homogeneized operator L_0 of the same kind as for non-random constant coefficients (see, e.g., [1-3]). The goal of this communication is to indicate some bounds for the error of approximation.

Everywhere below the dimension of the problem satisfies the restriction $d \geqslant 3$. The coefficient field has realizations continuous with probability 1, and for any $y \in R^d$, $\xi \in R^d$

$$A_1 |\xi|^2 \leqslant \Sigma a_{ij}(y) \xi^{(i)} \xi^{(j)} \leqslant A_2 |\xi|^2 , \quad |\xi|^2 = \Sigma (\xi^{(i)})^2 \tag{2}$$

while additionally assumed is the Cordes type condition

$$A_2/A_1 \leqslant 1 + A_3 , \tag{3}$$

where A_3 depends on the dimension of d and on the degree ρ in the bound (5) below.

The coefficient field is homogenuous with respect to integer shifts, i.e. the distribution of the field $\{ a_{ij}(y + z) \}$ does not depend on $z \in \mathbb{Z}^d$. It has a finite radius of dependence : if $\mathfrak{M}(B)$ is the σ-algebra, generated by the values of coefficients on the set $B \subset R^d$, then for any n

$$\mathfrak{M}(B_1), \dots, \mathfrak{M}(B_n) \text{ are independent if } \rho(B_i, B_j) \geqslant m_0 , \tag{4}$$

where $\rho(B, B')$ is the Euclidean distance between the sets.

The restrictions listed above allow us to estimate the approximation error, e.g., in the following case.

Let $Q \subset R^d$ be a bounded domain. Its boundary is smooth enough to ensure the validity of the a priori estimate

$$|D_i D_j u|_{Lp(Q)} \leq c\,|\Delta u|_{Lp(Q)} \tag{5}$$

for some $p > d + 1$, where Δ is the Laplace operator and $u(x) = 0$, $x \in \partial Q$. The non-random function f is also regular enough for the existence of a solution in the classe $C^{2+\alpha}$, $\alpha > 0$, for the Dirichlet problem

$$L_0 U = f, \quad x \in Q \;;\quad U = 0, \quad x \in \partial Q \tag{6}$$

with L_0 the homogeneized operator.

Theorem 1 - If conditions (1) - (5) hold, then the solution of the boundary problem

$$L_\varepsilon u_\varepsilon = f, \quad x \in Q \;;\quad u_\varepsilon = 0, \quad x \in \partial Q, \tag{7}$$

is connected to the solution of the homogeneized problem (6) by the relation

$$E \sup_Q |u_\varepsilon - U| \leq c\,\varepsilon^\beta$$

with the constant $\beta > 0$ determined by d, p and the constants in conditions (1), (2).

A similar result is also valid in the cas then the operator L_ε contains highly oscillating lower order terms with bounded coefficients. The boundary-value problem may also be replaced by the initial boundary-value one for the corresponding parabolic equation.

2 - METHOD OF PROOF.

Let η be a diffusion process with the generator (1), L_1 corresponding to a fixed realization of coefficient field ; P_y is the distribution in the space of trajectories of η corresponding to the initial condition $\eta(0) = y$, M_y is the integral for the measure P_y. Averaging with respect to the "environment" $\omega \in \Omega$ is, as above, denoted by Pr,

$$E(.) = \int_\Omega (.)\,Pr(d\omega), \quad EP_y(A) = EM_y I_A,$$

$$EM_y \varphi = \int_\Omega Pr(d\omega) \left(\int_C \varphi(\omega, u)\, P_y(du) \right).$$

Let k be one of the coefficients a_{ij},

$$\{k \mid T, U\} = (U - T)^{-1} \int_T^U k(\eta(s))\, ds \, .$$

Lemma 1 - For $T \geqslant 2$, $t \geqslant 1$, $y \in R^d$,

$$| EM_y \{k \mid T, T+t\} - EM_0 \{k \mid 0, t\} | \leqslant c(T^{-x_1} + t^{-x_1} + (T \ell n(2+T) \cdot t^{-d/2})^{x_2} \ell n\, t),$$

$$x_{1,2} > 0 \, .$$

Proof - Let $N[s, t]$ be the set of integer "numbers" $z \in \mathbb{Z}^d$ such that the diffusion path $\eta(s')$ for $s \leqslant s' \leqslant t$ "touches" the cube $z + [0, 1]^d$. Starting from the exponential bound

$$P_y \{ \max_{s \leqslant t \leqslant s+h} |\eta(t) - \eta(s)| \geqslant r \} \leqslant \exp\{-cr^2/h\} \, , \tag{8}$$

it is easy to verify that uniformly in "environments" $\omega \in \Omega$

$$M_y N[0, T] \leqslant c T \ell n(2 + T) \tag{9}$$

A smoothness estimate from [4] and the Markov property of η result in the equality

$$EM_y \{k \mid T, T+t\} = EM_y M_{\eta(T)} \{k \mid 0, t\} = E \sum_{L \subset \mathbb{Z}^d} \sum_{z \in \mathbb{Z}^d} P_y \{N[0,T] \tag{10}$$

$$= L, \eta(T) \in z + [0, 1]^d \}(M_z \{k \mid 0, t\} + \rho), |\rho| \leqslant ct^{-x_1}, x_1 > 0 .$$

In this equality $P_y \{N[0, T] = L, \ldots\}$ is determined by the values of a_{ij} on the set

$$V_L = \bigcup_{z \in L} (z + [0, 1]^d)$$

By means of (8) one verifies that with an error not exceeding $\exp\{-c(\ell n\, t)^2\}$ it is possible to recover $M_z \{k \mid 0, t\}$ from the values of a_{ij}, k on the ball $B = \{|x - z| \leqslant c t^{\frac{1}{2}} \ell n\, t\}$.

Due to (9) the Lebesgue measure of the set V_L has the order of magnitude $T \ell n(2 + T)$. Therefore, calculating $M_z \{k \mid 0, t\}$ one can change the coefficients a_{ij}, k in the neighbourhood of the set V_L, e.g., so that they do not depend on the σ-algebra $\mathfrak{M}(V_L)$. The "distorted mean" differs relatively little from the initial one :

$$|M_z \{k \mid 0,t\} - \widetilde{M}_z \{\widetilde{k} \mid 0,t\}| \leqslant c(((1 + \text{mes}\, V_L)\, t^{-d/2})^{x_2} \ell n\, t + \exp\{-c'(\ell n\, t)^2\} . \tag{11}$$

To obtain (11) we used inequalities of § 2, ch. 11 [5] and the a priori estimate of the form :

$$|D_i D_j u|_{Lp(R^d \times (0,t))} \leq c(d,p) \, |(\partial/\partial t + L_1) u|_{Lp(R^d \times (0,t))} .$$

Its validity follows from the Cordes condition (3) .

Due to construction and (4) the "distorted mean" $\tilde{M}_z \{\tilde{k} \mid 0, t\}$ and the probability $P_y \{N[0,T] = L, \eta(T) \in z + [0,1]^d\}$ are independent random variables on the probability space $(\Omega, \mathfrak{M}, Pr)$. By homogeneity and (11)

$$E\tilde{M}_z \{\tilde{k} \mid 0,t\} = EM_0 \{k \mid 0,t\} + \rho' ,$$

$$|\rho'| \leq c((1 + \text{mes}\, V_L)\, t^{-d/2})^{x_2} \, \ell n\, t + \exp\{-c'(\ell n\, t)^2\} .$$

Therefore, one can write (10) in the form :

$$EM_y \{k \mid T, T+t\} = EM_0 \{k \mid 0,t\} \times E \sum_{z \in \mathbb{Z}^d} \sum_{L \subset Z^d} P_y \{N[0,T] =$$

$$= L, \eta(T) \in z + [0,1]^d\} + \rho'' = EM_0 \{k \mid 0,t\} + \rho'' ,$$

where

$$|\rho''| \leq c\, EM_y (((1 + \text{mes}\, V_L)\, t^{-d/2})^{x_2} \, \ell n\, t + t^{-x_1})$$

is estimated by means of (9). To obtain the inequality of the lemma, one has to "round off" y by means of estimate in [4].

Lemma 2 - For any y

$$|EM_y \{k \mid 0,t\} - \langle k \rangle| \leq c\, t^{-\alpha_1} , \quad \alpha_1 > 0 ,$$

where

$$\langle k \rangle = \lim_{t \to \infty} EM_0 \{k \mid 0,t\} .$$

Proof - Let $K(t) = EM_0 \{k \mid 0,t\}$. From the equality

$$K(T) = \frac{1}{n} \sum_{j=0}^{n} EM_0 \{k \mid jt, jt+t\} + \mathcal{O}(1/n) , \tag{12}$$

where $n = [T/t]$, and from Lemma 1 it follows that for large t, T

$$K(T) - K(t) = \mathcal{O}(t/T + t^{-x_1} + T^{-x_1} + (T . \ln T . t^{-d/2})^{x_2} \ln t .$$

Hence the existence of the limit for the sequence $K_j = K(T_j)$, $T_j = e^{(1+\delta)^j}$ is easily obtained, if $\delta > 0$ is small enough ; besides,

$$| \langle k \rangle - K_j | \leqslant c\, T_j^{-\delta'} \quad , \quad \delta' > 0 .$$

Inequality of the lemma is deduced by interpolating this result on the interval $[T_j , T_{j+1}]$ by means of (12) since for $s \in [T_j , T_{j+1}]$ one of the ratios s/T_j , T_{j+1}/s is estimated from below by the value

$$(T_{j+1} / T_j)^{\frac{1}{2}} \geqslant s\, \delta'' \quad , \quad \delta'' > 0 .$$

Lemma 3

$$EM_y\, (\langle k \rangle - \{ k \mid 0 , t \})^2 < c\, t^{-\alpha_2} \quad , \quad \alpha_2 > 0 .$$

Proof - It is analogous to the derivation of the estimate of Lemma 1 . It is obvious that for $T = nt$, $\langle k \rangle = 0$

$$EM_y \{ k \mid 0, nt \}^2 = n^{-2} \sum_{\ell} EM_y \{ k \mid \ell t , \ell t + t \}^2 +$$

$$+ \frac{2}{n^2} \sum_{\ell < m} EM_y \{ k \mid \ell t , \ell t + t \} \{ k \mid mt , mt + t \} .$$

The first summand does not exceed c/n since the means $\{ k \mid \ldots \}$ are w.p. 1 bounded by the same constant as the corresponding coefficient of k . For $\ell < m$ one can use a representation similar to (10) :

$$\tilde{K}_{\ell m} = EM_y \{ k \mid \ell t , \ell t + t \} \{ k \mid mt , mt + t \} =$$

$$= E \sum_{L \subset \mathbb{Z}^d} \sum_{z \in \mathbb{Z}^d} M_y \{ \{ k \mid \ell t , \ell t + t \} ; N[0 , mt] = L ,$$

$$\eta(mt) \in z + [0 , 1]^d \} \, (M_z \{ k \mid 0 , t \} + \mathcal{O}(t^{-x_1})) .$$

Estimating the latter one can "break" the dependance with small error, distorting coefficients on the set V_L when calculating $M_z \{ k \mid 0, t \}$. After that the smallness of $\tilde{K}_{\ell m}$ follows from Lemma 2 since $\langle k \rangle = 0$.

An outline of proof of Theorem 1. The equation for the error term $v_\varepsilon = u_\varepsilon - U$ may be written in the form :

$$L_\varepsilon v_\varepsilon = \Sigma \; k_{ij} U_{ij} \zeta + (1 - \zeta) \Sigma \; k_{ij} U_{ij} \tag{13}$$

where $k_{ij} = -\frac{1}{2}(a_{ij}(x/\varepsilon) - \langle a_{ij} \rangle)$, $U_{ij} = D_i D_j U$, and $\zeta(x)$ is a non-random cutoff function for the domain Q, which equals 1 at a distance from the boundary greater than $\delta_1 = \delta_1(\varepsilon)$ while its first order derivatives do not exceed c/δ_1 in the absolute value.

The contribution r_1 of the second summand in the right-hand side of (13) is estimated by the value $c\,\delta_1^{\beta_0}$, $\beta_0 > 0$, using inequalities of Ch. 11, [5] since k_{ij}, U_{ij} are bounded.

To estimate the contribution of the first summand in the right-hand side of (13), one can write down the solution as a continuous integral. Let η^ε be the diffusion process with the generator (1) and $M_x = M_x^\varepsilon$ be the mean with respect to its distribution. Then

$$v_\varepsilon(x) = r_1(x) + \Sigma \; M_x \int_0^\tau (\zeta\, k_{ij} U_{ij})(\eta^\varepsilon(s))\, ds =$$

$$= r_1(x) + \sum_{i,j} \; \sum_{\ell=0}^{\infty} \; M_x M_{\eta^\varepsilon(\tau_\ell)} \int_0^{\tau_1} (\zeta\, k_{ij} U_{ij})(\eta^\varepsilon(s))\, ds \,, \tag{14}$$

where τ is the hitting time of the boundary of Q, $\tau_\ell = (\ell\delta) \wedge \tau$, and $\delta = \delta(\varepsilon)$, as δ_1 will be chosen later.

If the point $\eta^\varepsilon(\tau_\ell)$ is removed from the boundary for a distance no greater than δ_1^2, then the smallness of $\zeta(\eta^\varepsilon)$ in "typical" points of the trajectory of η^ε allows one to obtain the estimate

$$\left| M_{\eta^\varepsilon(\tau_\ell)} \int_0^{\tau_1} (\zeta\, k_{ij} U_{ij})(\eta^\varepsilon(\varepsilon))\, ds \right| \leq c(\delta_1^{\beta_1} + (\delta/\delta_1)^{\beta_2}) M_{\eta^\varepsilon(\tau_\ell)} \tau_1 \,. \tag{15}$$

In the case when $\eta^\varepsilon(\tau_\ell)$ is far from the boundary, the probability that τ_1 is not equal to the constant δ is negligible. Due to the smoothness of ζ, U_{ij}, one can substitute for these functions in the integrand of (13) their values at the point $\eta^{\varepsilon}(\tau_{\ell})$:

$$| M_{x}^{\varepsilon} M_{\eta^{\varepsilon}(\tau_{\ell})}^{\varepsilon} \int_{0}^{\tau_1} (\zeta k_{ij} U_{ij})(\eta^{\varepsilon}(s)\, ds | \leqslant$$

$$\leqslant c(\delta^{\beta_3} + (\delta/\delta_1)^{\beta_4}) \cdot M_{x}^{\varepsilon}(\tau_{\ell+1} - \tau_{\ell}) + \delta M_{x}^{\varepsilon} | M_{\eta^{\varepsilon}(\tau_{\ell})} \frac{1}{\delta} \int_{0}^{\delta} k_{ij}(\eta^{\varepsilon}(s))\, ds | . \quad (16)$$

The "environment" mean of expression (16) is estimated as in proofs of Lemma 1, 3, starting from equalities of type (10) (the position of $\eta^{\varepsilon}(\tau_{\ell})$ is no longer essential) :

$$E M_{x}^{\varepsilon} | M_{\eta^{\varepsilon}(\tau_{\ell})}^{\varepsilon} \frac{1}{\delta} \int_{0}^{\delta} k_{ij}(\eta^{\varepsilon}(s))\, ds | \leqslant c[(\varepsilon/\delta)^{\beta_5} + (\ell\, \varepsilon^{d-2}/\delta^{(d-2)/2})^{\beta_6}] . \quad (17)$$

Combining (15), (16) and (17), we obtain the inequality

$$E | M_{x}^{\varepsilon} | M_{\eta^{\varepsilon}(\tau_{\ell})}^{\varepsilon} \int_{0}^{\tau_1} (\zeta k_{ij} U_{ij})\, ds | \leqslant c\, \varepsilon_1(\ell)\, M_{x}^{\varepsilon}(\tau_{\ell+1} - \tau_{\ell}) , \quad (18)$$

where $\varepsilon_1(\ell) = (\varepsilon^2/\delta)^{\gamma_1} + \delta^{\gamma_2} + (\delta/\delta_1^2)^{\gamma_3} + (\ell\, \varepsilon^{d-2}\, \delta^{-(d-2)/2})^{\gamma_4}$, $\gamma_i > 0$.

This inequality is satisfactory for not too long "prehistories" ℓ. The remainder of the series in (14) is estimated by the formula

$$| \sum_{\ell \geqslant \ell_0} M_{x}^{\varepsilon} M_{\eta^{\varepsilon}(\tau_{\ell})}^{\varepsilon} \int_{0}^{\tau_1} (\zeta k_{ij} U_{ij})\, ds | \leqslant c(\ell\ \delta)^{-1} ,$$

since $M_{x}^{\varepsilon} \tau^2$ is bounded by a constant under the assumptions of the theorem.

Finally, for $\delta = \varepsilon^{\zeta_1}$, $\delta_1 = \delta^{\zeta_2}$, $\ell_0 \sim \delta^{-2}$, where $\zeta_i > 0$ are sufficiently small, we obtain the estimate

$$E | v_{\varepsilon}(x) | \leqslant c[\varepsilon_1(\ell_0) + (\ell_0\ \delta)^{-1} + \delta_1^{\beta_0}] \leqslant c\, \varepsilon^{\zeta_3} .$$

The inequality of the theorem follows from this "pointwise" estimate. Indeed, under the conditions of the conditions of the theorem, due to (3), (5), the solution and its first-order derivatives are bounded by a non-random constant, and, consequently,

$$E \sup_Q |v_\varepsilon(x)| \leqslant c\,(E \int_Q |v_\varepsilon(x)|\,dx)^\alpha, \quad \alpha > 0 .$$

REFERENCES

[1] Yurinskii V.V. On homogeneization of non-divergent second-order equations with random coefficients. Sib. Math. J., 1982, v. 23, N° 2, p. 176-188 (in Russian).

[2] Papanicolaou G.C., Varadhan S.R.S. Diffusions with random coefficients. – In : Statistics and Probability Essays in Honour of C.R.Rao. Amsterdam-New-York. North Holland, 1982, p. 547-552.

[3] Zhikov V.V., Sirazhudinov M.M. On G-compactness of a class of non-divergent elliptic operators of second order. Izv. AN SSSR, Ser. math., 1981, v. 45, N° 4, p. 718-734 (in Russian).

[4] Krylov N.V., Safonov M.V. A property of solutions of parabolic equations with measurable coefficients. Izv. AN SSSR, ser. math., 1980, v. 44, N° 1, p. 161-175 (in Russian).

[5] Krylov N.V. Controlled Processes of Diffusion Type, Moscow : "Nauka", 1977 (in Russian).

3. STOCHASTIC EQUATIONS. DIFFUSIONS

A NICE DISCRETIZATION FOR STOCHASTIC LINE INTEGRALS

J.M.C. Clark

Department of Electrical Engineering
Imperial College, London SW7 2BT
England

INTRODUCTION

Consider the following discretization problem of stochastic differential equations. A sequence of nested regular partitions of a time interval is given, with the partition mesh sizes converging to zero. For each partition, a solution at the final time of a stochastic differential equation is to be approximated by a function of the values of a driving Brownian motion at the points of the partition. Which discretization methods give, asymptotically, the best rate of convergence achievable by sequences of such "partition measurable" approximations, and also what is this best rate? The answer to these questions can be categorized as "semi-weak" or "strong" according to whether they are phrased in terms of an averaged measure of error, such as an L^2-norm or conditional L^2-norm, or as "almost sure" statements. There are many semi-weak results in the literature. Early references are McShane [5] and Mil'shtein [6]; the more recent work of Rümelin, Talay and others can be found in [8,9]. These are mainly results establishing that various methods give the best order of magnitude of error, which in the case of a scalar driving process is $O(h_n)$, where h_n is the partition mesh size, and in the vector case is $O(h_n^{\frac{1}{2}})$.

The discretization problem is essentially a question of the rate of convergence of convergent quasimartingales of approximations; so it is not surprising that more precise semi-weak rate results can be obtained from a central limit theorem. Conditional central limit theorems are given in [2,7] for the case of scalar driving processes, that give the correct scaling for the errors of particular methods and that establish the optimality of their rates of convergence.

The purpose of this note is to give both semi-weak and strong results for the case of a vector driving Brownian motion. To simplify matters we shall restrict our attention to approximation of stochastic line integrals, though this retains all the salient features of the discretization problem. It turns out that the best rate, in all senses, is achieved by the "trapezoidal rule" quadrature formula.

Semi-weak results based on a conditional central theorem are proved in the next section. The strong discretization problem is formulated, and solved, in the final section. Here the results are derived from a law of the iterated logarithm similar to that of Heyde for the tail sums of martingale differences [3,4], and also from a "strong lower bound" property of certain martingales.

A CONDITIONAL CENTRAL LIMIT THEOREM FOR THE ERRORS

Let $W_t(\omega) = \omega(t)$ be the d-dimensional coordinate process of the space $\Omega=C([0,1),\mathbb{R}^d)$ and F its Borel field generated by the uniform norm topology. The distribution P of (W_t) on F will be assumed to be Wiener measure. Then (W_t) is a continuous Brownian motion on the filtration (F_t), where F_t is the σ-algebra $\sigma(W_s:s\leq t)$. Let P_n the partition σ-algebra $\sigma(W_s:s\in\Pi_n)$ generated by (W_t) on the partition $\Pi_n:=\{0,1/2^n,2/2^n,\ldots,(2^n-1)/2^n,1)\}$, and let $h_n = 1/2^n$, the mesh size of Π_n.

Suppose I is the stochastic line integral

$$I = \sum_{i=1}^{d} \int_0^1 g_i(W_t)\circ dW_t^i \tag{1}$$

where the integrals are in Stratonovich form and the g_i are of class $C^2(\mathbb{R}^d)$ and satisfy a polynomial growth condition.

Let I_n be the "trapezoidal rule" approximation on $\Pi_n =: \{0=t_0<t_1<\ldots<1\}$

$$I_n = \sum_{k=0}^{2^n-1} \sum_i \tfrac{1}{2}(g_i(W_{t_k})+g_i(W_{t_{k+1}}))(W^i_{t_{k+1}}-W^i_{t_k}). \tag{2}$$

<u>Theorem 1.</u> Let $\phi_n(t,\omega)$ be a version, continuous in t, of the conditional characteristic function $E[\exp ith^{-\frac{1}{2}}(I-I_n)|P_n]$ of the normalized error of approximation $(I-I_n)h_n^{-\frac{1}{2}}$. Then for almost all ω

$$\lim_{n\uparrow\infty} \phi_n(t,\omega) = e^{-\frac{1}{2}\eta^2(\omega)t}$$

where

$$\eta^2 = \frac{1}{4}\sum_{i<j}\int_0^1 \left(\frac{\partial g_i}{\partial x^j}(W_t) - \frac{\partial g_j}{\partial x^i}(W_t)\right)^2 dt. \tag{3}$$

This theorem, besides giving the rate of convergence of the "trapezoidal rule" sequence I_n to I also shows that (I_n) is as fast, or as "asymptotically efficient", as any other (P_n)-adapted sequence. There are a number of ways this can be phrased, depending on the choice of measure of error, but the following corollary is representative of the sort of result that can be obtained:

Corollary: Let (Y_n) be a (P_n)-adapted sequence converging to I. If $\eta^2 > 0$ a.s., then for any $c > 0$

$$\underline{\lim}_{n\uparrow\infty} \frac{P(|I-Y_n|>ch_n^{\frac{1}{2}}|P_n)}{P(|I-I_n|>ch_n^{\frac{1}{2}}|P_n)} \geq 1 \text{ a.s.}$$

Proof: Since the normal distribution function $\Phi(n,\eta^2)$ with variance η^2 is continuous in x, by Theorem 1 the conditional distribution function of $h_n^{-\frac{1}{2}}(I-I_n)$ converges to $\Phi(x,\eta^2)$ uniformly in x for almost all ω. Hence for any $\varepsilon > 0$

$$|P(I-I_n > ch_n^{\frac{1}{2}}|P_n) - \Phi(-c,\eta^2| \leq \varepsilon \qquad \text{a.s.,}$$

and

$$|P(I-I_n <-ch_n^{\frac{1}{2}}|P_n) - \Phi(-c,\eta^2)| \leq \varepsilon \qquad \text{a.s.}$$

uniformly in c for all n sufficiently large. So by the unimodal symmetry of the normal distribution

$$\begin{aligned} P(|I-Y_n|>ch_n^{\frac{1}{2}}|P_n) &= P(I-I_n>ch_n^{\frac{1}{2}}+(Y_n-I_n)|P_n) \\ &\quad + P(I-I_n<-ch_n^{\frac{1}{2}}+(Y_n-I_n)|P_n) \\ &\geq 2\Phi(-c,\eta^2)-2\varepsilon \\ &\geq P(|I-I_n|>ch_n^{\frac{1}{2}}|P_n)-4\varepsilon. \end{aligned}$$

The convergence of the last probability to a positive value and the arbitrariness of

ε then establishes the result.

Remarks 1. Results similar to the corollary can be derived if the error is measured by conditional L^p norms. That is, if $p>0$ and $\eta^2>0$ a.s., for any $(\mathcal{P}_n)$-adapted (Y_n) $\underline{\lim}_n E[(I-Y_n)^2|\mathcal{P}_n]/E[(I-I_n)^2|\mathcal{P}_n] = 0$ a.s.

2. Theorem 1 can be extended to apply to processes (W_t) with distributions absolutely continuous with respect to Wiener measure (c.f.[2]), or to the trivial case where W_t is of the form $\psi(V_t)$ where (V_t) is a Brownian motion in $\mathbb{R}^d$ and ψ is a sufficiently smooth 1-1 transformation of $\mathbb{R}^d$. It is plausible that it also extends to the more interesting case where W_t is a non-degenerate diffusion of the form $dW_t=f(W_t)dV_t$, but the proof of this would require rather different techniques than those used below.

3. The inclusion of an ordinary integral $\int_0^1 g_0(W_t)dt$ in (1), with a corresponding modification of (2), would have no effect on the form of the limiting distribution of error. Also (1) is written in Stratonovich form simply for notational convenience. The Ito form suggests the alternative discretisation.

$$I_n^* = \sum_k \sum_i g_i(W_{t_k})(W^i_{t_{k+1}} - W^i_{t_k}) + \frac{1}{2} h_n \sum_k \sum_i \frac{\partial g_i}{\partial x^i}(W_{t_k}).$$

However, though (I_n^*) converges a.s. to I with errors of a similar order $h_n^{\frac{1}{2}}$, it is not in general as fast as I_n. For instance, it can be shown that (see e.g. [1]) if

$$I = \int_0^1 W_t^2 dW_t^1 \quad , \quad \text{then} \quad \lim_n \frac{E(I_n^*-I)^2}{E(I_n-I)} = 2.$$

4. Finally, it is possible to make plausible inferences for the discretization of stochastic differential equations. It is clear that any "sensible" second-order discretization scheme, such as the Mil'shtein scheme or a second-order Runge-Kutta scheme (see [6,8,9]) would converge as fast as any other partition-adapted scheme, provided that the equation to be discretized is driven in a nondegenerate way by more than one Brownian motion. "Nondegenerancy" here would be a non-commutativity condition for the Brownian vector fields similar to the non-integrability condition $\eta^2 > 0$.

Proof of Theorem 1. First recall that P is Wiener measure and (W_t) is a Brownian motion. Let t_k denote $k/2^n$ and let $W^i_{nk} = W^i(t_{k+1})-W^i(t_k)$. Consider the contribution to $I-I_n$ on the interval $[t_k,t_{k+1}]$. Expanding $g(W_t)$ in terms of Stratonovich integrals, we have

$$\sum_i\Big[\int_{t_k}^{t_{k+1}} g_i(W_t)\circ dW^i_t - \tfrac{1}{2}(g_i(W_{t_k})+g_i(W_{t_{k+1}}))\Delta W^i_{nk}\Big] = \sum_{ij}g_{ij}(W_{t_k})L^{ij}_{nk} + R_{nk}$$

where g_{ij} is short for $\frac{\partial g_i}{\partial x^j}$, L^{ij}_{nk} is the "stochastic area" integral of Lévy on the interval $[t_k,t_{k+1}]$.

$$L^{ij}_{nk} := \frac{1}{2}\int_{t_k}^{t_{k+1}} [(W^j_t-W^j_{t_k})\circ dW^i_t - (W^i_t - W^i_{t_k})\circ dW^j_t]$$

(notice that L^{ij}_{nk} expressed in Ito integrals takes the same form) and R_{nk} is a remainder term

$$R_{nk} := \frac{1}{2}\int_{t_k}^{t_{k+1}} \sum_{ij}\int_{t_k}^{t} [g_{ij}(W_s)-g_{ij}(W_{t_k})]\circ dW^j_s\circ dW^i_t$$
$$- \frac{1}{2}\int_{t_k}^{t_{k+1}} \sum_{ij} [g_{ij}(W_t)-g_{ij}(W_{t_k})](W^i_t-W^i_{t_k})\circ dW^j_t.$$

By reducing the Stratonovich integrals to Ito integrals it is straightforward to verify, given the polynomial growth conditions on the partial derivatives of g_i and the Markov property of (W_t), that:

$$E[|E[R_{nk}|F_{t_k}]|^2] \le C_1h^4_n$$
$$E[R^2_{nk}] \le C_2h^3_n$$

for some constants C_1, C_2. R_{nk} is $F_{t_{k+1}}$-measurable and the partial sums of R_{nk} over k form a quasimartingale on $(F_{t_{k+1}})$. If $R_n := \sum_{k=0}^{2^n-1}R_{nk}$ then $ER^2_n = O(h^2_n)$ and $\sum_0^\infty h^{-1}_n ER^2_n < \infty$, and by the Borel-Cantelli lemma $h^{-1/2}_n R_n \to 0$ a.s. . So the P_n-conditional characteristic function of $h^{-1/2}_n(I-I_n)$ is asymptotic to that of S_n, where

$$S_n := h_n^{-\frac{1}{2}} \sum_{k=0}^{2^n} \sum_{ij} g_{ij}(W_{t_k}) L_{nk}^{ij}$$
$$= \sum_{k=0}^{2^n-1} (h_n^{-\frac{1}{2}} \sum_{i<j} [g_{ij}(W_{t_k}) - g_{ji}(W_{t_k})] L_{nk}^{ij} =: \sum_k X_{nk} . \tag{4}$$

Now the following facts about the stochastic processes L_{nk}^{ij} will be required:

Conditional on P_n, L_{nk}^{ij} is independent of $L_{nk'}^{\ell m}$, if $k = k'$. (5a)

$$E[L_{nk}^{ij} | P_n = 0 \quad \text{a.s.} \tag{5b}$$

$$E[(L_{nk}^{ij})^2 | P_n] = \frac{h_n}{12}[\Delta W_{nk}^{i2} + \Delta W_{nk}^{j2} + h_n] \quad \text{a.s.} \tag{5c}$$

$$E[L_{nk}^{ij} L_{nk}^{\ell j} | P_n] = -E[L_{nk}^{ik} L_{nk}^{j\ell} | P_n] = \frac{h_n}{12} \Delta W_{nk}^{i} \Delta W_{nk}^{\ell} \text{ if } i \neq \ell \quad \text{a.s.} \tag{5d}$$

$$E[L_{nk}^{ij} L_{nk}^{\ell n} | P_n] = 0 \quad \text{a.s. if } \{i,j\} \cap \{\ell,m\} = \phi \tag{5e}$$

$$E[(L_{nk}^{ij})^4] \leq k_4 h_n^4 \qquad \text{for some constant } k_4 . \tag{5f}$$

Statements (5a,f) are obvious. A simple way of proving the remaining relations is to decompose $W_t^i - W_{t_k}^i$ into $h_n^{-1}(t-t_k)\Delta W_{nk}^i + \tilde{W}_{k,t-t_n}^i$. Then the process $(\tilde{W}_{kt}^i)_{t \leq h_n}$ is a Brownian bridge pinned to zero at 0 and h_n; that is, $\tilde{W}_{kt}^i$ is a continuous Gaussian semimartingale with $E\tilde{W}_{kt}^i = 0$ and $E(\tilde{W}_{ks}^i \tilde{W}_{kt}^i) = h_n^{-1} s(h_n - t)$ for $0 \leq s \leq t \leq h_n$. Furthermore the Brownian bridges $(\tilde{W}_{kt}^i)_{i,k}$ are independent of each other and P_n. So L_{nk}^{ij} can be expressed as the sum of three mutually orthogonal integrals

$$L_{nk}^{ij} = h_n^{-1} \Delta W_{nk}^i \int_0^{h_n} \tilde{W}_{kt}^j dt - h_n^{-1} \Delta W_{nk}^j \int_0^{h_n} \tilde{W}_{kt}^i dt + \int_0^{h_n} \tilde{W}_{kt}^j d\tilde{W}_{kt}^i .$$

It turns out that the first two integrals each have a second moment of $\frac{1}{12} h_n^3$ and the third, a second moment of $h_n^2/12$. The relations (5b,c,d,e) then follow by elementary computation.

As a consequence of (5a-f) the elements X_{nk} of the sum S_n are P_n-conditionally independent and have zero P_n-conditional mean. Furthermore

$$E[S_n^2|P_n] = \sum_k E[X_{nk}^2|P_n] = \sum_k (\sum_{\substack{i<j \\ \ell<m}} f_{ij\ell m}(W_{t_k})D_k^{ij\ell m})$$

where $f_{ij\ell m}(x) = (g_{ij}(x) - g_{ji}(x))(g_{\ell m}(x) - g_{m\ell}(x))$ and $D_k^{ij\ell m} := h_n^{-1}E[L_{nk}^{ij}\, L_{nk}^{\ell m}|P_n]$, as given by (5c,d,e). So it follows from the continuity of the $f_{ij\ell m}$, and the a.s. convergence of the sum $\sum_k \Delta W_{nk}^i \Delta W_{nk}^j$, $i \neq j$ and $\sum_k \Delta W_{nk}^{i2}$ to the respective mutual and quadratic variations 0 and 1 of independent Brownian motions on [0,1], that

$$\lim_n E[S_n^2|P_n] = \eta^2 \qquad \text{a.s.} \tag{6}$$

Finally, from (5f) and the Borel Cantelli lemma it follows that

$$\lim_n h_n^{-2} \sum_k E[L_{nk}^{ij^4}|P_n] = 0 \qquad \text{a.s.}$$

and with the uniform bound on $f_{ij\ell m}(W_{t_k})$ on each path ω, that

$$\lim_n \sum_k E[X_{nk}^4|P_n] = 0 \qquad \text{a.s.} \tag{7}$$

The conditional independence of the X_{nk} and the two relations (6,7) allow a central limit theorem of Hall and Heyde ([3], Corollary 3.1) for martingale difference arrays to be applied to show that the limiting characteristic function of S_n is $E[e^{-\frac{1}{2}\eta^2 t}]$. However we need to sharpen this theorem slightly to make a statement about conditional characteristic functions.

Let (Ω', F') be a replica of (Ω, F). For each n let $P_n(\omega, d\omega)$ be a conditional distribution of (W_t) conditional on P_n. In equations (5,6,7) replace all conditional expectations by integrations with respect to the respective P_n. Let $\Omega_0 \subset \Omega$ be a set with $P(\Omega_0) = 1$ on which all these equations hold. Then for each $\omega \in \Omega_0$ and each n the variables X_{nk} can be regarded as independent variables on $(\Omega', F', P_n(\omega,.))$. Since $\lim_n \sum_k E_n X_{nk}^2 = \eta^2(\omega)$ and $\lim_n \sum_k E_n X_{nk}^4 = 0$ it follows from a classical central limit theorem for arrays of independent random variables (or again, Corollary 3.1 in [3]), that

$$\lim_n \phi_n(t,\omega) = \lim_n \int e^{iS_n(\omega')t} P_n(\omega, d\omega') = e^{-\frac{1}{2}\eta^2(\omega)t}$$

for all t and all $\omega \in \Omega_0$. This completes the proof of the theorem.

A RATE OF CONVERGENCE THAT IS ALMOST SURELY BEST

From the point of view of someone who possesses a single path of $W_t(\omega)$ and who wants to know the rate of convergence of the discretization sequence (I_n) just for that one path, the preceding control limit theorem is not completely satisfactory as it is phrased in terms of probability laws for errors that have not occurred. A more appropriate measure of the rate of convergence would be the superior limit of the error sequence suitably normalized : $\overline{\lim}_n \frac{|I-I_n|}{N_n}$. (I_n) is a convergent quasimartingale, and it turns out that a law of the iterated algorithm of Heyde ([4] Corollary 2) for the tail sums of martingale difference sequences does provide a natural norming sequence that makes the limit equal to one. To establish that the rate of convergence of I_n is also a best rate, however, what is needed is an additional "strong lower bound" property for (I_n) of the form: for any P_n-adapted sequence (Y_n)

$$\overline{\lim}_n \frac{|I_n-Y_n|}{N_n} \overset{a.s.}{\geq} \overline{\lim}_n \frac{|I-I_n|}{N_n} \overset{a.s.}{\geq} = 1 \tag{8}$$

This raises a general question about the existence of a class of convergent martingales possessing such a strong lower bound property. During the conference at Luminy, H. Föllmer established that this class is at least not void with the following simple example for which an even stronger property holds. Suppose (Z_n) is an i.i.d sequence with $P(Z_k=1)=P(Z_k=-1)=\frac{1}{2}$. Let $I_n := \sum_{k=1}^n \frac{Z_k}{2^k}$. Then $I:=\lim_n I_n$ a.s. is uniform on $[-1,1]$. If P_n is now $\sigma(Z_1,Z_2,\dots,Z_n)$, then the lower bound property (8) holds for any (P_n)-adapted (Y_n) with $N_n = \frac{1}{2^n}$, and in addition for any $\varepsilon > 0$,

$$\overline{\lim}_n \left\{ \frac{|I-Y_n|}{|I-I_n|} \chi_{\{2^n|I-I_n|>1-\varepsilon\}} \right\} \geq 1 \quad \text{a.s. ;}$$

that is, "whenever I_n is a bad approximation of I, Y_n is equally bad infinitely often"

It turns out that a class of martingale covered by Heyde's iterated-logarithm law for tail sums also possesses the lower-bound property (8). One version of this result, phrased in terms of tail sums, is as follows.

Let $(X_n)_{n\in\mathbb{N}}$ be a sequence of martingale differences on an increasing filtration (F_n)(i.e. $E[X_n|F_{n-1}] = 0$) such that $\sum_1^\infty EX_n^2 < \infty$. Let S_n be the tail sums $\sum_{n+1}^\infty X_k$, which converges to zero a.s. and in L^2, and s_n^2 its variance $\sum_{n+1}^\infty EX_k^2$.

<u>Theorem 2.</u> If $s_n^2 > 0$ for all n and

i) $s_n^{-2}EX_n^2 \to 0$ as $n \to \infty$,

ii) $\sum_1^\infty s_n^{-4}EX_n^4 < \infty$,

(iii) For some random variable $\xi > 0$ a.s. $s_n^{-2}\sum_{n+1}^\infty E[X_k^2|F_{k-1}] \to \xi^2$ a.s.,

then for any (F_n)-adapted process (Z_n)

$$\overline{\lim}_{n\uparrow\infty} \frac{|S_n - Z_n|}{(2s_n^2 \log\log s_n^{-2})^{\frac{1}{2}}} \overset{\text{a.s.}}{\geq} \overline{\lim}_{n\uparrow\infty} \frac{|S_n|}{(2s_n^2 \log\log s_n^{-2})^{\frac{1}{2}}} \overset{\text{a.s}}{=} \xi . \tag{9}$$

The "equality" part of this theorem is essentially a corollary of Heyde's iterated-logarithm law [4]. The proof of the full result, and of a similar theorem for continuous martingales, will be published elsewhere.

To apply this theorem to the discretization problem of line integrals we need to introduce a slowly increasing sequence of partitions obtained by adding points one at a time. In what follows it is often convenient to replace the integer m by a pair (n,k) where $m = 2^n+k$, $k = 0,1,\ldots,2^n-1$. The dependence on m of (n,k) will be understood. Now let Π_m' be the partition of $m+1$ points:

$$\Pi_m' := \{i/2^{n+1}, \text{ for } 0 \leq i \leq 2k-1;\ j/2^n, \text{ for } k \leq j \leq 2^n\}.$$

Then the "average" mesh size of Π_m' is $1/m$. Let $P_m' = \sigma(W_t : t\in\Pi_m')$. The filtration (P_n) is a subsequence of (P_n') in that $P_n = P_{2^n}'$. Let I_m' be the "trapezoidal rule" discretization of I on Π_m' :

$$I_m' := \sum_{k=0}^{m-1} \frac{1}{2} \sum_i (g_i(W_{t_k}) + g_i(W_{t_{k+1}}))(W_{t_{k+1}}^i - W_{t_k}^i)$$

where $0 = t_i < t_1 \ldots < t_{m-1} < 1$ is a relabelling of the points of Π'_m. Then the strong lower bound property holds for I'_m in the following sense:

Theorem 3. Suppose that η^2 as given by (3) is strictly positive a.s. . Then for any (P'_m)-adapted sequence (Y_m),

$$\overline{\lim}_{m\uparrow\infty} \frac{|I-Y_m|}{(2m^{-1}\log\log m)^{\frac{1}{2}}} \overset{a.s.}{\geq} \lim_{m\uparrow\infty} \frac{|I-I'_m|}{(2m^{-1}\log\log m)^{\frac{1}{2}}} \overset{a.s.}{=} \eta. \quad (10)$$

Similar results can be established for approximations defined on more quickly converging subsequences of partitions, (though not for the subsequence P_n). The key condition is that the number of points that are added at each stage should be asymptotically negligible compared with the total number of points in the partition.

Proof. Many of the details are similar to parts of the proof of Theorem 1, and so the proof will be given in abbreviated form. Let b_m be the additional point in Π'_m that is not in Π'_{m-1}, and let a_m and c_m be the adjacent points with $a_m < b_m < c_m$. The difference

$$I'_m - I'_{m-1} = \frac{1}{2}\sum_i (\gamma_{im}\,\Delta W^i_m - \alpha^i_m \Delta g_{im}),$$

where $\Delta W^i_m = W^i_{c_m} - W^i_{a_m}$, $\Delta g_{im} = g_i(W_{c_m}) - g_i(W_{a_m})$, both P'_{m-1}-measurable, $\alpha^i_m = W^i_{b_m} - \frac{1}{2}(W^i_{c_m} + W^i_{a_m})$ and $\gamma_{im} = g(W_{b_m}) - \frac{1}{2}[g_i(W_{c_m}) + g_i(W_{a_m})]$. It can further be decomposed as $I'_m - I'_{m-1} =: X_m + R_m$, where

$$X_m = \frac{1}{2}\sum_{ij}(g_{ij}(W_{a_m})\alpha^j_m \Delta W^i_m - \alpha^j_m \Delta g_{im})$$

is a martingale difference on (P'_m) and R_m is a remainder

$$R_m = \frac{1}{2}\sum_{ij}\Delta W^i_m \left(\int_{a_m}^{b_m} \operatorname{sgn}(b_m - t)(g_{ij}(W_t) - g_{ij}(W_{a_m}))\circ dW^j_t\right.$$

that turns out to be negligible. Note that, for $2^n \leq m < 2^{n+1}$, (R_m) is a quasi-martingale on (F_{c_m}), and it is not difficult to verify that $E|E[R_m|P_{c_{m-1}}]|^2 = 0(2^{-4n})$ and $E[R_m^2] = 0(2^{-3n})$. It follows from the Doob inequality that

$K_n = E[\max_{2^n \le m < 2^{n+1}} |\sum_m^{2^{n+1}-1} R_k|] = O(2^{-n})$ and $E|\sum_{2^n}^{\infty} R_k| \le \sum_n K_n = O(2^{-n})$. Consequently $\lim_m m^{\beta} \sum_m^{\infty} R_k = 0$ a.s. for any $\beta < 1$. But the norming in (10) increases more slowly than, say, $m^{2/3}$, and so the contribution of $\sum_m^{\infty} R_k$ to the limit is negligible.

Now consider (X_m) as a martingale difference sequence on (P'_m). The α_m^i are normal and independent with zero means and variances $2^{-(n+2)}$. So for $2^n \le m < 2^{n+1}$,

$$E[X_m^2|P'_{m-1}] = 2^{-(n+4)}\sum_k(\sum_i g_{ik}(W_{a_m})\Delta W_m^i - \Delta g_{km})(\sum_j g_{jk}(W_{a_m})\Delta W_m^j - \Delta g_{km}).$$

With further expansion, and taking into account the convergence of the mutual variation process to zero, we have as $n \to \infty$

$$\sum_{m=2^n}^{2^{n+1}-1} E[X_m^2|P'_{m-1}] \sim 2^{-(n+3)}\sum_{m=2^n}^{2^{n+1}-1} (\sum_{i<k} g_{ik}(W_{a_m}) - g_{ki}(W_{a_m}))^2 \Delta W_m^{i2} \quad \text{a.s.}$$

$$\sim 2^{-(n+1)}\eta^2 \quad \text{a.s.}\ .$$

Consequently $\sum_{2^n}^{\infty} E[X_m^2|P'_{m-1}] \sim 2^{n}\eta^2$, and if $s_m^2 = \sum_{m+1}^{\infty} EX_k^2$, as $m \to \infty$

$s_m^2 \sim m^{-1}E\eta^2$ and $s_m^{-2}\sum_{m+1}^{\infty} E[X_k^2|P'_{k-1}] \sim \eta^2/E\eta^2$. Now apply Theorem 2 with $F_n = P'_n$, $S_n = I - I_n$ and $Z_n = Y_n - I_n$. Conditions (i) and (ii) are easily verified. Condition (iii) is clearly satisfied with $\xi^2 = \eta^2/E\eta^2$. Finally we note that $s_m^2 \log\log s_m^{-2} \sim E\eta^2 m^{-1} \log\log m$ and the result follows.

REFERENCES

[1] J.M.C. CLARK, R.J. CAMERON. The maximum rate of convergence of discrete approximations for stochastic differential equations. B. Grigelionis (Ed.) Stochastic Systems : Proc. IFIP-WG 7/1 Working Conference Vilnius 1978. Lect. Notes Contr. Inf. Sciences No. 25, Springer-Verlag, Berlin 1980, pp 162-171.

[2] J.M.C. CLARK. An efficient approximation scheme for a class of stochastic differential equations. W.H. Fleming, L.G. Gorostiza (Eds.) Advances in Filtering and Optimal Stochastic Control:Proc. IFIP-WG 7/1 Working Conference Cocoyoc, Mexico 1982. Lect. Notes Contr. Inf. Sciences, No. 42. Springer-Verlag, Berlin 1982 pp 69-78.

[3] P. HALL, C.C. HEYDE. Martingale Limit Theory and its Application, Academic Press, New York, 1980..

[4] C.C. HEYDE. On Central limit and iterated logarithm supplements to the martingale convergence theorem. J. Appl. Prob. 14 1977, pp 758-775.

[5] E.J. McSHANE. Stochastic Calculus and Stochastic Models. Adademic Press, New York 1974.

[6] G.N. MIL'SHTEIN. Approximate integration of stochastic differential equations. Theory Prob. Appl., 19, 1974, pp. 557-562.

[7] N.J. NEWTON. Discrete Approximations for Markov-chain filters, Ph.D. Thesis, University of London 1983.

[8] E. PARDOUX, D. TALAY. Discretization and simulation of stochastic differential equations. To appear in Acta Applicandae Mathematicae.

[9] W. RÜMELIN. Numerical treatment of stochastic differential equations. SIAM J. Numer. Anal. 19(3) 1982, pp 604-613.

ON ONE-DIMENSIONAL STOCHASTIC DIFFERENTIAL EQUATIONS WITH GENERALIZED DRIFT

H.J. Engelbert and W. Schmidt
Friedrich-Schiller-Universität
Sektion Mathematik, 6900 Jena, GDR

1. INTRODUCTION

In the present paper we consider a type of one-dimensional stochastic differential equations which includes equations with "ordinary" drift as well as equations with singular drift describing for instance the so-called skew Brownian motion (cf. [7]). In a certain limit case (which, however, will not be treated here) diffusions with reflecting barriers can also be constructed (cf. [11]). The essential point consists in driving the equation not by a usual drift function but by a measure. We deal with this problem in Section 3 and give necessary and sufficient conditions for the existence and uniqueness of solutions. Moreover, we state a sufficient condition for the existence of a pathwise unique strong solution improving several known results. The general solution will be constructed by space transformation from a solution without drift. For this reason, Section 2 is devoted to a profound study of the existence and uniqueness of solutions of one-dimensional stochastic differential equations without drift, improving the results presented by us at the foregoing meeting (cf. [4]).

Let be b a universally measurable real function and v a signed measure on the real line $\mathbb{R}$. We shall consider the one-dimensional stochastic differential equation

$$(1) \qquad X_t = X_0 + \frac{1}{2} \int_{\mathbb{R}} L^X(t,y)\, v(dy) + \int_0^t b(X_s)\, dB_s$$

where B is a Wiener process and L^X denotes the right (or left) local time of the semimartingale X. If b^2 0 and v is given by the density $2a\,b^{-2}$ where a is any universally measurable real function then (1) reduces to the stochastic differential equation with "ordinary" drift

$$(1') \qquad X_t = X_0 + \int_0^t a(X_s)\, ds + \int_0^t b(X_s)\, dB_s .$$

If b=1 and v is concentrated in the point 0 with $|v(\{0\})|<1$, equation (1) describes the skew Brownian motion which was treated in [7].

Now let us introduce some basic notions. Let $(\Omega, \underline{\underline{F}}, P)$ be a complete probability space equipped with an increasing family $\mathbb{F}=(\underline{\underline{F}}_t)$ of sub-

σ-algebras of $\underline{\underline{F}}$. We always suppose that $\mathbb{F}$ satisfies the usual conditions, i.e. $\mathbb{F}$ is right-continuous and $\underline{\underline{F}}_o$ contains all $\underline{\underline{F}}$-sets of P-measure zero. For a process $X=(X_t)$ we write $(X, \mathbb{F})$ to indicate that X is $\mathbb{F}$-adapted. For any continuous local martingale $(X, \mathbb{F})$ we denote by $\langle X\rangle$ the unique continuous increasing process such that $(X^2-\langle X\rangle, \mathbb{F})$ is a local martingale and $\langle X\rangle_o=0$.

Let now S be an $\mathbb{F}$-stopping time. A process $(X, \mathbb{F})$ defined on the stochastic interval $[0,S)$ is called a semimartingale (resp. a local martingale) up to time S if there exists an increasing sequence (S_n) of $\mathbb{F}$-stopping times such that $S_n<S$ for all n and $\lim_n S_n=S$ and the process $(X^n, \mathbb{F})$ obtained from $(X, \mathbb{F})$ by stopping at S_n is a semimartingale (resp. a local martingale) for all n. If $(X, \mathbb{F})$ is a semimartingale up to time S then we can find a decomposition

(2) $$X_t = X_o + M_t + V_t \qquad \text{for } t<S$$

on $[0,S)$ where $(M, \mathbb{F})$ is a local martingale up to time S with $M_o=0$ and V is a right-continuous process with paths of bounded varation on $[0,t]$ for every $t<S$ and with $V_o=0$. If X is continuous on $[0,S)$ then there exists a decomposition (2) such that M and V are continuous on $[0,S)$ and this decomposition is unique on $[0,S)$.

The following facts are well-known for semimartingales (cf. P.A. Meyer [12], J. Azéma and M. Yor [1]). Their extension to semimartingales up to time S is obvious. Let $(X, \mathbb{F})$ be a continuous semimartingale up to time S. Then there exists the (right) local time L^X up to time S which is a function of $[0,S)\times\mathbb{R}$ into $[0,+\infty)$ such that for every convex function f

(IF) $$f(X_t) = f(X_o) + \int_0^t D^+f(X_s)\, dX_s + \frac{1}{2}\int_{\mathbb{R}} L^X(t,y)n(dy)$$

for all $t<S$ P-a.s. where D^+f denotes the right derivative of f and the measure n on $\mathbb{R}$ is the second derivative of f (in the sense of distributions). Formula (IF) is the generalized Itô's formula.

For the problem which will be treated by us it is convenient to consider solutions of equation (1) admitting explosions. Let $\overline{\mathbb{R}}=\mathbb{R}\cup\{-\infty,+\infty\}$ be the extended real line. By $\underline{\underline{B}}(\overline{\mathbb{R}})$ and $\underline{\underline{B}}(\mathbb{R})$ we denote the σ-algebras of Borel subsets of $\overline{\mathbb{R}}$ and $\mathbb{R}$, respectively.

DEFINITION 1. A <u>continuous</u> $(\overline{\mathbb{R}},B(\overline{\mathbb{R}}))$-valued stochastic process X defined on a probability space $(\Omega,\underline{\underline{F}},P)$ with a filtration $\mathbb{F}=(\underline{\underline{F}}_t)$ is called <u>a solution of equation</u> (1) if

(i) X_o is real-valued,

(ii) $X_t(w)=\pm\infty$ for all $t\geq S_\infty(w)$

where $S_\infty=\inf\{s\geq 0:\ X_s=-\infty \text{ or } +\infty\}$,

(iii) $(X, \mathbb{F})$ is a semimartingale up to time S_∞,

(iv) there exists a Wiener process $(B, \mathbb{F})$ such that equation (1) holds for all $t<S_\infty$ P-a.s.

The $\mathbb{F}$-stopping time S_∞ is called the explosion time of X.

We say that the solution of (1) is unique (in law) if any two solutions $(X^1, \mathbb{F}^1)$ and $(X^2, \mathbb{F}^2)$, the initial distributions of which coincide, possess the same image law on the space of continuous functions over $\bar{\mathbb{R}}$.

A solution $(X, \mathbb{F})$ of (1) is called a strong solution if X is $\mathbb{F}^B$-adapted where $\mathbb{F}^B$ is the filtration generated by B and completed in the usual way.

We say that for equation (1) the pathwise uniqueness holds if for any two solutions X^1 and X^2, defined on the same probability space $(\Omega, \underline{\underline{F}}, P)$ with a filtration $\mathbb{F}$ and the same Wiener process $(B, \mathbb{F})$, such that $X_0^1 = X_0^2$ P-a.s. it follows $X_t^1 = X_t^2$ for all $t \geq 0$ P-a.s.

Finally, a solution $(X, \mathbb{F})$ of (1) is called trivial if $X_t = X_0$ for $t \geq 0$ P-a.s.

2. EQUATIONS WITHOUT DRIFT

In this section we consider the stochastic differential equation without drift

$$(3) \qquad Y_t = Y_0 + \int_0^t b(Y_s)\, dB_s .$$

Solutions of (3) were already investigated by the authors in [4] and [5]. However, in the following we shall improve the results which were obtained there.

Let b be any universally measurable real function. By N we denote the set $N = \{x \in \mathbb{R}: b(x) = 0\}$ of zeros of b. We put $b^{-2}(x) = +\infty$ for $x \in N$ and introduce the set

$$M = \left\{ x \in \mathbb{R}: \int_G b^{-2}(y)\, dy = +\infty \text{ for every open neighbourhood } G \text{ of } x \right\}$$

We notice that M is a closed subset of $\mathbb{R}$.

It can easily be verified that every solution of (3) has the explosion time $+\infty$ (cf. proof of Proposition 1). Since we do not give a proof here, we shall include this property in the definition of a solution of (3).

THEOREM 1. For every initial distribution there exists a solution of equation (3) if and only if $M \subseteq N$, i.e. if and only if the following condition (E) is satisfied:

(E) If $b(y) \neq 0$ then b^{-2} is integrable in a neighbourhood of y.

P r o o f . The proof is based on the well-known method of random time change. First we suppose that (E) is satisfied. Let μ be an arbitrary probability measure on $\mathbb{R}$. We consider a Wiener process $(W, \mathbb{H})$ on a probability space $(\Omega, \underline{\underline{F}}, P)$ with initial distribution μ and define

$$T_t = \int_0^{t+} b^{-2}(W_s)\, ds \qquad \text{for all } t \geq 0.$$

Let A be the inverse of the increasing process T defined by $A_t = \inf\{s \geq 0: T_s > t\}$. Furthermore, let U(M) be the first entry time of W in M:

$$U(M) = \inf\{s \geq 0: W_s \in M\}.$$

Now we state the following lemma.

LEMMA 1. (i) $\int_0^t b^{-2}(W_s)\, ds < +\infty$ <u>for all</u> $t < U(M)$ P-a.s.

(ii) $\int_0^{U(M)+} b^{-2}(W_s)\, ds = +\infty$ P-a.s.

(iii) $A_\infty = U(M)$ P-a.s.

P r o o f . Statement (iii) precisely expresses (i) and (ii) together. For proving (i), we notice that b^{-2} is integrable over every compact subset K of $M^c = \mathbb{R} - M$. Now the assertion follows exactly as in the proof of Theorem 1 of 4 . It remains to verify (ii). If $M = \emptyset$ then $U(M) = +\infty$ and the equality stated in (ii) follows from [4], Theorem 3. Finally, we assume that $M \neq \emptyset$. Then $U(M) < +\infty$ P-a.s. and

$$\int_0^{U(M)+t} b^{-2}(W_s)\, ds \geq \int_0^t b^{-2}(W_{U(M)} + \overline{W}_s)\, ds$$

where $\overline{W}_s = W_{U(M)+s} - W_{U(M)}$ is a Wiener process. Since M is closed we get $W_{U(M)} \in M$ P-a.s. and the latter integral is equal to $+\infty$ P-a.s. in view of the Lemma in [4]. □

Now we continue the proof of Theorem 1. Obviously, $A = (A_t)$ is a continuous increasing family of $\mathbb{H}$-stopping times. Because of $T_\infty = +\infty$ P-a.s. (cf. [4], Theorem 3), A_t is finite P-a.s. for every $t \geq 0$. Now we put [1)]

$$Y = W \circ A, \qquad \mathbb{F} = \mathbb{H} \circ A.$$

Then $(Y, \mathbb{F})$ is a continuous local martingale with $\langle Y \rangle = A$ (cf. [9]). Let us compute the explicite form of A. The definition of M implies that $M^c \cap N$ has Lebesgue measure zero. Hence

$$W \in M \cup N^c \qquad L \times P\text{-a.e.}$$

where L is the Lebesgue measure on $[0, +\infty)$. Using Lemma 1 this gives

$$b^2(W_s) > 0 \qquad \text{for all } s < A_\infty \qquad L \times P\text{-a.e.}$$

Lemma 1 also yields $W_{A_\infty} \in M$ on $\{A_\infty < +\infty\}$ and, because of condition (E), we get

$$b^2(W_{A_\infty}) = 0 \qquad \text{on } \{A_\infty < +\infty\} \qquad P\text{-a.s.} \tag{4}$$

Using these properties and the definition of T we obtain

1) $(\mathbb{H} \circ A)_t = \underline{\underline{H}}_{A_t}$.

$$A_t = \int_0^{A_t} b^2(W_s)\, dT_s = \int_0^{(T\circ A)_t} b^2(Y_s)\, ds$$

where the last equality follows by time change in the integral. (Note that T is not necessarily finite. Cf., for example, Lemma (1.6) of [5].) Consequently,

$$A_t = \begin{cases} \int_0^t b^2(Y_s)\, ds & \text{if } A_t < A_\infty \\ \int_0^\infty b^2(Y_s)\, ds & \text{if } A_t = A_\infty. \end{cases}$$

If $A_t = A_\infty$ then $Y_s = W_{A_\infty}$ for all s t and (4) implies $b^2(Y_s)=0$ for all $s \geq t$. Consequently,

$$A_t = \int_0^t b^2(Y_s)\, ds \qquad \text{for all } t \geq 0 \qquad P\text{-a.s.}$$

From this follows that there exists a Wiener process $(B, \mathbb{F})$ (on a, possibly, enlarged probability space) such that (3) holds (cf., for example, Theorem 7.1' of [8]) and hence $(Y, \mathbb{F})$ is a solution of (3) with initial distribution μ.

Now we are going to prove the necessity of condition (E). For this we suppose that $(Y, \mathbb{F})$ is any solution of (3). Then the increasing process $A' = \langle Y \rangle$ is given by

$$(5) \qquad A'_t = \int_0^t b^2(Y_s)\, ds \qquad \text{for all } t \geq 0 \qquad P\text{-a.s.}$$

Let T' be the inverse of A' defined by $T'_t = \inf\{s \geq 0 : A'_s > t\}$. We put

$$W' = Y \circ T', \qquad \mathbb{H}' = \mathbb{F} \circ T'$$

where $Y_\infty = \lim_{t \to +\infty} Y_t$ on $\{A'_\infty < +\infty\}$. Then $(W', \mathbb{H}')$ is a continuous local martingale such that $\langle W' \rangle_t = t \wedge A'_\infty$, with initial value $W'_0 = Y_0$. This means that $(W', \mathbb{H}')$ is a Wiener process stopped at A'_∞ (cf. [2], p.330). By time change in the integral we obtain

$$(6) \qquad \begin{aligned} \int_0^{t \wedge A'_\infty} b^{-2}(W'_s)\, ds &= \int_0^{(A' \circ T')_t} b^{-2}(W'_s)\, ds \\ &= \int_0^{T'_t} b^{-2}(Y_s)\, dA'_s = \int_0^{T'_t} b^{-2}(Y_s) b^2(Y_s) ds \leq T'_t. \end{aligned}$$

Now the proof of the necessity of condition (E) is easily accomplished. Let $(Y, \mathbb{F})$ be a solution of (3) satisfying $Y_0 = y_0$ where $y_0 \in \mathbb{R}$ is such that $b(y_0) \neq 0$. Then $(Y, \mathbb{F})$ cannot be trivial and hence $P(\{0 < A'_\infty\}) > 0$. This yields that there exists $t>0$ such that $P(\{t < A'_\infty\}) > 0$. Since $T'_t < +\infty$ on $\{t < A'_\infty\}$, from inequality (6) we observe that

$$P(\{\int_0^t b^{-2}(W'_s)\, ds < +\infty,\ t < A'_\infty\}) > 0.$$

Now we can apply the Lemma of [4] and conclude that there exists an open neighbourhood G of y_0 such that b^{-2} is integrable over G. This establishes condition (E), completing the proof of Theorem 1. ■

COROLLARY. *If b^{-2} is locally integrable then there is a solution of* (3) *for every initial distribution.*

REMARKS. (i) Under the condition of the above Corollary there exists not only any solution but also a nontrivial solution. The converse is also true: If there is a nontrivial solution of (3) for every initial distribution then b^{-2} is locally integrable (cf. 4 , Theorem 4).

(ii) Let b be a Lipschitz function and $b(y_o)=0$ for some $y_o \in \mathbb{R}$. Then b^{-2} is not integrable in any open neighbourhood of y_o. But condition (E) of Theorem 1 is satisfied. However, if we start at y_o there is only the trivial solution $Y_t=y_o$ for all $t\geq 0$. This makes clear that it is not justified to exclude trivial solutions.

In general, the solution of (3) is not unique. However, we have the following result which improves Theorem (3.2) of [5].

THEOREM 2. *For every initial distribution there exists a unique solution of equation* (3) *if and only if* M=N, *i.e. if and only if the following condition* (E+U) *is satisfied:*

(E+U) $b(y)\neq 0$ *if and only if* b^{-2} *is integrable in a neighbourhood of* y

P r o o f . The necessity of condition (E) follows from Theorem 1. Now we assume that there exists a point $y_o \in N\cap M^c$ and show that the solution of (3) starting at y_o is not unique. Let $(W, \mathbb{H})$ be a Wiener process starting at y_o. As in the proof of Theorem 1 we construct a solution $(Y, \mathbb{F})$ of (3) with $Y_o=y_o$. For $A=\langle Y\rangle$, by Lemma 1(iii) we have $A_\infty=U(M)$ where U(M) is the first entry time of W in M. This yields $P(\{0<A_\infty\})=1$ and hence the solution $(Y, \mathbb{F})$ is not trivial. On the other side, $\overline{Y}_t=y_o$ for every $t\geq 0$ is a trivial solution of (3) starting at y_o. Consequently, uniqueness fails. This proves the necessity of the condition $N\cap M^c=\emptyset$ and hence the necessity of (E+U). Now we assume that (E+U) is satisfied Then the existence of a solution to (3), for every initial distribution is guaranteed by Theorem 1 and it remains to verify the uniqueness. To this end, let $(Y, \mathbb{F})$ be an arbitrary solution of (3). We adopt the notations introduced in the second part of the proof of Theorem 1 before formula (6) and first prove the following lemma.

LEMMA 2. *Suppose* $N\subseteq M$. *We then have:*

(i) $T'_t = \int_0^t b^{-2}(W'_s)\, ds$ *for all* $t < A'_\infty$ P-a.s.

(ii) A'_∞ *is an* $\mathbb{F}^{W'}$*-previsible stopping time where* $\mathbb{F}^{W'}$ *is the filtration generated by* W' *and completed in the usual way.*

P r o o f . Setting $S(M) = \inf\{s\geq 0:\ Y_s\in M\}$ we observe that $b^2(Y_s)>0$ for all $s\leq T'_t$ on $\{T'_t<S(M)\} = \{t<A'_{S(M)}\}$ and therefore inequality (6) be-

comes an equality on this set. For proving (i), thus it remains to establish that $A'_\infty = A'_{S(M)}$. To this end we define $U'(M)=\inf\{s\geq 0:\ W'_s\in M\}$. In view of inequality (6) we get

$$T'_{U'(M)} \int_0^{U'(M)+} b^{-2}(W'_s)\,ds \qquad \text{on } \{U'(M)< A'_\infty\} \qquad P\text{-a.s.}$$

Applying Lemma 1(ii) to W' we see that the right member of this inequality is equal to $+\infty$ on $\{U'(M)< A'_\infty\}$. Since the left member is finite on this set we conclude that $\{U'(M) < A'_\infty\}$ has probability zero and hence

(7) $\qquad A'_\infty \leq U'(M) \qquad\qquad P\text{-a.s.}$

On the other side, it can easily be verified that

(8) $\qquad U'(M) = A'_{S(M)} \qquad \text{on } \{S(M)<+\infty\} \qquad P\text{-a.s.}$

From (7) and (8) we obtain the desired equality $A'_\infty = A'_{S(M)}$, proving (i). It remains to verify that A'_∞ is an $\mathbb{F}^{W'}$-previsible stopping time. Since $U'(M)$ is $\mathbb{F}^{W'}$-previsible this will follow from $A'_\infty = U'(M)$ P-a.s. In view of (7) and (8) this equality is valid on $\{S(M)< +\infty\}$ and, because of (7), it remains to show that

(9) $\qquad U'(M) \leq A'_\infty \qquad \text{on } \{S(M)=+\infty\} \qquad P\text{-a.s.}$

On $\{S(M)=+\infty\}$ we have $A'_t<A'_\infty$ for all $t\geq 0$. Indeed, on $\{A'_t=A'_\infty\}$ the process Y is constant on $[t,+\infty)$ and from (5) follows $b^2(Y_t)=0$ and hence $Y_t\in M$ on this set. Thus $S(M)\leq t$ on $\{A'_t=A'_\infty\}$ P-a.s., proving the assertion. Now we apply (i) and obtain

$$(T'\circ A')_t = \int_0^{A'_t} b^{-2}(W'_s)\,ds \qquad \text{on } \{S(M)=+\infty\} \qquad P\text{-a.s.}$$

Since the left member converges to $+\infty$ as $t\to+\infty$ we conclude

$$\int_0^{A'_\infty} b^{-2}(W'_s)\,ds = +\infty \qquad \text{on } \{S(M)=+\infty\} \qquad P\text{-a.s.}$$

Applying Lemma 1(i) to W' we observe that (9) is fulfilled. □

After these preparations the proof of the uniqueness is easily accomplished. Noting that $T'_t=+\infty$ for $A'_\infty\leq t$, Lemma 2 yields that T' is $\mathbb{F}^{W'}$-adapted and, moreover, A'_∞ is an $\mathbb{F}^{W'}$-previsible stopping time. From Theorem 2 and Proposition 7 of [2] we now obtain that $(Y, \mathbb{F}^Y)$ possesses the representation property for continuous local martingales (cf. [2]). Thus every solution of (3) possesses the representation property. Precisely as in [3] (proof of Corollary 1 to Theorem 6), from this follows that for any two solutions $(Y^1, \mathbb{F}^1)$ and $(Y^2, \mathbb{F}^2)$ with the same initial distribution their image laws on the space of continuous functions coincide, proving the uniqueness. ■

REMARK. Without the condition (E), the condition that $N\subseteq M$ is sufficient but not necessary for the uniqueness of the solution of equation (3).

3. EQUATIONS WITH GENERALIZED DRIFT

In this section we deal with the general equation (1). We shall always assume that the signed measure v is locally finite and that

(10) $\quad v(\{x\}) > -1 \qquad$ for all $x \in \mathbb{R}$.

In a certain sense this jump condition is a natural restriction since there is, in general, no solution of (1) if $v(\{x\}) < -1$ for some $x \in \mathbb{R}$ (cf. [7], [10]). The case $v(\{x\})=-1$ is excluded to avoid technical complications. Let g denote the unique solution of the equation

$$g(x) = \begin{cases} 1 - \int_{[0,x]} g(y)\, v(dy) & \text{for } x \geq 0 \\ 1 + \int_{(x,0)} g(y)\, v(dy) & \text{for } x < 0. \end{cases}$$

which can be written explicitely by

$$g(x) = \begin{cases} \exp(-v([0,x])) \prod_{0 \leq y \leq x} (1+v(\{y\}))^{-1} \exp(v(\{y\})), & x \geq 0, \\ \exp(v((x,0))) \prod_{x<y<0} (1+v(\{y\}))\ \exp(v(\{y\})), & x<0 \end{cases}$$

(cf. [17], Theorem (3-4) and Theorem (2-1)). We notice that g is of locally bounded variation and strictly positive. The same properties hold for g^{-1}. We put

$$G(x) = \int_0^x g(y)\, dy, \qquad x \in \overline{\mathbb{R}}.$$

Obviously, G is strictly increasing and continuous. Let H be the inverse of G which is defined on $(G(-\infty), G(+\infty))$ and is also strictly increasing and continuous. We extend H, g, and the diffusion coefficient b by setting

$$H(x) = -\infty \text{ for } x \in [-\infty, G(-\infty)],\ H(x) = +\infty \text{ for } x \in [G(+\infty), +\infty],$$
$$g(-\infty) = g(+\infty) = b(-\infty) = b(+\infty) = 0.$$

The following proposition shows that solving equation (1) is equivalent to solving an equation of type (3).

PROPOSITION 1. The process $(X, \mathbb{F})$ is a (resp. strong) solution of equation (1) if and only if $(Y, \mathbb{F})$ defined by $Y = G \circ X$ is a (resp. strong) solution of the equation

(11) $$Y_t = Y_0 + \int_0^t (g \circ H)(Y_s)\,(b \circ H)(Y_s)\, dB_s, \qquad t \geq 0.$$

P r o o f . Let S_∞ be the explosion time of X (cf. Definition 1).

1° Since G (restricted to $\mathbb{R}$) is the difference of two convex function we can apply Itô's formula (IF) and for $t < S_\infty$ we obtain P-a.s.

$$G(X_t) = G(X_0) + \int_0^t g(X_s) b(X_s)\, dB_s + \frac{1}{2}\int_0^t g(X_s) \int_{\mathbb{R}} L^X(ds,y)\, v(dy) - \frac{1}{2}\int_{\mathbb{R}} L^X(t,y) g(y)\, v(dy)$$

The difference of the last two terms is equal to zero and because of $X = H \circ Y$ it follows

(12) $$Y_t = Y_o + \int_0^t (g\circ H)(Y_s)(b\circ H)(Y_s)\,dB_s \quad \text{on } \{t < S_\infty\} \quad P\text{-a.s.}$$

Now we prove that the equality remains valid on the set $\{S_\infty \le t\}$. For this it suffices to show that

(13) $$Y_{S_\infty} = Y_o + \int_0^{S_\infty} (g\circ H)(Y_s)(b\circ H)(Y_s)\,dB_s \quad \text{on } \{S_\infty < +\infty\} \quad P\text{-a.s.}$$

Indeed, if $S_\infty \le t$ then $Y_t = Y_{S_\infty}$ by definition of Y and, on the other side, the integrand vanishes for all s such that $S_\infty \le s \le t$. For proving (13), let S_n be the first exit time of X from $[-n,n]$. Then (M_n) defined by $M_n = Y_{S_n}$ where $Y_\infty = \lim_{t\to\infty} Y_t$ on $S_n = +\infty$ (which exists) is a (if Y_o is not integrable, generalized) martingale of class C (see [6]) and according to Corollary 4 to Theorem 1 of [6] we get P-a.s. <u>either</u> $\liminf M_n = -\infty$ and $\limsup M_n = +\infty$ <u>or</u> $\lim M_n$ exists and is <u>finite</u>. Using the continuity of X and G we observe that $M_n = G(X_{S_n})$ <u>converges</u> to the finite or infinite points $G(-\infty)$ or $G(+\infty)$ on $\{S_\infty < +\infty\}$ P-a.s. This excludes the first case and we see that $\lim Y_{S_n}$ exists and is <u>finite</u> on $\{S_\infty < +\infty\}$ P-a.s. Now we replace t by S_n in (12) and pass to the limit as $n \to +\infty$, proving the asserted equality (13). Consequently, $(Y, \mathbb{F})$ is a solution of (3) (with explosion time $+\infty$).

2°. Conversely, suppose $(Y, \mathbb{F})$ is a solution of (11). Using

(14) $$H(x) = \int_0^x (g\circ H)^{-1}(y)\,dy, \qquad x \in \mathbb{R},$$

an application of Itô's formula (IF) gives

(15) $$X_t = H(Y_t) = H(Y_o) + \int_0^t (g\circ H)^{-1}(Y_s)(g\circ H)(Y_s)(b\circ H)(Y_s)\,dB_s + C_t \quad \text{on } \{t < S_\infty\}$$

P-a.s. where C is a process of locally bounded variation. Once more applying Itô's formula and using (15) we conclude

(16) $$Y_t = G(X_t) = Y_o + \int_0^t g(X_s)b(X_s)\,dB_s + \int_0^t g(X_s)\,dC_s - \frac{1}{2}\int_{\mathbb{R}} L^X(t,y)g(y)v(dy)$$

on $\{t < S_\infty\}$ P-a.s. Since $(Y, \mathbb{F})$ is a solution of (11) this implies

$$\int_0^t g(X_s)\,dC_s = \frac{1}{2}\int_{\mathbb{R}} L^X(t,y)g(y)v(dy) = \int_0^t g(X_s)\frac{1}{2}\int_{\mathbb{R}} L^X(ds,y)v(dy)$$

and consequently

$$C_t = \frac{1}{2}\int_{\mathbb{R}} L^X(t,y)\,v(dy) \quad \text{on } \{t < S_\infty\} \quad P\text{-a.s.}$$

Now (15) becomes

$$X_t = X_o + \frac{1}{2}\int_{\mathbb{R}} L^X(t,y)\,v(dy) + \int_0^t b(X_s)\,dB_s \quad \text{on } \{t < S_\infty\} \quad P\text{-a.s.}$$

and the proof is finished. □

Now we come to our main result concerning the existence of a solution to equation (1).

THEOREM 3. <u>For every initial distribution there exists a solution of equation</u> (1) <u>if and only if the condition</u> (E) <u>for the diffusion coef-</u>

<u>ficient</u> b <u>is satisfied.</u>

P r o o f . By Proposition 1, for all initial distributions there exists a solution of (1) if and only if for all initial distributions there is a solution of (11). According to Theorem 1 we have to show that the condition (E) for b and for $(g \circ H)(b \circ H)$ are equivalent. Since the latter function vanishes outside of $(G(-\infty),G(+\infty))$, condition (E) is fulfilled there automatically. Let $(x-a,x+a) \subseteq (G(-\infty),G(+\infty))$. Because of (14) we obtain

$$\int_{(x-a,x+a)} g^{-2}(H(y))b^{-2}(H(y))\,dy = \int_{(H(x-a),H(x+a))} g^{-1}(z)b^{-2}(z)dz. \tag{17}$$

Let $x \in (G(-\infty),G(+\infty))$ be fixed and choose $a, a_1, a_2 > 0$ such that $(H(x)-a_1,H(x)+a_1) \subseteq (H(x-a),H(x+a)) \subseteq (H(x)-a_2,H(x)+a_2) \subseteq (G(-\infty),G(+\infty))$. The functions g and g^{-1} are locally bounded and hence bounded by a constant $C>0$ on $(H(x)-a_2,H(x)+a_2)$. Using (17), from this we conclude

$$C^{-1} \int_{(-a_1,a_1)} b^{-2}(H(x)+y)dy \le \int_{(x-a,x+a)} g^{-2}(H(y))b^{-2}(H(y))dy \le C \int_{(-a_2,a_2)} b^{-2}(H(x)+y)dy$$

and the equivalence of the conditions follows. ■

In an analogous way but now using Theorem 2 one can prove the following result on the uniqueness of solutions to (1).

THEOREM 4. <u>For every initial distribution there exists a unique solution of equation</u> (1) <u>if and only if the condition</u> (E+U) <u>for the diffusion coefficient</u> b <u>is satisfied.</u>

REMARKS. (i) Note that the conditions (E) and (E+U) do not depend on the measure v. The only conditions on v are its local finiteness and the jump condition (10).

(ii) The local finiteness of v can be weakened considerably. It is sufficient to suppose the following condition: For all $x \in \mathbb{R}$, if $b(x) \neq 0$ then v is finite in some open neighbourhood of x.

(iii) Using results of the authors [4] J.A. Yan [16] obtained similar statements for equations with "ordinary" drift under the assumption that b^{-2} is locally integrable.

(iv) The solution $(X, \mathbb{F})$ of equation (1) <u>does not explode</u> if and only if the solution $(Y, \mathbb{F})$ of equation (11) with initial value $Y_0=G(X_0)$ does not reach the boundary points $G(-\infty)$ and $G(+\infty)$. In particular, this is the case if $G(-\infty)=-\infty$ and $G(+\infty)=+\infty$. A very simple sufficient condition is the following: $v^+((0,+\infty))<+\infty$ and $v^-((-\infty,0))<+\infty$.

(v) As in Section 2 the condition $N \subseteq M$ (without condition (E)) on b is sufficient but not necessary for the uniqueness of the solution to (1) if it exists.

(vi) In case of nonuniqueness it is possible to give a complete description of the set of all solutions. If b^{-2} is locally integrable this can be done without any difficulties as in [5] where we characterized the set of solutions to an equation without drift by time delays in the set of zeros $N=\{x\in\mathbb{R}: b(x)=0\}$.

(vii) If $(X, \mathbb{F})$ is the unique solution to (1) then it is a strong Markov process. However, in case of nonuniqueness there are many strong Markov solutions and it is also possible to characterize them completely. Moreover, there are also many solutions that are neither strong Markov nor Markov. In case of equations without drift see [5].

(viii) Analogous results can be obtained by using the left local time L^X in equation (1). The jump condition (10) changes into $v(\{x\})<1$ for all $x\in\mathbb{R}$ and we have to work with the unique solution g of the equation considered at the beginning of this section except for g(y) is replaced by g(y-). If v has no atoms both approaches are the same. In general, the behaviour of solutions of (1) is another if we use the left local time. For including both cases in a more general equation it would be of interest to investigate a certain modification of equation (1) where the solution is driven through the left and right local times by two measures separately.

In case of the equation (1') with "ordinary" drift we obtain the following simple conditions for existence and uniqueness improving some known results (cf. [8]).

COROLLARY. Each of the following conditions is sufficient for the existence of a solution to (1') for every initial distribution:

(i) a is locally bounded a.e. (with respect to the Lebesgue measure) and b^{-2} is locally integrable.

(ii) a is locally integrable and b^{-2} is locally bounded a.e.

(iii) There exists $p>1$ such that a^p and b^{-2q} are locally integrable where q denotes the conjugate exponent.

If, additionally, $b(x)\neq 0$ for all $x\in\mathbb{R}$ then uniqueness holds.

Finally, we state a theorem on the pathwise uniqueness of solutions to (1) that implies in connection with Theorem 3 the existence of a pathwise unique strong solution. The proof is omitted.

THEOREM 5. Suppose that there exist functions f and h of $\mathbb{R}$ into $[0,+\infty]$ such that the following condition (PU) is satisfied:

(PU)
(i) fb^{-2} is locally integrable in M^c.
(ii) $\int_{(-a,a)} h^{-1}(y)\,dy = +\infty$ for every $a>0$.
(iii) There exists $a>0$ such that
$(b(x+y)-b(x))^2 \le f(x)\cdot h(y)$ for all $x\in\mathbb{R}$ and $y\in(-a,a)$.

(iv) If $b(x)=0$ then $f(x)< +\infty$, for all $x \in \mathbb{R}$.
Then pathwise uniqueness holds for equation (1). If, additionally, condition (E) is satisfied then for every given Wiener process $(B, \mathbb{F})$ and for every $\underline{\underline{F}}_0$-measurable random variable X_0 there exists a pathwise unique strong solution $(X, \mathbb{F})$ of equation (1) with initial value X_0.

REMARKS. (i) The pathwise uniqueness of the solution to equation (1) was already obtained by D.W. Stroock and M. Yor [14] in the case that b=1 and the singular part of v is finite. J.F. Le Gall [10] proved the pathwise uniqueness of the solution to this equation under the conditions that b is of bounded variation with $0<c\leq b(x)\leq C$ and v is finite.

(ii) Using a similar but stronger condition than (PU) J.F. Le Gall [10] proved the statement of Theorem 5 for equations (1') with "ordinary" drift under the conditions that a and b are bounded and $0<c\leq |b(x)|$ for every $x\in\mathbb{R}$. He also proved the interesting fact that equations of type (1) occur in a certain limit procedure from equations with "ordinary" drift.

(iii) We notice that (PU) implies the condition $N\subseteq M$ which is sufficient for the uniqueness (in law) of the solution to equation (1) (cf. Remark (v) to Theorem 4). Hence there is no need for using additional assumptions ensuring the uniqueness in law, contrary to other papers.

(iv) The statement of Theorem 5 on the strong existence already follows from the pathwise uniqueness and the existence (in law), the latter is satisfied in view of Theorem 3. This is a general result due to T. Yamada and S. Watanabe [15] (see also [8]).

(v) Condition (PU) is a modification of conditions used by E. Perkins [13] and J.F. Le Gall [10] improving a condition of T. Yamada and S. Watanabe [15]. Note that we do not assume the strict monotony of h on the half lines. Hence b need not be continuous, contrary to other papers. Also b need not be neither right-continuous nor left continuous. Furthermore, since we suppose that h is a function defined on $\mathbb{R}$ (and not only on $[0,+\infty)$ where y has to be replaced by $|y|$ on the right hand side of inequality (PU)(iii)) the condition (PU)(iii) is only an one-sided condition on the behaviour of b in a right or left neighbourhood of $x\in\mathbb{R}$.

(vi) If b is continuous then, in connection with (PU)(iii), (PU)(i) already follows if we only assume that f is locally integrable in M^c. J.F. Le Gall [10] supposed that f is locally integrable but on the right hand side of inequality (PU)(iii) the additional factor $b^2(x)$. This excludes, except for the trivial case b=0, zeros of b, contrary to (PU).

(vii) Since b^{-2} is locally integrable in M^c we can formulate (PU) choosing f=1 and the condition simplifies considerably in this case.

REFERENCES

1. J. Azéma and M. Yor (Editors), Temps locaux, Astérisque 52-53(1978).
2. H.J. Engelbert and J. Hess, Stochastic Integrals of Continuous Local Martingales I, Math.Nachr. 97, 325-343 (1980).
3. H.J. Engelbert and J. Hess, Stochastic Integrals of Continuous Local Martingales II, Math.Nachr. 100, 249-269 (1981).
4. H.J. Engelbert and W. Schmidt, On the Behaviour of Certain Functionals of the Wiener Process and Applications to Stochastic Differential Equations, Proceedings of the 3rd IFIP-WG 7/1 Working Conference on Stochastic Differential Systems, Visegrád (Hungary), September 15 - 20, 1980, Lecture Notes in Control and Information Sciences 36, 47-55, Springer-Verlag Berlin 1981.
5. H.J. Engelbert and W. Schmidt, On Solutions of One-Dimensional Stochastic Differential Equations without Drift, Z.Wahrscheinlichkeitstheorie verw.Geb., to appear.
6. H.J. Engelbert and A.N. Shiryaev, On the Sets of Convergence of Generalized Submartingales, Stochastics 2, 155-166 (1979).
7. J.M. Harrison and L.A. Shepp, On Skew Brownian Motion, Ann.Probab. 9, 309-313 (1981).
8. N. Ikeda and S. Watanabe, Stochastic Differential Equations and Diffusion Processes, North Holland, Amsterdam 1981.
9. N. Kazamaki, Changes of Time, Stochastic Integrals, and Weak Martingales, Z.Wahrscheinlichkeitstheorie verw.Geb. 22, 25-32 (1972).
10. J.F. Le Gall, Temps locaux et équations différentielles stochastiques, Thèse 3e cycle, Université de Paris VI, 1982.
11. H.P. McKean, A. Skorohod's Stochastic Integral Equation for a Reflecting Barrier Diffusion, J.Math.Kyoto Univ. 3-1, 85-88 (1963).
12. P.A. Meyer, Un cours sur les intégrales stochastiques, Séminaire de Probabilités X, Lecture Notes in Mathematics 511, 245-400, Springer-Verlag Berlin 1976.
13. E. Perkins, Local Time and Pathwise Uniqueness for Stochastic Differential Equations, Séminaire de Probabilités XVI, Lecture Notes in Mathematics 920, 201-208, Springer-Verlag Berlin 1982.
14. D.W. Stroock and M. Yor, Some Remarkable Martingales, Séminaire de Probabilités XV, Lecture Notes in Mathematics 850, 590-603, Springer-Verlag Berlin 1981.
15. T. Yamada and S. Watanabe, On the Uniqueness of Solutions of Stochastic Differential Equations, J.Math.Kyoto Univ. 11, 155-167 (1971).
16. J.A. Yan, Une remarque sur les solutions faibles des équations différentielles stochastiques unidimensionelles, Séminaire de Probabilités XVII, Lecture Notes in Mathematics 986, 78-80, Springer-Verlag Berlin 1983.
17. Ch. Yoeurp, Decompositions des martingales locales et formules exponentielles, Séminaire de Probabilités X, Lecture Notes in Mathematics 511, 432-480, Springer-Verlag Berlin 1976.

AN ENTROPY APPROACH TO THE TIME REVERSAL OF DIFFUSION PROCESSES

H. FÖLLMER
Mathematikdepartement
ETH Zentrum, CH-8092 Zürich

Abstract: We introduce an entropy technique which allows to treat some infinite-dimensional extensions of the classical duality equations for the time reversal of diffusion processes.

1. Introduction

Consider the time reversal of a process X_t $(0 \leq t \leq 1)$ with stochastic differential equation

$$dX_t = dW_t + b_t dt \; . \tag{1.1}$$

In the Markovian case we have $b_t = b(X_t, t)$, and under some regularity assumptions the drift of the time-reversed process $\hat{X}_t = X_{1-t}$ is given by the duality equation

$$\hat{b}(x, 1-t) = - b(x,t) + \nabla \log p(x,t) \tag{1.2}$$

where $p(.,t)$ denotes the density of X_t ; cf. [9] , [11] and also [1] , [7] , [12] , [3] , [17] , [10] for some recent developments and applications, in particular for the role of (1.2) in Stochastic Mechanics.

One possible approach to (1.2) is to calculate the time-reversed drift as a stochastic backward derivative of the process in the sense of [11] . The purpose of this note is to show that finite entropy of (X_t) with respect to Wiener measure is a condition which allows us to make this approach rigorous. Moreover, the entropy technique can be used to extend (1.2) to certain infinite-dimensional situations where other time reversal methods, which have been developed in the theory of Markov processes, do not readily apply. One such situation is equation

(1.1) in the non-Markovian case where b_t depends on the past history. As another example, we mention an extension of (1.2) to diffusion processes on an infinite-dimensional state space which arise in connection with Gibbs states; this part refers to joint work with A. WAKOLBINGER.

2. A remark on entropy and the Girsanov transformation

Let W_t $(0 \leq t \leq 1)$ be a Wiener process on $(\Omega, \underline{\underline{F}}, \underline{\underline{F}}_t, P)$, and let $\tilde{P}$ be a probability measure which is absolutely continuous with respect to P on $\underline{\underline{F}}_1$. It follows from the general theory of the (Cameron-Martin-Maruyama-)Girsanov transformation that there exists an adapted process b_t $(0 \leq t \leq 1)$ with

$$(2.1) \qquad \int_0^1 b_t^2 dt < \infty \qquad \tilde{P}\text{-a.s.}$$

such that

$$(2.2) \qquad W_t^b \equiv W_t - \int_0^t b_s ds \qquad (0 \leq t \leq 1)$$

is a Wiener process under $\tilde{P}$; cf. [8]. We say that $\tilde{P}$ has <u>finite energy</u> if not only (2.1) but also

$$(2.3) \qquad \tilde{E}\left[\int_0^1 b_t^2 dt\right] < \infty$$

holds where $\tilde{E}$ denotes expectation with respect to $\tilde{P}$. This condition includes the diffusions with singular drift which arise in Stochastic Mechanics [3]; it is also used in [13] . In [16] it is shown that condition (2.3) allows to identify the drift (b_t) as a stochastic forward derivative of (W_t) with respect to $\tilde{P}$, i.e.,

$$(2.4) \qquad b_t = \lim_{h \downarrow 0} \frac{1}{h} \tilde{E}\left[W_{t+h} - W_t \middle| \underline{\underline{F}}_t\right] \qquad \text{in } L^2(\tilde{P}).$$

Let us now introduce the <u>relative entropy</u>

$$(2.5) \qquad H(\tilde{P}|P) = \tilde{E}\left[\log \frac{d\tilde{P}}{dP}\right]$$

of $\tilde{P}$ with respect to P. By Jensen's inequality we have $0 \leq H(\tilde{P}|P) \leq \infty$, and $H(\tilde{P}|P) = 0$ if and only if $\tilde{P} = P$. The following lemma shows that finite entropy implies finite energy, and therefore also (2.4).

(2.6) Lemma: $E[\int_0^1 b_t^2 dt] \leq 2\, H(\tilde{P}|P)$

Proof: Put $T_n = \inf\{t > 0 \mid \int_0^t b_s^2 ds > n\} \wedge 1$ and

$$Z_n = \exp\left[\int_0^{T_n} b_s dW_s - \frac{1}{2}\int_0^{T_n} b_s^2 ds\right] = \exp\left[\int_0^{T_n} b_s dW_s^b + \frac{1}{2}\int_0^{T_n} b_s^2 ds\right].$$

Since $E[Z_n] = 1$ by Novikov's criterion, $dP_n = Z_n dP$ defines a probability measure P_n which is equivalent to P. We have

$$\begin{aligned} H(\tilde{P}|P) &= H(\tilde{P}|P_n) + \tilde{E}[\log Z_n] \\ &\geq \tilde{E}\left[\int_0^{T_n} b_s dW_s^b + \frac{1}{2}\int_0^{T_n} b_s^2 ds\right] = \frac{1}{2}\tilde{E}\left[\int_0^{T_n} b_s^2 ds\right], \end{aligned}$$

and since $\sup T_n = 1$ $\tilde{P}$-a.s. by (2.1) we get

$$\tilde{E}\left[\int_0^1 b_s^2 ds\right] = \sup_n \tilde{E}\left[\int_0^{T_n} b_s^2\, ds\right] \leq 2\, H(\tilde{P}|P).$$

We are going to use the lemma in the following manner: Since finite entropy is preserved under time reversal, (2.5) and (2.6), applied to the distribution of the time-reversed process, will allow to calculate the time-reversed drift as a stochastic backward derivative.

3. Time reversal on C[0,1]

Let X_t $(0 \leq t \leq 1)$ denote the coordinate process on $\Omega = C[0,1]$, $(\underline{\underline{F}}_t)$ the canonical filtration and P_x the Wiener measure with starting point x. We consider a probability measure $Q = \int \mu(dx) Q_x$ on $C[0,1]$ which is absolutely continuous with respect to Wiener measure in the sense that

(3.1) $$Q \ll P_\mu \equiv \int \mu(dx) P_x$$

This implies that there is an adapted process (b_t) with $\int_0^1 b_t^2\,dt < \infty$ Q-a.s. such that

$$W_t^b \equiv X_t - X_0 - \int_0^t b_s ds \qquad (0 \leq t \leq 1)$$

is a Wiener process with respect to Q . Moreover, we have

$$(3.2) \qquad \frac{dQ}{dP_\mu} = \exp\left[\int_0^1 b_t dW_t^b + \frac{1}{2} \int_0^1 b_t^2 dt \right] \qquad \text{Q-a.s.}$$

cf.[8] CH.7, (7.30), and this implies

$$(3.3) \qquad H(Q|P_\mu) = \frac{1}{2} E_Q\left[\int_0^1 b_t^2 dt \right] .$$

Now let R denote the pathwise time reversal on $C[0,1]$ given by $X_t \circ R = X_{1-t}$, and let $\hat{Q} = Q \circ R^{-1}$ denote the <u>time reversal of Q</u> . In the sequel we denote by E and $\hat{E}$ the expectation with respect to Q and $\hat{Q}$.

(3.4) <u>Theorem</u>: If Q has finite energy then its time reversal $\hat{Q}$ has locally finite energy, i.e., there is an adapted process $(\hat{b}_t)$ such that $d\hat{W} \equiv dX - \hat{b}_t dt$ is a Wiener process with respect to $\hat{Q}$ and

$$(3.5) \qquad \hat{E}\left[\int_0^t \hat{b}_s^2 ds \right] < \infty \qquad (0 \leq t < 1).$$

The time-reversed drift $(\hat{b}_t)$ can be calculated, for almost all $t \in [0,1]$ as

$$(3.6) \qquad \hat{b}_t = \lim \frac{1}{h} \hat{E}\left[X_{t+h} - X_t \mid \underline{\underline{F}}_t \right]$$

$$= \lim \frac{1}{h} E\left[X_{1-t-h} - X_{1-t} \mid \hat{\underline{\underline{F}}}_{1-t} \right] \circ R$$

where $(\hat{\underline{\underline{F}}}_t)$ is the canonical filtration of $\hat{X}_t = X_{1-t}$ $(0 \leq t \leq 1)$.

We refer to [6] for a detailed proof. The main point is that finite energy of Q implies via (3.3) that $H(\hat{Q}|\hat{P}_\mu) = H(Q|P_\mu)$ is finite. But $\hat{P}_\mu = P_\mu \circ R^{-1}$ is a well-known object, namely the distribution of Brownian motion conditioned to have distribution μ at time 1, and we have an explicit formula for the process (c_t) such that $dW = dX - c_t dt$

is a Wiener process under $\hat{P}_\mu$. Applying lemma (2.6) with $P = \hat{P}_\mu$ and $\tilde{P} = \hat{Q}$, we obtain (3.5).

(3.7) Remarks: 1) Already the simplest case $Q = P_0$, where $\hat{Q}$ is Brownian motion with initial distribution N(0,1) pinned to 0 at time 1, shows that we can expect finite energy of $\hat{Q}$ only in the local form (3.5), not globally.

2) The argument becomes even simpler if we assume that the initial distribution has finite variance σ^2 and finite entropy

$$-\log\sqrt{2\pi\sigma^2 e} \leq H(\mu|\lambda) = \int \log \frac{d\mu}{d\lambda}\, d\lambda < \infty$$

with respect to Lebesgue measure λ. Then we can use the σ-finite measure $P = \int P_x \, dx$ as reference measure. But P is reversible, and so the finite energy condition on Q implies that $H(\hat{Q}|P) = H(Q|P) = H(\mu|\lambda) + H(Q|P_\mu)$ is finite. This allows to conclude that $\hat{Q}$ has finite energy, now even globally.

3) The method can also be applied to diffusion processes of the form $dX = \sigma(X_t)dW + b_t dt$ if we proceed as in 2) and use as reference model the reversible σ-finite measure on $C[0,1]$ which corresponds to a solution of $dX = \sigma(X)dW$ started off with its reversible equilibrium distribution.

If Q has finite energy then (3.4) shows that the following standard argument, which is usually carried out under additional regularity assumptions (see, e.g., [3], [1]) , is in fact rigorous. For a smooth bounded function f and for almost all $t \in [0,1]$ we get, using (2.5), (3.6), Itô's formula and some straight forward estimates,

(3.8)
$$\hat{E}\left[\hat{b}_{1-t} f(X_{1-t})\right] = \lim \frac{1}{h} \hat{E}\left[(X_{1-t+h} - X_{1-t}) f(X_{1-t})\right]$$
$$= -\lim \frac{1}{h} E\left[(X_t - X_{t-h})\left[f(X_{t-h}) + \int_{t-h}^{t} f'(X_s)\,dX_s + \frac{1}{2}\int_{t-h}^{t} f''(X_s)\,ds\right]\right]$$

$$= - E\left[b_t f(X_t)\right] - E\left[f'(X_t)\right].$$

The absolute continuity (3.1) implies that $\mu_t \equiv Q \circ X_t^{-1}$ is given by a density function $p(x,t)$ so that

$$E\left[f'(X_t)\right] = \int f'(x)p(x,t)dx = -\langle f, p'(\cdot,t)\rangle \tag{3.9}$$

But (3.8) implies that the distributional derivative is in fact a function, and that it satisfies the duality equation

$$p'(x,t) = p(x,t)\left(E\left[\hat{b}_{1-t} \circ R \,\middle|\, X_t = x\right] + E\left[b_t \,\middle|\, X_t = x\right]\right). \tag{3.10}$$

If Q is the distribution of a Markov process then (3.10) reduces to the classical duality equation (1.2). In the non-Markovian case, (3.10) does show that <u>the densities of a process with finite energy are smooth in x.</u> But (3.10) does not yet determine the time-reversed drift, only its projections on the present states. In order to identify (b_t) completely, we have to replace the function f in (3.8) by a smooth functional F on $C[0,1]$ which depends on the past up to time t. In particular, we have to replace the integration by parts on R^1 in (3.9) by an integration by parts on Wiener space. This can be made exact if we assume smoothness of the functionals b_t and use Bismut's version [2] of the Malliavin calculus. Denoting by $D_\omega b_t(ds)$ the derivative of b_t at $\omega \in C[0,1]$, viewed as a measure on $[0,1]$, we obtain the following path-space formula for the time-reversed drift; cf. [6] for details.

(3.11) <u>Theorem:</u> $\hat{b}_{1-t} = - E[a_t \mid \hat{\underline{\underline{F}}}_t] \circ R$

where

$$a_t \equiv b_t + \frac{1}{t}\int_0^t \left(1 - \int_0^t DB_s[r,1]\,dr\right)dW_s^b + \int_t^1 \left(Db_s[t,1] - \frac{1}{t}\int_0^t Db_s[r,1]\,dr\right)dW_s^b. \tag{3.12}$$

4. <u>Time reversal of infinite-dimensional diffusions</u>

Consider an infinite-dimensional diffusion process $X_t = (X_t^i)_{i\in I}$ $(0 \leq t \leq 1)$ with state space $S \subset R^I$ and stochastic differential equation

(4.1) $$dX_t^i = dW_t^i + b^i(X_t,t)dt \qquad (i \in I),$$

where I is a countable index set and (W^i) a collection of independent Brownian motions. Under some regularity conditions on the drift, there is a unique strong solution on $S = L^2(\gamma)$ with some finite measure γ on I; cf. [5], [14]. In this situation the natural analogue of (1.2) is

(4.2) $$\hat{b}^i(x,1-t) = - b^i(x,t) + \frac{\partial}{\partial x_i} \log p_t^i(x^i | x^j (j \neq i)) \; ,$$

where $p_t^i(.|.)$ is the conditional density of the i-th coordinate, given the coordinates $j \neq i$. If the drift is time-homogeneous and if μ is a reversible equilibrium distribution then (4.2) implies that μ is a Gibbs measure in the sense of Statistical Mechanics; cf. [5] . A rigorous derivation of (4.2) involves some additional technical problems which do not yet arise in the finite-dimensional case; cf. [5], [14] for a discussion of the reversible case. To begin with, after time reversal the coordinate processes X_t^i $(0 \leq t \leq 1)$ may not even be semimartingales; cf. [16] for an explicit example (the reason is the same as in the non-Markovian example [15]: after time reversal, we may know too much). Here again, a rigorous approach can be based on entropy conditions. Let Q denote the distribution of the process on $C([0,1],S)$, and let Q^i denote the distribution which coincides with Q on $\underline{\underline{F}}^i = \sigma(X_0^i, X_t^j \; (j \neq i, 0 \leq t \leq 1))$ and which makes $X_t^i - X_0^i$ a Wiener process which is independent of $\underline{\underline{F}}^i$. Using lemma (2.6) and arguments similar to the proof of (3.4) we obtain the following result (joint work [16] with A. WAKOLBINGER):

(4.3) <u>Theorem</u>: If $H(Q|Q^i) < \infty$ for all $i \in I$ then the duality equations (4.2) do hold.

It is shown in [16] that the entropy conditions $H(Q|Q^i) < \infty$ are satisfied under assumptions which are very close to the usual conditions for existence and uniqueness of a strong solution of (4.1).

<u>References</u>

[1] B.D.O. ANDERSON: Reverse time diffusion equation models.
Stoch. Proc. and Appl. 12, 313-326 (1982)

[2] J.M.BISMUT: Martingales, the Malliavin Calculus and Hypoellipticity under general Hörmander conditions. Z.Wahrscheinlichkeitstheorie verw. Geb. 56, 469-505 (1981)

[3] E.A.CARLEN: Conservative Diffusions: A constructive approach to Nelson's stochastic mechanics. Thesis, Princeton Univ. (1984)

[4] D.A.CASTANON: Reverse time diffusion processes. IEEE Trans.Inf.Theory, vol IT-28, 953-956 (1982)

[5] H.DOSS,G.ROYER: Processus de diffusion associés aux mesures de Gibbs. Z. Wahrscheinlichkeitstheorie verw.Geb.46,125-158 (1979)

[6] H.FÖLLMER: Time reversal of absolutely continuous measures on C 0,1 To appear.

[7] U.HAUSSMANN: On the drift of a reversed diffusion. In: Proc. 4th IFIP Conf. on Stochastic Diff.Systems (this volume)

[8] R.S.LIPTSER, A.N.SHIRYAEV: Statistics of Random Processes I. Springer (1977)

[9] M.NAGASAWA: Time reversion of Markov Processes. Nagoya Math.J.24, 117-204 (1964)

[10] M.NAGASAWA: Segregation of a population in an environment. J.Math. Biology 9, 213-235 (198o)

[11] E.NELSON: Dynamical theories of Brownian motion. Princeton UP (1967)

[12] E.NELSON: Quantum Fluctuations. Princeton UP (1984)

[13] D.OKONE: To appear in Proc.4th IFIP Conf. on Stoch.Diff.Systems (this volume)

[14] T.SHIGA,A.SHIMIZU: Infinite dimensional stochastic differential equations and their applications. J.Math.Kyoto Univ.20,395-416 (1980)

[15] J.WALSH: A non reversible semi-martingale. Sém. Prob.XVI, Springer Lecture Notes in Math.920, 212 (1982)

[16] A.WAKOLBINGER, H.FÖLLMER: Time reversal of infinite-dimensional diffusions. To appear.

[17] W.ZHENG, P.A.MEYER: Quelques resultats de mécanique stochastique. Sém.Prob. XVIII, Springer Lecture Notes in Math. 1059, 223-244 (1984)

ON THE DRIFT OF A REVERSED DIFFUSION*

U. G. Haussmann
Mathematics Department
University of British Columbia
Vancouver, Canada, V6T 1Y4

1. Introduction. Let $\{X_t: 0 \leq t \leq 1\}$ be a diffusion process in R^d, solution of

(1.1) $$dX_t = b(t,X_t)dt + \sigma(t,X_t)dw_t$$

where $\{w_t: 0 \leq t \leq 1\}$ is a standard Brownian motion in R^d. Let $\bar{X}_t = X_{1-t}$ be the reversed diffusion. Under suitable hypotheses it is shown in [3] that $\bar{X}_t$ satisfies

(1.2) $$d\bar{X}_t = \bar{b}(t,\bar{X}_t)dt + \bar{\sigma}(t,\bar{X}_t)d\bar{w}_t$$

for some Brownian motion $\{\bar{w}_t: 0 \leq t \leq 1\}$ with $\bar{\sigma}(t,x) = \sigma(1-t,x)$ and

$$\bar{b}^i(t,x) = -b^i(1-t,x) + p(1-t,x)^{-1}[a^{ij}(1-t,x)p(1-t,x)]_{x_j}.$$

Here $\bar{b}^i$ is the ith component of $\bar{b}$, a^{ij} is the (i,j) entry of the matrix $a \equiv \sigma\sigma^*$ (σ^* is the transpose of σ), $p(t,\cdot)$ is the probability density of X_t, $f_{x_j} = \partial f/\partial x_j$, and we have used the convention that repeated indices are summed. Sometimes it is of interest to know when $|\bar{b}(t,x)| \leq K(1+|x|)$ or when it is bounded [5]. We study here the equivalent question for $\nabla p/p$ (with $\nabla p = px$). To show that $\nabla p/p$ is bounded we show that the Cauchy problem for the system of quasilinear parabolic equations formally satisfied by $\nabla p/p$, has a bounded classical solution. This is carried out in the appendix - the method is a standard application of the maximum principle and the Leray-Schauder fixed point theorem. If u is this solution we show in §2 that a solution $\tilde{p}$ of the Kolmogorov forward equation can be constructed so that $\nabla\tilde{p}/\tilde{p} = u$. By uniqueness $\tilde{p} = p$. The major restrictions are regularity of the coefficients b, σ and the requirement that σ be independent of x. The case with linear growth is then reduced to the bounded case and thus solved.

2. The Main Result. The density p satisfies

$$p_t(t,x) = L^*p(t,x) \equiv \frac{1}{2}[a^{ij}(t,x)p(t,x)]_{x_i x_j} - [b^i(t,x)p(t,x)]_{x_i}$$

$$p(0,x) = p_o(x)$$

where p_o is the initial density. Differentiating (2.1) formally, setting $u(t,x) = \nabla p(t,x)/p(t,x)$, and using the fact that $u^i_{x_j} = u^j_{x_i}$ gives

---

*This research was supported by NSERC under grant A8051, and was in part carried out at the Laboratoire de Probabilités, Université Pierre et Marie Curie, Paris.

(2.2) $$u^k_t = \frac{1}{2} a^{ij} u^k_{x_i x_j} - (b^i - a^{ij}_{x_j} - a^{ij}u^j)u^k_{x_i} - b^i_{x_k} u^i - b^i_{x_i x_k} + a^{ij}_{x_i x_j} u^k + \frac{1}{2} a^{ij}_{x_i x_j x_k} + a^{ij}_{x_k}(u^i u^j + u^i_{x_j}).$$

To establish the existence of a solution of (2.2) - c.f. the appendix - we need to apply the maximum principle, which is not valid for (2.2) because of the quadratic term $u^i u^j$. To eliminate this term we assume that $a(t,x) = a(t)$ so that (2.2) becomes

$$L^k u \equiv u^k_t - \frac{1}{2} a^{ij}(t) u^k_{x_i x_j} + \left[b^i(t,x) - a^{ij}(t)u^j\right]u^k_{x_i} + \left[b^i_{x_i x_k}(t,x) + b^i_{x_k}(t,x)u^i\right] = 0.$$

We make the following hypotheses. Let $\|f\|$ denote the supremum of $|f(x)|$ over $\mathbf{R}^d$ or of $|f(t,x)|$ over $[0,1] \times \mathbf{R}^d$, and write $\phi(x)$ for $\nabla p_o/p_o$. $\alpha \in (0,1)$ is fixed.

(H) i) p_o is bounded, positive and differentiable, $\|\phi\| < \infty$.

ii) $b: [0,1] \times \mathbf{R}^d \to \mathbf{R}^d$ is twice differentiable in x,

$$|b^i(t,x)| \le K(1 + |x|), \quad |b^i_{x_k}(t,x)| + |b^i_{x_i x_k}(t,x)| \le K,$$

and for any open bounded $G \subset \mathbf{R}^d$, b^i, $b^i_{x_k}$, $b_{x_i x_k}$ are in $H^{\alpha,\alpha/2}(\overline{Q})$ with $Q = (0,1) \times G$.

iii) $\sigma:[0,1] \to \mathbf{R}^d \otimes \mathbf{R}^d$ is Holder continuous (order $\alpha/2$) and $a(t) \equiv \sigma(t)\sigma(t)^* > 0$ $\forall t$.

For the definition of $H^{\alpha,\alpha/2}(\overline{Q})$ see the appendix. Observe that (H iii) implies that $a(t) \ge \nu I$ for some $\nu > 0$. In the appendix we prove the following theorem.

Theorem 2.1. Assume (H) and

(2.3) $$\phi \in H^{2+\alpha}(\overline{G}) \text{ for every bounded } G \subset \mathbf{R}^d .$$

Then there exists a bounded continuous function u which satisfies

(2.4) $$L^k u = 0 \quad \text{on} \quad (0,1) \times \mathbf{R}^d, \; k = 1, 2, \dots, d,$$
$$u(0,x) = \phi(x) \qquad x \in \mathbf{R}^d$$

and which lies in $H^{2+\alpha,1+\alpha/2}(\overline{Q})$ for any bounded Q such that $\overline{Q} \in (0,1) \times \mathbf{R}^d$. Moreover ∇u is locally bounded and $u^i_{x_j} = u^j_{x_i}$.

Let $Q_\varepsilon = (\varepsilon,1) \times G$ for G open, bounded, and set

(2.5) $$\Psi(t,x) = \frac{1}{2} a^{ij}(t)\left[u^i(t,x)u^j(t,x) + u^i_{x_j}(t,x)\right] - b^i(t,x)u^i(t,x) - \operatorname{div} b(t,x)$$

with u as given by theorem 2.1. Then Ψ is bounded and continuous on $Q \equiv (0,1) \times G$ and is $H^{1+\alpha,\frac{1+\alpha}{2}}(\overline{Q}_\varepsilon)$, so that Ψ is integrable and $\tilde{p}$ is well defined if

(2.6) $$\tilde{p}(t,x) = p_o(x) \exp \int_o^t \Psi(s,x)ds.$$

Theorem 2.2. Assume (H) and (2.3). Then $\tilde{p}_{x_k} = \tilde{p}\, u^k$.

Proof: Define

$$\tilde{p}_\varepsilon(t,x) = \begin{cases} p_o(x), & t \leq \varepsilon \\ p_o(x)\exp\int_\varepsilon^t \Psi(s,x)ds, & t > \varepsilon. \end{cases}$$

Since Ψ_{x_k} exists on $\overline{Q}_\varepsilon$ then for $t > \varepsilon$

$$\left(\int_\varepsilon^t \Psi(s,x)ds\right)_{x_k} = \int_\varepsilon^t \Psi_{x_k}(s,x)ds$$

and

$$\Psi_{x_k}(s,x) = \frac{1}{2}a^{ij}(s)(u^i_{x_k}u^j + u^i u^j_{x_k} + u^i_{x_j x_k}) - b^i_{x_k}u^i - b^i u^i_{x_k} - b^i_{x_i x_k} = u^k_t$$

because $u^i_{x_j} = u^j_{x_i}$. Hence for $t > \varepsilon$

$$(2.7)\qquad \begin{aligned} \tilde{p}_\varepsilon(t,x)_{x_k} &= p_o(x)_{x_k}\exp\int_\varepsilon^t \Psi(s,x)ds + \tilde{p}_\varepsilon(t,x)\int_\varepsilon^t u^k_t(s,x)ds \\ &= \tilde{p}_\varepsilon(t,x)[\phi^k(x) + u^k(t,x) - u^k(\varepsilon,x)]. \end{aligned}$$

The local boundedness of Ψ implies that $\tilde{p}_\varepsilon \to \tilde{p}$ pointwise as $\varepsilon \to 0$, and the continuity of u on $[0,1] \times R^d$ implies that $(\tilde{p}_\varepsilon)_{x_k} \to \tilde{p}\,u^k$ pointwise, c.f. (2.6), (2.7).

Moreover for $(t,x) \in Q$

$$|\tilde{p}_\varepsilon(t,x)_{x_k}| \leq p_o(x)\exp\int_0^t |\Psi(s,x)|ds(\,\|\phi^k\| + 2\,\|u^k\|) \leq c(Q)$$

for some constant $c(Q)$. By dominated convergence for appropriate x^o.

$$\begin{aligned} \tilde{p}(t,x) &= \lim_{\varepsilon\to 0}\tilde{p}_\varepsilon(t,x) = \lim_{\varepsilon\to 0}\left[\int_{x_k^o}^{x_k}\tilde{p}_\varepsilon(t,x)_{x_k}dx_k + \tilde{p}_\varepsilon(t,x^o)\right] \\ &= \int_{x_k^o}^{x_k}\tilde{p}(t,x)u^k(t,x)dx_k + \tilde{p}(t,x^o) \end{aligned}$$

from which the result follows.

<u>Corollary</u> 2.1. Assume (H) and (2.3). Then $\nabla p/p$ is bounded.

Proof: Since $[\ell n\,\tilde{p}(t,x)]_{x_k} = u^k$ is bounded, then

$$|\tilde{p}(t,x)| \leq |\tilde{p}(t,0)|\exp(\,\|u\|\,|x|) \leq A\exp(B|x|^2).$$

Moreover

$$\tilde{p}_{x_i x_j} = (\tilde{p}u^i)_{x_j} = \tilde{p}_{x_j}u^i + \tilde{p}\,u^i_{x_j} = \tilde{p}(u^i u^j + u^i_{x_j})$$

so that (2.5), (2.6) imply

$$\tilde{p}_t(t,x) = \tilde{p}(t,x)\Psi(t,x) = L^* \tilde{p}.$$

Thus $\tilde{p}$ is a classical solution of (2.1) satisfying an exponential bound. According to [2], ch.6, corollary 4.2, such solutions of the Cauchy problem are unique. Since $\tilde{p}(0,x) = p_o(x)$, then $\tilde{p} = p$, the density of $\{X_t\}$, and thus $\nabla p/p = u$ is bounded.

Corollary 2.2. Assume (H). Then $\nabla p/p$ is bounded.

Proof: Let ρ^n be a non-negative smoothing kernel in $H^{2+\alpha}(\mathbf{R}^d)$ with support in $\{x: |x| < \frac{1}{n}\}$. Let $p_o^n = p_o * \rho^n$ (convolution). Then $p_o^n \in H^{2+\alpha}(\bar{G})$ for any bounded G, and if $\lambda = \inf\{p_o(x): x \in G\}$ (so $\lambda > 0$) then it can be shown that

$$|\phi^n|_G^{(2+\alpha)} \leq \|\phi\| \; \|p_o\| \; |\rho^n|_{\mathbf{R}^d}^{(2+\alpha)} \int_G dx(\lambda^{-1} + \|p_o\| \lambda^{-2})$$

so $\phi^n \in H^{2+\alpha}(\bar{G})$ if $\phi^n \equiv \nabla p_o^n/p_o^n$. See the appendix for the definition of $|\cdot|_G^{(2+\alpha)}$. Moreover $\|\phi^n\| \leq \|\phi\|$. Since ϕ^n satisfies (2.3) then $|\nabla p^n(t,x)|/p^n(t,x) \leq k$, where k depends on $\|\phi\|$ but not on n. Here p^n is the density of X_t when the initial density is p_o^n. By [4], ch. I, theorem 2.5, there exists a constant M_o such that $\|p^n\| \leq M_o \; \|\phi^n\| \leq M_o \|\phi\|$. Let $\hat{Q}_n = (\frac{1}{n}, 1) \times \{x: |x| < n\}$, $Q_n = (0,1) \times \{x: |x| < n\}$. By [4], ch.IV, theorem 10.1, if $n > m + 1$

$$|p^n|_{\hat{Q}_m}^{(2+\alpha)} \leq c_2 \, M_o \; \|\phi\|$$

where c_2 depends on m but not n, and by [4], ch. III, theorem 10.1 $|p^n|_{Q_m}^{(\gamma)} \leq c_3$ for some $\gamma > 0$. Again γ and c_3 are independent of n. The last two inequalities, Ascoli's lemma and the usual diagonalization procedure allow us to extract a subsequence which converges (together with the t-derivatives and the first two x-derivatives) to a solution of (2.1), hence to p. Then $|\nabla p|/p$ is also bounded by k.

The condition $\|\phi\| < \infty$, c.f. (Hi) is not met if p_o is a normal density. Observe that if $p_o(x) = q(x)\exp(-x^*Qx)$ with $Q \geq 0$, $q(x) > 0$, then

$$\phi(x) = \nabla p_o(x)/p_o(x) = \nabla q(x)/q(x) - 2Qx.$$

This remark motivates the next theorem.

Theorem 2.3. Assume (H iii) and

(2.8) $\quad b(t,x) = B(t)x + \beta(t,x)$ with $B^{ij} \in H^{\alpha/2}([0,1])$,
$\quad \beta$ satisfies (H ii) as well as $|x||\nabla\beta(t,x)| \leq K$,

(2.9) $\quad p_o$ is positive, bounded and differentiable,
$$\nabla p_o(x)/p_o(x) \equiv \phi(x) = \psi(x) + \Psi x$$
$\quad$ with $\Psi \leq 0$, symmetric, and $\|\psi\| < \infty$.

Then there exists k such that $|\nabla p(t,x)|/p(x) \leq k(1 + |x|)$.

Proof: Note that (2.8) implies that $|\beta(t,x)| \leq K + |\beta(t,0)|$, i.e. is bounded. Again we must find a solution of (2.4). We set $u(t,x) = P(t)x + v(t,x)$ so that (2.4) becomes

$$P_t^{ki}x_i + v_t^k = \frac{1}{2}a^{ij}v^k_{x_ix_j} - [(B^{ij} - a^{i\ell}P^{\ell j})x_j + \beta^i - a^{ij}v^j]v^k_{x_i}$$

$$(2.10) \qquad - [\beta^i_{x_ix_k} + P^{ki}\beta^i + (B^{ik} + \beta^i_{x_k} - P^{kj}a^{ij})v^i + \beta^i_{x_k}P^{ij}x_j]$$

$$- [B^{ik}P^{ij} + P^{ki}(B^{ij} - a^{i\ell}P^{\ell j})]x_j$$

$$P(0)x + v(0,x) = \Psi x + \psi(x).$$

Observe that the last term on the right of (2.10) is the kth component of

$$-[B(t)^*P(t) + P(t)B(t) - P(t)a(t)P(t)]x.$$

From the theory of the linear regulator, [1], it is well known that the optimal value of the problem

$$\max_{\mu}\{y(1)^*\Psi\, y(1) - \int_{1-t}^{1} |\mu(s)|^2 ds: \frac{dy}{ds} = -B(s)y - \sigma(s)\mu,\ y(1-t) = x\}$$

can be written as $x^*\bar{P}(t)\,x$ where

$$\frac{d\bar{P}}{dt} = \bar{P}\,a(t)\bar{P} - B(t)^*\bar{P} - \bar{P}\,B(t),\ \bar{P}(0) = \Psi.$$

Now set $P = \bar{P}$ so $P \geq 0$, symmetric and bounded. Then $u = Px + v$ is a solution of (2.4) if and only if

$$v_t^k = \frac{1}{2}a^{ij}v^k_{x_ix_j} - (b^i - a^{ij}v^j)v^k_{x_i} - (\hat{b}^i_{x_k}v^i + \hat{b}^i_{x_ix_k} + (x^*P\,\beta)_{x_k}),$$

$$v(0,x) = \psi(x)$$

with $\hat{b}(t,x) = [B(t) - a(t)P(t)]x + \beta(t,x)$. A bounded solution of this problem exists by the same proof as for theorem 2.1, provided q is locally $H^{2+\alpha}$. The result follows as in theorem 2.2 and corollary 2.1. Note that we still have $|\tilde{p}| \leq A\exp(B|x|^2)$ since $|[\ell n\,\tilde{p}]_{x_k}| \leq k(1 + |x|)$. Finally the proof of corollary 2.2 allows us to eliminate the condition that q is locally in $H^{2+\alpha}$. The result follows as in theorem 2.2 and corollary 2.1. Note that we still have $|\tilde{p}| \leq A\exp(B|x|^2)$ since $|[\ell n\tilde{p}]_{x_k}|$ $\leq k(1 + |x|)$. Finally the proof of corollary 2.2 allows us to eliminate the condition that q is locally in $H^{2+\alpha}$.

Appendix. Our aim is to prove theorem 2.1. We begin with a maximum principle. Let $Q = (0,1) \times G$, G open, bounded, connected, and let $\partial^*Q = (\{0\} \times G) \cup ([0,1]) \times \partial G)$. A classical solution refers to a function u which has continuous partials u_t, u_{x_i}, $u_{x_ix_j}$. $\bar{Q}$ is the closure of Q.

Theorem A.1. Assume a^{ij}, a_i, A^i, B^{ij} are bounded on $\bar{Q}$ for each fixed $u \in R^d$, $a^{ij}(t,x)\xi_i\xi_j \geq 0 \quad \forall\,\xi$, and u is a classical solution of

$$u^k_t = a^{ij}(t,x)u^k_{x_i x_j} + a_i(t,x,u)u^k_{x_i} + A^k(t,x) + B^{ki}(t,x)\, u^i \text{ on } Q$$

$$u = \phi \quad \text{on } \partial^*Q$$

Then

$$\sup_Q |u(t,x)| \le \max\{\sup_{\partial^*Q}|\phi(t,x)|,\ \inf_{\lambda>c_o} e^{\lambda}\sqrt{\frac{c_{oo}}{\lambda-c_o}}\}$$

where

$$c_0 \ge \frac{1}{2} + \sup_Q \sqrt{\Sigma_j\, B^{ij}(t,x)^2}\,,\quad c_{oo} \ge \frac{1}{2}\Sigma\, A^i(t,x)^2.$$

Proof: If $v = e^{\lambda t}u$, then

$$v^k_t = -\lambda v^k + a^{ij}(t,x)v^k_{x_i x_j} - a_i(t,x,e^{-\lambda t}v)v^k_{x_i} - e^{-\lambda t}A^k(t,x) - \sum_i B^{ki}(t,x)v^i.$$

Set $\psi = \sum_k (v^k)^2 = e^{-2\lambda t}|u|^2$. Then

$\psi_t = -2\lambda\psi + a^{ij}\psi_{x_i x_j} - 2a^{ij}v^k_{x_i}v^k_{x_j} - a_i\psi_{x_i} - 2e^{-\lambda t}A^k v^k - 2B^{ki}v^k v^i$. Since $\psi \ge 0$, then a non-negative maximum over $\bar{Q}$ is attained either on ∂^*Q or at some point $(t,x) \in [0,1] \times G$. In the latter case we have $\psi_t(t,x) \ge 0$, $\psi_{x_i}(t,x) = 0$, $a^{ij}(t,x)\psi_{x_i x_j}(t,x) \le 0$, and $a^{ij}(t,x)\, v^k_{x_i}(t,x)v^k_{x_j}(t,x) \ge 0$, so that

$$0 \le -2\lambda\psi - 2e^{-\lambda t}A^k v^k - 2B^{ki}v^k v^i,$$

and hence $0 \ge \lambda\psi - c_0\psi - c_{oo}$.
It follows that $\psi(t,x) \le c_{oo}/(\lambda-c_0)$ if $\lambda > c_0$. The result follows.

We define now some Sobolev spaces. Let $Q = (0,1) \times G$, G open, connected. $H^{\ell,\ell/2}(\bar{Q})$ is the Banach space of functions which together with all derivatives of the form $(\frac{\partial}{\partial t})^r(\frac{\partial}{\partial x})^s u$, $2r + s \le \ell$, are continuous on $\bar{Q}$. Note that $(\frac{\partial}{\partial x})^s u$ denotes any partial derivative in the x-variables of order s. We need only the cases $\ell = i+\alpha$, $i = 0, 1, 2$, $0 < \alpha < 1$. We define

$$\langle u\rangle^{(\alpha)}_Q = \sup_Q |u| + \sup_{(t,x)\in Q,(s,x)\in Q} \frac{|u(t,x) - u(s,x)|}{|t-s|^\alpha},$$

$$|u|^\alpha_Q = \langle u\rangle^{(\alpha/2)}_Q + \sup_{\substack{(t,x)\in Q,(t,y)\in Q\\ |x-y|<1}} \frac{|u(t,x) - u(t,y)|}{|x-y|^\alpha},$$

$$|u|^{(1+\alpha)}_Q = \langle u\rangle_Q^{(\frac{1+\alpha}{2})} + \sum_i |u_{x_i}|^{(\alpha)}_Q$$

$$|u|^{(2+\alpha)}_Q = \sup_Q |u| + \sum_i \langle u_{x_i}\rangle_Q^{(\frac{(1+\alpha)}{2})} + \sum_{ij} |u_{x_i x_j}|^{(\alpha)}_Q + |u_t|^{(\alpha)}_Q.$$

Note that our symbol $\langle u\rangle^{(\alpha)}_Q$ is not the same as that in [4]. For $G \subset R^d$, $H^\alpha(\bar{G})$ is defined in the obvious way.

We proceed now with the proof of theorem 2.1.

The method of proof is standard: we solve the first boundary value problem on a bounded domain and then let the domain converge to $(0,1) \times R^d$. Let $G_n = \{x: |x| < n\}$, $Q_n = (0,1) \times G_n$. We must first introduce boundary conditions on $\partial^* Q_n$ which satisfy a compatibility condition on $\{0\} \times \partial G$; moreover this must be done so as to preserve the symmetry $u^i_{x_j} = u^j_{x_i}$.

Define

$$\text{(A1)} \qquad \zeta(x) = \begin{cases} \sin(2\pi x)/2\pi + 1 - x, & 0 \leq x \leq 1 \\ 0 & , x \geq 1 \\ \zeta(-x) & , x < 0 . \end{cases}$$

and with $\mu = (2 + 2\alpha)^{-1}$ let $\gamma(s,x) = \zeta(x/s^\mu)/s^\mu$. Then

i) $\int_{-s_\mu}^{s_\mu} \gamma(s,x)dx = 1$, $\gamma(s,x) = 0$ if $|x| \geq s^\mu$,

ii) $0 \leq \gamma(s,x) \leq s^{-\mu}$, $|\gamma_x(s,x)| \leq 2s^{-2\mu}$, $|\gamma_{xx}(s,x)| \leq 2\pi s^{-3\mu}$.

Let

$$\psi(t,x) = \phi(x) + \int_0^t \int_{R_d} \chi(x-y) \prod_{i=1}^{d} \gamma(s, y_i)dy\, ds$$

where

$$\chi^k(x) = \frac{1}{2} a^{ij}(0)\, \phi^k_{x_i x_j}(x)$$

It can be shown that $\psi \in H^{2+\beta,1+\beta/2}(\overline{Q}_n)$ for $\beta = \alpha/(1+\alpha)$, $||\psi|| < \infty$, $\psi^i_{x_i} = \psi^j_{x_i}$, and $\psi_t(0,x) \equiv \lim_{s \to 0} \psi_t(s,x) = \chi(x)$. Let us now define

$$L^k_n u = u^k_t - \frac{1}{2} a^{ij}(t)\, u^k_{x_i x_j} + \delta_n(t)\left[(b^i(t,x) - a^{ij}(t)u^j)u^k_{x_i} + (b^i_{x_i x_k}(t,x) + b^i_{x_k}(t,x)u^i)\right]$$

where $\delta_n(t)$ is smooth, $0 \leq \delta_n(t) \leq 1$, $\delta_n(0) = 0$, $\delta_n(t) = 1$ if $t \geq 1/n$.

We shall now solve

$$\text{(A2)} \qquad L^k_n u = 0 \text{ on } Q_n,\ k = 1,\dots,d,$$

$$u = \psi \text{ on } \partial^* Q_n.$$

It follows from the fact that $\psi_t(0,x) = \chi(x)$ and $\delta_n(0) = 0$, that

$$L^k_n \psi(0,x) = 0 \quad \text{for } |x| = n,$$

so that the compatibility condition is satisfied. We apply the Leray-Schauder fixed point theorem much as in [4], ch V, §6. Let δ^{ij} be the Kronecker delta, i.e. the entries of the identity matrix, and define, for fixed n and for $\tau \in [0,1]$,

$$L^k_\tau u = \tau L^k_n u^k + (1 - \tau)[u^k_t - u^k_{x_i x_i} - \psi^k_t + \psi^k_{x_i x_i}]$$

$$= u^k_t - [\tau \tfrac{1}{2} a^{ij} + (1-\tau)\delta^{ij}]u^k_{x_i x_k} + \tau\, \delta_n(t)[(b^i - a^{ij}u^j)u^k_{x_i}$$

$$+ (b^i_{x_i x_k} + b^i_{x_k} u^i)] - (1-\tau)[\psi^k_t - \psi^k_{x_i x_i}].$$

Suppose that u is a classical solution of

$$\text{(A3)} \qquad L^k_\tau u = 0 \quad \text{on } Q_n$$

$$u = \psi \quad \text{on } \partial^* Q_n,$$

then we can apply theorem A.1 with c_o given in terms of K, c.f. (H ii), and c_{oo} in terms of K and $|\psi|^{(\beta)}_{Q_n}$. It follows that

$$\text{(A4)} \qquad \sup_{Q_n} |u(t,x)| \leq k_n$$

for some constant k_n independent of τ. For $\tau = 1$ i.e. for solution of (A2) we have moreover

$$\text{(A5)} \qquad \sup_{Q_n} |u(t,x)| \leq k_\infty.$$

We obtain further a priori estimates as follows. Let

$$N^w_\tau v = v_t - [\tau.\, \tfrac{1}{2} a^{ij} + (1-\tau)\, \delta^{ij}]v_{x_i x_j} + \tau\, \delta_n(t)[b^i - a^{ij}w^j]v_{x_i},$$

$$f^{w,k}_\tau = \tau\, \delta_n(t)[b^i_{x_i x_k} - b^i_{x_k} w^i] - (1-\tau)[\psi^k_t - \psi^k_{x_i x_i}].$$

Then N^w_τ is a scalar linear operator and

$$L^k_\tau u = N^u_\tau u^k + f^{u,k}_\tau.$$

Suppose that u is a classical solution of (A3), then u^k is a solution of

$$N^u_\tau v + f^{u,k}_\tau = 0 \quad \text{on } Q_n$$

$$v = \psi^k \quad \text{on } \partial^* Q_n.$$

By [4], ch V, theorem 1.1 (with ν, μ, μ, independent of τ, $\phi_o = \phi_1 = 0$, ϕ_2 depending on the bounds on b, b_x, b_{xx}, a, ψ_t, ψ_{xx} and k_n but not on τ, r = 2, q = ∞) there exist γ, c_1 depending on n but not τ such that $|u|^{(\gamma)}_{Q_n} \leq c_1$.

By [4], ch IV, theorem 5.2 (since now u in the coefficient of N^u_τ is smooth)

$$\text{(A6)} \qquad |u|^{2+\gamma}_{Q_n} \leq c(|f^{u,k}_\tau|^{(\gamma)}_{Q_n} + |\psi|^{(2+\gamma)}_{\partial^* Q_n}) \leq c_2$$

where c_2 depends on n but not τ.

Now let B be the set

$$\{w: \bar{Q}_n \to R_d,\ w,\ w_{x_i} \text{ continuous},\ w^i_{x_j} = w^j_{x_i},\ |w|_B \equiv \sum_k |w^k|^{(\gamma)}_{Q_n} + \sum_{ki} |w^k_{x_i}|^{(\gamma)}_{Q_n} < \infty\}$$

Let $\Phi(w,\tau)$ be the solution of the linear problem

$$v^k_t - [\tau.\frac{1}{2}.a^{ij} + (1-\tau)\delta^{ij}]v^k_{x_i x_j} + \tau\,\delta_n(t)[b^i - a^{ij}w^j)w^k_{x_i} + (b^i_{x_i x_k} + b^i_{x_k}w^i)]$$

$$\text{(A7)}\qquad - (1-\tau)(\psi^k_t - \psi^k_{x_i x_j}) = 0 \quad \text{on } Q_n$$

$$v = \psi \quad \text{on } \partial^* Q_n$$

For $w \in B$ such solutions exists and are unique, c.f. [4], ch.III. Moreover by [4], ch.IV, theorem 5.2 $\Phi(\cdot,\tau): B \to H^{2+\gamma,1+\gamma/2}(\bar{Q}_n)$. To show that $\Phi(\cdot,\tau): B \to B$ requires us to establish that $v^i_{x_j} = v^j_{x_i}$ if v solves (A7).

Let us smooth ψ to

$$\psi_m(t,x) = \int \psi(t,x-y) \prod_{i=1}^{d} \gamma(1/m, y_i)dy$$

and similarly w to w_m, b to b_m. Since ψ_t is bounded it follows that

$$\frac{\partial}{\partial t}\psi_m(t,x) = \int \psi_t(t, x-y) \prod_{i=1}^{d} \gamma(1/m, y_i)dy.$$

Define

$$Nv = v_t - [\tau\frac{1}{2}a^{ij} + (1-\tau)\,\delta^{ij}]v_{x_i x_j}$$

$$g^k(w) = \tau\,\delta_n(t)\,[(b^i - a^{ij}w^j)w^k_{x_i} + (b^i_{x_i x_k} + b^i_{x_k}w^i)] - (1-\tau)(\psi^k_t - \psi^k_{x_i x_i})$$

and similarly define $g^k_m(w)$ to be $g^k(w)$ but with b, ψ replaced by b_m, ψ_m. Let v^k_m be the unique solution of

$$Nv^k + g^k_m(w_m) = 0 \quad \text{on } Q_n,$$

$$v = \psi_m \quad \text{on } \partial^* Q_n.$$

Let $v^{k\ell}_m = (v^k_m)_{x_\ell}$. It satisfies

$$Nv^{k\ell} + [g^k_m(w_m)]_{x_\ell} = 0 \quad \text{on } Q_n,$$

$$v^{k\ell} = \psi^{k\ell}_m = \psi^{\ell k}_m \quad \text{on } \partial^* Q_n.$$

Note that ψ_m inherits the symmetry properties of ψ. Now (suppressing some subscripts momentarily)

$$[g^k_m(w_m)]_{x_\ell} = \tau\,\delta_n(t)[(b^i_{x_\ell} - a^{ij}w^j_{x_\ell})w^k_{x_i} + (b^i - a^{ij}w^j)w^k_{x_i x_\ell}$$

$$+ b^i_{x_i x_k x_\ell} + b^i_{x_k x_\ell}w^i + b^i_{x_k}w^i_{x_\ell}] - (1-\tau)(\psi^k_{tx_\ell} - \psi^k_{x_i x_i x_\ell})$$

$$= [g^\ell_m(w_m)]_{x_k},$$

since $(\psi^k_m)_{x_\ell} = (\psi^\ell_m)_{x_k}$, and

$$(\psi_m^k)_{tx_\ell} = \int \psi_t^k(t,y) \prod_{i\neq\ell} \gamma(\tfrac{1}{m}, x_i-y_i)\gamma_x(\tfrac{1}{m}, x_\ell - y_\ell)dy$$

$$= \frac{\partial}{\partial t}\int \psi^k(t,y) \prod_{i\neq\ell} \gamma(\tfrac{1}{m}, x_i-y_i)\gamma_x(\tfrac{1}{m}, x_\ell - y_\ell)dy$$

$$= \frac{\partial}{\partial t}\int \psi_{x_\ell}^k(t,y) \prod_i \gamma(\tfrac{1}{m}, x_i - y_i)dy$$

is symmetric in k, ℓ, and since $(w_m^k)_{x_\ell} = (w_m^\ell)_{x_k}$ and

$$b_{x_\ell}^i w_{x_i}^k + b_{x_k}^i w_{x_\ell}^i = b_{x_\ell}^i w_{x_i}^k + b_{x_k}^i w_{x_i}^\ell$$

is also symmetric in k and ℓ. By uniqueness $(v_m^k)_{x_\ell} = (v_m^\ell)_{x_k}$. Finally the usual estimate gives (A8)

$$|v^k - v_m^k|_{Q_n}^{(2+\gamma)} \leq c\{|\psi^k - \psi_m^k|_{\partial^* Q_n}^{(2+\gamma)} + |g^k(w) - g_m^k(w_m)|_{Q_n}^{(\gamma)}\},$$

and it also holds with γ replaced by, say $\gamma/2$. We shall show that $w_m \to w$ in $H^{\gamma/2,\gamma/4}(\bar{Q}_n)$. The same proof applied to b_m, $(b_m)_x$, $(b_m)_{xx}$, ψ_t, ψ_{xx}, w_x implies that the right side of (A8) converges to 0 i.e. $(v_m^k)_{x_\ell} \to v_{x_\ell}^k$ so that $v_{x_\ell}^k = v_{x_k}^\ell$.

First we have

$$\sup_{Q_n} |w_m(t,x) - w(t,x)| = \sup_{Q_n} |\int [w(t,x-y) - w(t,x)] \prod_i \gamma(\tfrac{1}{m}, y_i)dy|$$

(A9) $$\leq |w|_{Q_n}^{(\gamma)} \int |y|^\gamma \prod_i \gamma(\tfrac{1}{m}, y_i)dy \to 0.$$

Next consider

$$\{[w_m(t,x+h) - w(t,x+h)] - [w_m(t,x) - w(t,x)]\}h^{-\gamma/2}$$

$$= \left(\frac{w_m(t,x+h) - w_m(t,x)}{h^{\gamma/2}}\right) - \left(\frac{w(t,x+h) - w(t,x)}{h^{\gamma/2}}\right)$$

$$= w_m^h(t,x) - w^h(t,x)$$

if we define $w^h(t,x) = [w(t,x+h) - w(t,x)]h^{-\gamma/2}$. Now

$$|w^h(t,x) - w^h(t,y)| = \left|\frac{w(t,x+h) - w(t,x) - w(t,y+h) + w(t,y)}{|h|^{\gamma/2}}\right|$$

$$\leq \min\{2|w|_{Q_n}^{(\gamma)}|x-y|^\gamma/|h|^{\gamma/2},\ 2|w|_{Q_n}^{(\gamma)}|h|^\gamma/|h|^{\gamma/2}\}$$

$$\leq 2|w|_{Q_n}^{(\gamma)}|x-y|^{\gamma/2}.$$

It follows as in (A9) that $w_m^h \to w^h$ uniformly in h, t, x. Finally

$$\sup_{h,t,x} \{[w_m(t+h,x) - w(t+h,x) - [w_m(t,x) - w(t,x)]\} h^{-\gamma/4} \to 0$$

follows by a similar argument. Hence $|w_m - w|^{(\gamma/2)}_{Q_n} \to 0$, and $\Phi(\cdot,\tau)$: $B \to B$.

Now any classical solution of (A3) is a fixed point of $\Phi(\cdot,\tau)$, and it lies in B by (A6). Conversely if $u \in B$ is a fixed point $|u|^{(\gamma)}_{Q_n} < \infty$ and so (A6) holds again, i.e. the fixed point is a $H^{\gamma,\gamma/2}(\overline{Q}_n)$ solution of (A3). Moreover from (A4) (A6) we know that all the fixed points lie in the interior of the convex set

$$\{w \in B: \sup_{Q_n} \sum_k |w^k| \le k_n + \varepsilon,\ \sup_{Q_n} \sum_{ik} |w^k_{x_i}| \le c_2 + \varepsilon,\ |w|_B \le c_2 + \varepsilon\}$$

for any $\varepsilon > 0$. The proof now proceeds exactly as in [4], pp.454, 455 (for us β of the text is 1, α is γ), to establish that $\Phi(\cdot,\tau)$ has at least one fixed point for each τ, specifically for $\tau = 1$. Note that the unique fixed point of $\Phi(\cdot,0)$ is ψ which is in B. Finally one can argue as on p.455 that solutions of (A3) are not just in $H^{2+\gamma\ 1+\gamma/2}(\overline{Q}_n)$ but rather $H^{2+\beta\ 1+\beta/2}(\overline{Q}_n)$. Note that $\beta = \frac{\alpha}{\alpha+1}$ unlike γ is independent of n.

Let u_n now denote a solution of (A2). From (A5) we have that $\|u_n\| \le k_\infty$. By [4], ch VII, theorem 6.1 we have for $n > m+1$

(A10)
$$\max_{Q_{m+1}} |(u^k_n)_{x_i}| \le c_{3,m}$$

where $c_{3,m}$ depends on m, k_∞, and $\max_{G_{m+2}} |\phi_{x_j}(x)|$ but not n. Now [4], ch VII, theorem 5.1, gives that

$$|u_n|^{(2+\alpha)}_{\hat{Q}_m} \le c_{4,m}$$

where $c_{4,m}$ depends on m, k_∞ but not n, and $\hat{Q}_m = (1/m,1) \times G_m$. Note that δ_n is not Holder uniformly in n at $t = 0$, so we use $\hat{Q}_m$. The Ascoli theorem and diagonalization allows us to extract a subsequence, again denoted u_n, converging to a function $u \in H^{2+\alpha,1+\alpha/2}(\overline{Q})$ for any finte cylinder Q with $\overline{Q} \subset (0,1) \times R^d$. By [4], ch.VII, theorem 6.2, there exist γ, $c_{5,m}$ independent of n such that

$$|u_n|^{(\gamma)}_{Q_m} \le c_{5,m}.$$

Thus we may take $u_n \to u$ uniformly on Q_m, i.e. $u \in C(\overline{R}_T)$. Moreover (A10) implies that u_x is bounded on finite $\overline{Q}$. Of course u also inherits the symmetry $u^i_{x_j} = u^j_{x_i}$ on $(0,1) \times R^d$, and u solves (2.4).

References

[1] W.H. Fleming, R.W. Rishel, Deterministic and Stochastic Optimal Control, Springer-Verlag, New York, 1975.

[2] A. Friedman, Stochastic Differential Equations and Applications, Academic Press, New York 1975.

[3] U.G. Haussmann, E. Pardoux, Time reversal of diffusions, these proceedings.

[4] O.A. Ladyzenskaja, V.A. Solonnikov, N.N. Ural'ceva, Linear and Quasilinear Equations of Parabolic Type, A.M.S., Providence, R.I. 1968.

[5] E. Pardoux, Time reversal of diffusion processes and non-linear smoothing, Proc. Twente Workshop on Systems and Optimization, to appear in Lecture Notes in Control and Information Sciences, Springer-Verlag.

TIME REVERSAL OF DIFFUSION PROCESSES

U.G. HAUSSMANN+ and E. PARDOUX++

\+ University of British Columbia, Vancouver, Canada.

++ Université de Provence, Marseille, France; and INRIA.

Abstract -

We give conditions under which a diffusion process, solution of an Ito stochastic differential equation, remains a diffusion after time reversal. The method of proof, which is only sketched here, is to check that the original process solves a time-reversed martingale problem. We thus identify the coefficients of the stochastic differential equation for the reversed process.

1. Introduction -

Let $\{X_t, t \in [0,1]\}$ be a diffusion process with values in $\mathbb{R}^d$, solution of the Ito stochastic differential equation :
(1.1) $dX_t = b(t,X_t)dt + \sigma(t,X_t)dW_t$ where $\{W_t, t \in [0,1]\}$ is a standard Wiener process with values in $\mathbb{R}^\ell$, independent of the random vector X_o. Define, again for $t \in [0,1]$ the reversed process :

$$\overline{X}_t = X_{1-t}$$

clearly, $\{\overline{X}_t\}$ is a Markov process, since $\{X_t\}$ is . The aim of this note is to give conditions on b, σ and the law of X_o, under which $\{\overline{X}_t\}$ again solves a stochastic differential equation :
(1.2.) $d\overline{X}_t = \overline{b}(t,\overline{X}_t)dt + \overline{\sigma}(t,\overline{X}_t)d\overline{W}_t$
where $\{\overline{W}_t\}$ is a standard Wiener process independent of $\overline{X}_o = X_1$. We also seek to identify $\overline{b}$, $\overline{\sigma}$ and possibly $\{\overline{W}_t\}$.

The problem has been of interest to physicists-see in particular NELSON [12] and to control theoretists. In ANDERSON [1] and in PARDOUX [13], rather unverifiable conditions on the solution of the Fokker-Planck

----

This work was carried out while the first author was visiting the Université de Provence .

equation associated with (1.1) are given to guarantee the time-reversibility of the diffusion property . Another approach, which is in fact related to the theory of enlargement of filtration("grossissement de filtration") is used in CASTANON [3] and ELLIOTT and ANDERSON [5] , but unfortunately either with unverifiable assumptions or with lacunae in the proofs .Finally FÖLLMER [6] has interesting results concerning the same problem (with $\sigma = I$) in the non-Markov case and for an infinite dimensional example.

In section 2 we formulate a set of hypotheses which imply that $\{\bar{X}_t\}$ is a diffusion, and which permit us to compute $\bar{b}$ and $\bar{\sigma}$. Part of those hypotheses concern existence and smoothness of a density of the law of $X_t, t \in [0,1]$. In section 3 we give three sets of conditions on b, σ and the law of X_o, which imply the hypotheses of section 2 . In section 4, we discuss the identification of $\{\bar{W}_t\}$,and a related problem of enlargement of a filtration . The proofs, which are only sketched here, will be published elsewhere .

2 - A time reversed martingale problem.

Consider the following Ito stochastic differential equation :

(2.1) $\quad dX_t = b(t,X_t)dt + \sigma(t,X_t)d\,W_t$

where $\{W_t,\ t \in [0,1]\}$ is an $\mathbb{R}^\ell$-valued standard Wiener process defined on a probability space with filtration $(\Omega, F, \{F_t\}, P)$.

We suppose that the random vector X_o is F_o measurable .

(A.1) $\quad b: [0,1] \times \mathbb{R}^d \to \mathbb{R}^d$ and $\sigma: [0,1] \times \mathbb{R}^d \to \mathbb{R}^d \otimes \mathbb{R}^\ell$ are Borel measurable, locally bounded , and locally Lipschitz in x uniformly in t .

(A.2) $\quad \dfrac{\partial^2 \sigma^{im}}{\partial x_j\, \partial x_k}$ is a locally bounded function on $[0,1] \times \mathbb{R}^d$, for $i,j,k = 1,\dots,d$; $m = 1,.,\ell$

(A.3) The process $\{ X_t, t \in [0,1]\}$ is non exploding a.s.

(A.4) $\forall\, t > 0$, the law of X_t prossesses a density $p(t,x)$, and $\forall t_o > 0$, $p \in L^2(t_o,1;H^1_{loc})$.

The last part of (A.4) means that for any open bounded domain $D \subset \mathbb{R}^d$, $p \in L^2(t_o,1;\ H^1(D))$ i.e. $p, \frac{\partial p}{\partial x_i} \in L^2(\,]t_o,1[\, \times D)$; $i = 1,\dots,d$ (the partial derivatives are taken in the distributional sense) .

(A.1) + (A.3) imply that (2.1) has a unique solution on $[0,1]$, which is then a Markov process with generator L_t defined by :

$$L_t\, v(x) = \frac{1}{2}\, a^{ij}(t,x)\, \frac{\partial^2 v}{\partial x_i\, \partial x_j}(x) + b^i(t,x)\, \frac{\partial v}{\partial x_i}\,(x)$$

$$= \frac{1}{2}\, \frac{\partial}{\partial x_j}\, [a^{ij}\, \frac{\partial v}{\partial x_i}]\,(t,x) + \tilde{b}^i(t,x)\, \frac{\partial v}{\partial x_i}(x)$$

where $a = \sigma\sigma^*$ (σ^* denotes the transposed of σ), $\tilde{b}^i = b^i - \frac{1}{2}\frac{\partial a^{ij}}{\partial x_j}$,
and we use here and in the sequel the convention of summation over repeated indices .

We now introduce what we want to prove to be the infinitesimal generator of the Markov process $\{\overline{X}_t\}$. In the expressions which follow , any term involving p^{-1} is taken to be zero at any point (t,x) where $p(t,x) = 0$. Let us define $\overline{L}_t$, $t \in [0,1]$,by :

$$\overline{L}_t\, v(x) = \frac{1}{2}\, \overline{a}^{ij}(t,x)\, \frac{\partial^2 v}{\partial x_i\, \partial x_j}(x) + \overline{b}^i(t,x)\, \frac{\partial v}{\partial x_i}\,(x)$$

where :

$$\overline{b}^i(t,x) = -\, b^i(1-t,x) + p^{-1}(1-t,x)\, \frac{\partial}{\partial x_j}[a^{ij}p]\ (1-t,x) ; i=1,\dots,d$$

$\overline{a}^{ij}(t,x) = a^{ij}(1-t,x)$;similarly $\overline{\sigma}(t,x) = \sigma(1-t,x)$

We have the following result :

Theorem 2.1. Assume (A.1,2,3,4). Then $\{\overline{X}_t, t \in [0,1]\}$ is a solution of the martingale problem associated to $\overline{L}_t$;i.e. it is a diffusion process with generator $\overline{L}_t$.

Corollary 2.2. Under the assumptions of theorem 2.1 ; there exists a standard Wiener process $\{\ \overline{W}_t, t \in [0,1]\ \}$ which is a $\overline{F}_t$ martingale $\{\ \overline{F}_t = \sigma(\overline{X}_s, 0 \leq s \leq t)\ \}$ with :

$$(2.2)\quad \overline{X}_t = \overline{X}_o + \int_o^t \overline{b}(s,\overline{X}_s)ds + \int_o^t \overline{\sigma}(s,\overline{X}_s)d\,\overline{W}_s, t \in [0,1]$$

Remark 2.3. We do not know a priori that $\overline{b}$ is locally bounded. HAUSSMANN [7] gives conditions under which $\overline{b}$ is bounded or has linear growth . Nevertheless,(2.2) makes sense , provided the process $p^{-1}(t,X_t)\, \frac{\partial}{\partial x_j}\ [a^{ij}\ p\ ]\ (t,X_t)$ is a.s. Lebesgue integrable. This follows from the hypotheses . Inded, to show that :

$$\int_o^1 |\, p^{-1}(t,X_t)\, \frac{\partial}{\partial x_j}\ [a^{ij}p\ ](t,X_t)\ |\ dt < \infty \qquad \text{a.s.}$$

we need only show that, $\forall\ f \in C_c^\infty\ (\mathbb{R}^d)$, the set of C^∞ functions from $\mathbb{R}^d$ into $\mathbb{R}$ with compact support , $f \geq 0$,

$$E \int_o^1 f(X_t)\ p^{-1}(t,X_t)\ |\frac{\partial}{\partial x_j}\ [\ a^{ij}\ p\ ](t,X_t)\ |\ dt < \infty$$

But $\frac{\partial}{\partial x_j}\ [\ a^{ij}p\ ] = \frac{\partial a^{ij}}{\partial x_j}\ p + a^{ij}\ \frac{\partial p}{\partial x_j}$,and since the distributional derivative $\frac{\partial p}{\partial x_j}$ equals a.e. the usual derivative (See NECAS [11] ,Thm 2.2.,p.61) , which is zero at any point of $\{(t,x);p(t,x)=0\}$ where it does exist, since at such a point $p(t,x)$ attains its minimum.

Therefore the above quantity equals :

$$\int_0^1 \int_{\mathbb{R}^d} f(x) \left|\frac{\partial}{\partial x_j}[a^{ij}p](t,x)\right| dx\, dt$$

which is finite .

□

<u>Sketch of the proof of theorem 2.1</u> :

We weed to show that for any $f \in C_c^\infty(\mathbb{R}^d)$, and $0 < s < t < 1$,

$$E\{f(\overline{X}_t)-f(\overline{X}_s)- \int_s^t \overline{L}_u f(\overline{X}_u)du \mid \overline{X}_r,\ 0 \leq r \leq s\}= 0$$

Since $\overline{X}$ is markovian , this amounts to show that for any $g \in C_c^\infty(\mathbb{R}^d)$,

$$E\{[f(\overline{X}_t)-f(\overline{X}_s)- \int_s^t \overline{L}_u f(\overline{X}_u)du]\, g(\overline{X}_s)\} = 0$$

or, in other words, that $\forall f,g \in C_c^\infty(\mathbb{R}^d)$, $0 < s < t < 1$,

$$E\{[f(X_t)-f(X_s)- \int_s^t \hat{L}_u f(X_u)du]\, g(X_t)\} = 0$$

where $\hat{L}_u = - L_{1-u}$. Define :

$$v(u,x) = E[g(X_t) / X_u = x] ; \quad s \leq u \leq t,\ x \in \mathbb{R}^d$$

Formally, v satisfies the Kolmogorov backward equation :

$$(2.3) \quad \begin{cases} \dfrac{dv}{du} + L_u v = 0 , & s \leq u \leq t \\ v(t) = g & \end{cases}$$

Now :

$$\begin{aligned} E\{[f(X_t)-f(X_s)]\, g(X_t)\} &= E[f(X_t)v(t,X_t)] - E[f(X_s)v(s,X_s)] \\ &= (f\, p(t),v(t))- (f\, p(s),v(s)) \end{aligned}$$

where $(h,k) \triangleq \int_{\mathbb{R}^d} h(x)k(x)dx$

Taking into account (2.3) and the Kolmogorov forward equation satisfied in a weak sense by p :

$$(2.4) \qquad \frac{dp}{du} = L_u^* p$$

we obtain formally :

$$\begin{aligned} (fp(t),v(t))-(fp(s),v(s)) &= \int_s^t [(f\frac{dp}{du}(u),v(u))+(fp(u), \frac{dv}{du}(u))]\, du \\ &= \int_s^t [(L_u^*(fp(u))+p(u)\hat{L}_u f,v(u))-(fp(u),L_u v(u))]du \\ &= E[g(X_t)\int_s^t \hat{L}_u f(X_u)du] \end{aligned}$$

□

3 - <u>Sufficient conditions for assumptions (A.3),(A.4) to hold</u>

Let us note that there exist well-known sufficient conditions for (A3) to hold, see e.g. STROOCK -VARADHAN [14] or IKEDA-WATANABE [9] .

We now concentrate on sufficient conditions for (A.4)to hold. Since smoothness of p implies local boundedness, clearly a sufficient condition for (A 4), regardless of the nature of the law of X_o, is :

(B) $\frac{\partial}{\partial t} + L_t$ is hypoelliptic

We don't want to describe the well-known assumptions of HÖRMANDER's theorem, which imply (B), and refer the interested reader to e.g. [8]. Note that these assumptions include in particular the fact that b and σ be C^∞ in t and x. On the other hand, a version of the needed HÖRMANDER's theorem, in case of coefficients not necessarily smooth in t can be found in MICHEL and CHALEYAT-MAUREL [4] .

We now formulate two other sets of hypotheses, where we don't assume C^∞ regularity of the coefficients, but where we do restrict their growth as well as assume the existence of an initial density .

(C.1) $\exists k > 0$ s.t. $\forall\, t \in [0,1]$; $x,y \in \mathbb{R}^d$,
$|\tilde{b}(t,x)| + |\sigma(t,x)| \leqslant k\ (1 + |x|)$

(C.2) $\exists\, \alpha > 0$ s.t. $\forall\, t \in [0,1], x \in \mathbb{R}^d$,
$a(t,x) \geqslant \alpha I$

(C.3) The law of X_o has a density $p_o(x)$, and the exists $\lambda < 0$ s.t.
$p_o \in L^2(\mathbb{R}^d, (1+|x|^2)^\lambda\, dx)$

(D.1) $\exists\, k > 0$ s.t. $\forall\, t \in [0,1]$, $x \in \mathbb{R}^d$,
$|b(t,x)| + |\sigma(t,x)| + \sum_{i,j,k} |\frac{\partial a^{ij}}{\partial x_k}(t,x)| \leqslant k(1+|x|)$

(D.2) $\frac{\partial b^i}{\partial x_k}, \frac{\partial^2 a^{ij}}{\partial x_k \partial x_j}, \frac{\partial(\mathrm{div}\, \tilde{b})}{\partial x_k} \in L^\infty(\mathbb{R}^d)$; $i,j,k = 1,\ldots,d$

(D.3) The law of X_o has a density $p_o(x)$, and $\exists\, \lambda < 0$
s.t. $p_o, \frac{\partial p_o}{\partial x_i} \in L^2(\mathbb{R}^d; (1+|x|^2)^\lambda\, dx)$, $i = 1\ldots d$.

<u>Theorem 3.1.</u> Each of the set of hypotheses (C.1)-(C.2)-(C.3) and (D.1)-(D.2)-(D.3) implies (A.4).

The proof of this result uses techniques for estimating the solution of a PDE with unbounded coefficients in weighted Sobolev spaces, as in BENSOUSSAN-LIONS [2] and MENALDI [10] .

4. Identification of $\overline{W}$

From (2.2), we get a new equation for $\{X_t\}$:

(4.1) $X_t = X_o + \int_0^t \hat{b}(s,X_s)ds + \int_0^t \sigma(s,X_s) \oplus d\,\hat{W}_s$

where $\hat{b}(s,x) = -\bar{b}(1-s,x)$, and $\hat{W}_s = \bar{W}_{1-s}$.

Define the " backward filtration"$\{H^t, t \in [0,1]\}$ (the family of σ-algebras H^t is decreasing) by :

$$H^t \triangleq \sigma\{X_u,\ t \leq u \leq 1\} = \bar{F}_{1-t}$$

The"backward Wiener process " $\hat{W}_t$, starting from 0 at $t = 1$, is a " backward H^t martingale ", and the sign $\oplus$ means that the integral is a " backward Ito integral ", i.e. it is the limit of sums of the type :

$$\sum_i \sigma_i\,(X_{s_{i+1}})\,(\hat{W}_{s_{i+1}} - \hat{W}_{s_i})$$

($\sigma_i(x)$ denoting an approximation of $\sigma(s,x)$ on the interval(s_i, s_{i+1})).
After having rewritten both (1.1) and (4.1) in the "Stratonovitch language ", we get by comparison :

$$\sigma(t,X_t)\circ d\hat{W}_t = \sigma(t,X_t)\circ dW_t + p^{-1}(t,X_t)\ \sigma(t,X_t)\nabla.(p\sigma)(t,X_t)dt$$

where o means that the corresponding stochastic integral has to be taken in the Stratonovitch sense, and $\nabla.(p\sigma)(t,x)$ denotes the vector in $\mathbb{R}^\ell$ whose i-th component equals $\frac{\partial}{\partial x_j}(p\,\sigma_{ji})(t,x)$. If $\ell = d$ and $\sigma(t,x)$ is invertible $\forall (t,x)$, it then follows :

(4.2) $\hat{W}_t = W_t - W_1 - \int_t^1 p(s,X_s)^{-1}\ \nabla.(p\sigma)(s,X_s)ds$

We recall the convention that the term involving p^{-1} is taken to be zero whenever p is zero, and the argument of Remark 2.3. In particular, $\hat{W}_t$, giben by (4.2), is a H^t - " backward Wiener process " . Suppose now that we can show directly that the just mentioned result holds. We would then get the result of §2 as a consequence, with the identification (4.2), even in case σ is degenerate . It is possible to show such a result, under hypotheses similar to those in(C)or(D). Let us now indicate the connection with a problem of enlargement of a filtration (" grossissement de filtration " in the French probability litterature). Proving the above mentioned result is equivalent to showing that $\{W_t - W_1, t \in [0,1]\}$ is a H^t semi-martingale, and identifying its bounded variation part. It would be enough to do the same after having replaced $\{H^t\}$ by the larger filtration $\{G^t\}$, with :

$$G^t = F^t \vee \sigma(X_t) = F^t \vee \sigma(X_1)$$

where $F^t = \sigma(W_u - W_t, t \leq u \leq 1)$.

But $\{W_t - W_1, t \in [0,1]\}$ is a F^t"backward Wiener process",and G^t consits of F^t, enlarged by $\sigma(X_1)$. Our question is then typically what the theory of enlargement of filtration is about .

References

[1] B.D.O. ANDERSON - Reverse time diffusion equation models. Stock-Proc.and Appl.12 ,313-326 (1982)

[2] A.BENSOUSSAN, J.L. LIONS - Applications des inéquations variationnelles en contrôle stochastique. Dunod, Paris (1978)

[3] D.A. CASTANON - Reverse time diffusion process. IEEE . Trans.Inf.Theory,IT 28,953-956 (1982)

[4] M. CHALEYAT-MAUREL , D.MICHEL - Hypoellipticity theorems and conditional laws. Zeit.für Wahrschein. 65 , 573-597 (1984)

[5] R.J.ELLIOTT, B.D.O.ANDERSON - Reverse time diffusions, preprint.

[6] H. FÖLLMER - An entropy approach to the time reversal of diffusion processes, these Proceedings

[7] U.HAUSSMANN - On the drift of a reversed diffusion, these Proceedings .

[8] L. HÖRMANDER - Hypoelliptic second order differential equations . Acta Math. 117, 147-171 (1967)

[9] N.IKEDA,S.WATANABE - Stochastic differential equations and diffusion processes North Holland(1981)

[10] J.L. MENALDI - Optimal impulse control problems, SIAM J. Control & Optimization, 18 , (1980)

[11] J. NECAS - Les méthodes directes en théorie des équations elliptiques -Masson, Paris and Academia, Prague (1967).

[12] E.NELSON - Dynamical theories of Brownian motion Princeton Univ. Press (1967)

[13] E.PARDOUX - Smoothing of a diffusion conditioned at final time, in Stochastic Differential Systems, M.Kohlmann, N.Christopeit Eds, Lecture Notes in Control and Info. Sci. 43 , 187-196, Springer-Verlag (1982)

[14] D.W. STROOCK, S.R.S. VARADHAN-Multidimensional diffusion processes Springer-Verlag (1979) .

DIVERGENCE, CONVERGENCE
AND MOMENTS OF SOME INTEGRAL FUNCTIONALS OF DIFFUSIONS

M. MUSIELA

Laboratoire IMAG-TIM 3
BP : 68
38402 SAINT-MARTIN D'HERES CEDEX
FRANCE

In this paper we study some integral functionals of diffusions. We present, without proofs, criteria for their divergence, their convergence and the existence of their moments. An extended version of this paper with proofs, completed by a study of some particular cases will be published elsewhere.

1. INTRODUCTION

Let D be an open, connected subset of $\mathbb{R}^d$, $d \geq 2$, and let $D_\delta = D \cup \{\delta\}$ with $\delta = \infty(D)$ be the one point compactification of D.
Let Ω be the set of all continuous functions ω of $\mathbb{R}_+$ into D which are stopped at the first time of hitting δ : if $\omega(t) = \delta$, then $\omega(t') = \delta$ for $t' \geq t$. Denote, $(X_t)_{t\geq 0}$ the coordinate process, $F_t = \sigma(X_s ; s \leq t)$, $F = F_\infty$. Moreover, let $W = \mathbb{R}_+ \times \Omega$, $Y_t(c, \omega) = (c+t, \omega(t))$, $G_t = \sigma(Y_s ; s \leq t)$ $(= \sigma(Y_0, X_s, s \leq t))$, $G = G_\infty$.
If U is an open (for the relative topology) subset of $[c,\infty[\times D$ for some $c \in \mathbb{R}_+$ we set $S(U) = \inf\{s \geq 0 ; Y_s \notin U\}$ ($\inf \emptyset = \infty$). S(U) is a stopping time of (G_t). The lifetime of (Y_t) is $S = S(\mathbb{R}_+ \times D) = \inf\{s \geq 0 ; X_s = \delta\}$.
Assume that the functions $b : \mathbb{R}_+ \times D \to \mathbb{R}^d$ and $\sigma : \mathbb{R}_+ \times D \to \mathbb{R}^d \otimes \mathbb{R}^r$ are continuous and that for any $R > 0$ there exists a constant $c_R > 0$ such that :

$$|b(t,x) - b(t,y)| + |\sigma(t,x)-\sigma(t,y)| \leq c_R |x-y|$$

for all $t \in \mathbb{R}_+$ $x,y \in D$ provided $|x| + |y| \leq R$.
For the vectors (matrices) x,y the symbols $|x|$ and $x\cdot y$ stand for the Euclidean norm and the Euclidean scalar product, respectively. The symbol x^* stands for the transpose of x.
Let $a = \sigma\sigma^*$ and let Λ be the second order differential operator on $\mathbb{R}_+ \times D$:

$$\Lambda = \mathbb{D}_t + L, \quad L = \frac{1}{2} a\cdot\mathbb{D}_x^2 + b\cdot\mathbb{D}_x .$$

For $y \in \mathbb{R}_+ \times D$, let Q_y be the probability on (W,G) such that $Q_y\{Y_0 = y\} = 1$ and $(g(Y_t) - \int_0^t \Lambda g(Y_s)ds)$ is a (G_t, Q_y)-local martingale on $[0, S[$ for each $g \in C^{1,2}(\mathbb{R}_+ \times D)$. (W, G, G_t, X_t, S, Q_y) is the canonical realization of an inhomogeneous strongly Markov diffusion process on D generated by the differential operator Λ. If the functions $a(t,\cdot)$ and $b(t,\cdot)$ do not depend on t, then (W,G,G_t,X_t,S,P_x) with $P_x = Q_{0,x}$, $x \in D$, is a homogeneous diffusion on D with infinitesimal generator L.

Existence and uniqueness of Q_y, for all $y \in \mathbb{R}_+ \times D$, follow from the classical results on stochastic differential equations (cf. Ikeda and Watanabe [5] and Narita [11] for example) and Doob's representation of stopped diffusions (cf. Dynkin [4]).

Let the function $f : \mathbb{R}_+ \times D \to \mathbb{R}_+$ be Borel measurable. In this paper we investigate the integral :

$$(1.1) \qquad I_t = \int_0^{t\wedge S} f(Y_s)\, ds \quad , \quad 0 \le t \le \infty .$$

We obtain criteria ensuring its a.s. divergence or its a.s. convergence. We give also sufficient conditions for the existence of exponential moments of the integral. Criteria which are formulated in terms of differential inequalities generalize those of Stroock and Varadhan [14] and Narita [11-13] and relate the problem to a purely analytical one. The radial criteria generalize the classical Khas'minskii's [8, 9] tests of non explosion or explosion (see also Azencott [1], Bhattacharya [2, 3] Mc Kean Jr [10]).

We shall use the following notation. For a Q_y-integrable function ξ on W the symbol $Q_y\xi$ stands for the expectation $\int_W \xi \, d\, Q_y$. For a real function f defined on a non empty set B $\sup_B f$ and $\inf_B f$ stand, respectively for the supremum and the infinimum of f over B.

2. GENERAL CRITERIA

In this section we discuss a few simple criteria which relate the problem to a purely analytical one.

Let Δ be an open connected non empty subset of $[c, \infty[\times D$ for some $c \in \mathbb{R}_+$. Moreover, let U be a compact subset of Δ such that $\inf_U f > 0$ ($\inf \emptyset = \infty$). We have the following

Theorem 2.1.

Assume that there exist $\lambda > 0$ and positive function $u \in C^{1,2}(\Delta)$ such that $\Lambda u \le \lambda f \phi\circ u$ on $\Delta - U$, where $\phi : \mathbb{R}_+ \to \mathbb{R}_+$ is increasing (= nondecreasing), differentiable, such that $\int_0^{\infty} \frac{dt}{1+\phi(t)} = \infty$.

Then for $y \in \Delta$ one has Q_y a.s.

$$\{\overline{\lim_{t\uparrow S(\Delta)}}\, u(Y_t) = \infty\} \subset \{I_{S(\Delta)} = \infty\} .$$

In particular if $\lim_{\Delta \ni z \to \infty(\Delta)} u(z) = \infty$, then

$$Q_y\{I_{S(\Delta)} = \infty\} = 1.$$

Now we discuss the convergence of the integral $I_{S(\Delta)}$.

Theorem 2.2. : Assume that there exist $\lambda > 0$ and $u \in C^{1,2}(\Delta)$ such that $u \geq 0$ and $\Lambda u \geq \lambda f u$ on Δ. Then for $y \in \Delta$

$$Q_y\{I_{S(\Delta)} < \infty\} \geq \frac{u(y)}{\sup_\Delta u}$$

Concerning the moments we have (see also Khas'minski [7])

Theorem 2.3.

(i) If there exist $\lambda > 0$ and $u \in C^{1,2}(\Delta)$ such that $\Lambda u \geq \lambda f$, then

$$Q_y I_{S(\Delta)} \leq \lambda^{-1} (\sup_\Delta u - u(y)),\ y \in \Delta .$$

(ii) If $\alpha = \sup_\Delta Q_y\, I_{S(\Delta)} < 1$, then

$$Q_y \exp(I_{S(\Delta)}) \leq \frac{1}{1-\alpha}\ (< \infty),\ y \in \Delta .$$

Let π be the projection of $\mathbb{R}_+ \times D$ on D.

Corollary 2.1. : If $\pi(\Delta)$ is bounded and if there exists $z \in \mathbb{R}^d$ such that $\frac{1}{2}\, z \cdot az + z \cdot b \geq \lambda$ on Δ for some $\lambda > 0$, then $Q_y\, S(\Delta) < \infty$, $y \in \Delta$.

We use the strong Markov property of (Y_t) under Q_y to prove the following result.

Theorem 2.4. : Let $q(y) = Q_y\{I_{S(\Delta)} < \infty\}$, $y \in \Delta$. Assume that

a) $Q_y\{S(U) < \infty\} = 1$ for each $y \in U$ and each open subset U of Δ such that $\pi(U)$ is bounded and $\overline{U} \subset \Delta$,

b) $\lim_{\pi(\Delta) \ni \pi(y) \to \infty(\pi(\Delta))} q(y) = 1.$

Then $q \equiv 1$ on Δ.

We finish this section by criteria of divergence or convergence obtained via change of measure. Let $u \in C_t^{1,2}(\mathbb{R}_+ \times D)$ be real non vanishing. Let $c = \frac{\mathbb{D}_x u}{u}$ and $Z_t = \frac{u(Y_t)}{u(Y_0)} \exp\left(-\int_0^t \frac{\Lambda u}{u}(Y_s)\, ds\right)$. Under Q_y one has a.s.

$$Z_t = \exp\left(\int_0^t c(Y_s) \cdot dM_s - \frac{1}{2}\int_0^t c \cdot ac(Y_s)\, ds\right),\ t < S,$$

where $(M_t = X_t - \int_0^t b(Y_s)\, ds)$ is a local martingale on $[0,S[$ with quadratic variation $(\int_0^t a(Y_s)\, ds)$. Let $\Delta_n = [0,n[\times D_n$, where D_n is an increasing sequence of bounded open subsets of D such that $\bigcup_n \overline{D_n} = D$. It follows from Girsanov's theorem that the process $(X_t - \int_0^t (b+ac)(Y_s)\, ds)$ stopped at $S(\Delta_n)$ is a bounded martingale under

$Z_{S(\Delta_n)} Q_y$.

This proves that the diffusion measures Q_y^u associated with the operator $\Lambda + ac \cdot \mathbb{D}_x$ satisfy

$$Q_y^u = Z_{S(\Delta_n)} Q_y \text{ on } G_{S(\Delta_n)} .$$

Note that if we set $Z_S = \overline{\lim_{t \uparrow S}} Z_t$ we have the following decomposition :

$$Q_y^u = Z_\tau Q_y + Q_y^u\{\{Z_\tau = \infty\} \cap \cdot\} \text{ on } G_\tau$$

for each stopping time $\tau \le S$. Moreover, in this decomposition we can replace $\{Z_\tau = \infty\}$ by $\{\int_0^\tau c \cdot ac(Y_t)dt = \infty\}$ (see Kabanov, Liptser, Shiryaev [6] for details).

Theorem 2.5. : Suppose that $\frac{\Lambda u}{u} \le \lambda f$ on Δ for some $\lambda > 0$ and that $\underline{\lim}_{n\to\infty} Q_y^u \frac{1}{u}(Y_{S_n}) = 0$ for some sequence (S_n) of stopping times such that $S_n < S(\Delta)$, $S_n \uparrow S(\Delta)$. Then $Q_y\{I_{S(\Delta)} = \infty\} = 1$, $y \in \Delta$.

Suppose that $\frac{\Lambda u}{u} \ge \lambda f$ on Δ for some $\lambda > 0$. Then for any $y \in \Delta$ and any sequence (S_n) of stopping times such that $S_n < S(\Delta)$, $S_n \uparrow S(\Delta)$, we have :

$$Q_y\{I_{S(\Delta)} < \infty\} \ge \overline{\lim_{n\to\infty}} Q_y^u \frac{u(y)}{u(Y_{S_n})} .$$

3. RADIAL CRITERIA

Now we shall use results of Section 2 in order to obtain more explicit criteria of divergence or convergence of the integral I_S, where $S = \inf\{t \ge 0 ; X_t = \delta\}$. We shall study also the integral $I_{T_a \wedge T_b}$, where $T_a = \inf\{t \ge 0 ; |X_t - z| = (2a)^{1/2}\}$, $z \in \mathbb{R}^d$, $0 \le a < b \le \infty$.
We consider only the case $D = \mathbb{R}^d$. We assume that the continuous functions $b : \mathbb{R}_+ \times \mathbb{R}^d \to \mathbb{R}^d$ and $\sigma : \mathbb{R}_+ \times \mathbb{R}^d \to \mathbb{R}^d \otimes \mathbb{R}_+^r$ are uniformly in t locally Lipschitz continuous in x and that for all $x \in \mathbb{R}^d$ the functions $a(\cdot,x)$ and $b(\cdot,x)$ are bounded ($a = \sigma\sigma^*$).
Let $z \in \mathbb{R}^d$, $r > 0$. We assume also that there are continuous functions:

$$\bar{\alpha} : [r, \infty[\to]0,\infty[$$
$$\bar{\beta} : [r, \infty[\to \mathbb{R}$$
$$\underline{\gamma} : [r, \infty[\to \mathbb{R}_+$$

such that for all $|x| \ge (2r)^{1/2}$, $t \in \mathbb{R}_+$

$$x \cdot a(t,x+z)x \le \bar{\alpha}\left(\frac{|x|^2}{2}\right)$$

$$\text{(3.1)} \quad \operatorname{tr} a(t,x+z) + 2x \cdot b(t,x+z) \le x \cdot a(t,x+z)x\, \bar{\beta}\left(\frac{|x|^2}{2}\right),$$

$$\underline{\gamma}\left(\frac{|x|^2}{2}\right) \le f(t,x+z).$$

Define for $t \geq r$

$$\bar{e}(t) = \exp\left(-\int_r^t \bar{\beta}(u)du\right), \quad \bar{m}(t) = 2\int_r^t \frac{\underline{\gamma}}{\bar{e}\,\underline{\alpha}}(u)\,du,$$

$$\bar{s}(t) = \int_r^t \bar{e}(u)\,du, \quad \bar{k}(t) = \int_r^t \bar{e}\,\bar{m}(u)\,du.$$

Let $B(z,r) = \{x : |x-z| < (2r)^{1/2}\}$. Our following result, which reduces to Khas'minskii's test of non explosion (cf. Stroock, Varadhan [14] for example) if $f \equiv 1$, is a consequence of Theorem 2.1.

Theorem 3.1. : If $\bar{k}(\infty) = \infty$ and $\inf_{\mathbb{R}_+ \times B(z,r)} f > 0$, then for $y \in \mathbb{R}_+ \times \mathbb{R}^d$ $Q_y\{I_S = \infty\} = 1$.

As before we fix $z \in \mathbb{R}^d$, $r > 0$ and we assume that there are continuous functions :

$$\underline{\alpha} : [r,\infty[\to]0,\infty[,$$
$$\underline{\beta} : [r,\infty[\to \mathbb{R},$$
$$\bar{\gamma} : [r,\infty[\to \mathbb{R}_+$$

such that for all $|x| \geq (2r)^{1/2}$, $t \in \mathbb{R}_+$,

$$\underline{\alpha}\left(\frac{|x|^2}{2}\right) \leq x \cdot a(t,x+z)x,$$

$$(3.2) \quad x \cdot a(t,x+z)x\,\underline{\beta}\left(\frac{|x|^2}{2}\right) \leq \operatorname{tr} a(t,x+z) + 2x \cdot b(t,x+z),$$

$$f(t,x+z) \leq \bar{\gamma}\left(\frac{|x|^2}{2}\right).$$

We define for $t \geq r$ the functions

$$\underline{e}(t) = \exp\left(-\int_r^t \underline{\beta}(u)du\right), \quad \underline{m}(t) = 2\int_r^t \frac{\bar{\gamma}}{\underline{e}\,\underline{\alpha}}(u)\,du$$

$$\underline{s}(t) = \int_r^t \underline{e}(u)\,du, \quad \underline{k}(t) = \int_r^t \underline{e}\,\underline{m}(u)\,du,$$

and

$$\bar{\delta}(t) = \frac{\bar{e}\,\bar{m}}{\underline{e}}(t), \quad \underline{\delta}(t) = \frac{\underline{e}\,\underline{m}}{\bar{e}}(t).$$

Now we introduce an additional assumption which will be used in the following study.

(A) For each open subset U of $\mathbb{R}_+ \times \mathbb{R}^d$ such that $\pi(U)$ is bounded and for each $y \in U$ $Q_y\, S(U) < \infty$.

A simple sufficient condition for (A) to hold is given in Corollary 2.1. Therefore if the functions a and b do not depend on t and a(x) is positive definite for all $x \in \mathbb{R}^d$, then (A) holds. The next result, which for $f \equiv 1$ reduces to Khas'minskii's test of explosion (cf. Mc Kean [10] for example), can be proved using Theorem 2.4.

Theorem 3.2. : Assume (A) holds. If $\underline{k}(\infty) < \infty$, then for $y \in \mathbb{R}_+ \times \mathbb{R}^d$ $Q_y\{I_S < \infty\} = 1$.

From now until the end of this section we suppose that the continuous functions $\underline{\alpha}$, $\underline{\beta}$, $\overline{\gamma}$ and $\overline{\alpha}$, $\overline{\beta}$, $\underline{\gamma}$ satisfy (3.1) and (3.2), respectively, on the interval $]0, \infty[$. We shall study the integral $I_{T_a \wedge T_b}$, where $T_a = \inf\{t \geq 0 \,;\, |X_t - z| = (2a)^{1/2}\}$, $z \in \mathbb{R}^d$, $0 \leq a < b \leq \infty$. Note that $T_\infty = \lim_{b\to\infty} T_b = S$ and $T_0 = \inf\{t \geq 0 \,;\, X_t = z\}$.

Theorem 3.3. : Assume (A) holds. Denote $y = (t,x)$, where $t \in \mathbb{R}_+$ and $x \in B(z,b) - \overline{B(z,a)}$.

1) If $\overline{k}(a) = \infty$, $\overline{k}(b) = \infty$, then $Q_y\{I_{T_a \wedge T_b} = \infty\} = 1$.

2) If $\overline{k}(a) < \infty$, $\overline{k}(b) = \infty$, then $\{I_{T_a \wedge T_b} < \infty\} = \{T_a < T_b\}$ Q_y a.s.

If additionally :

a) $\underline{s}(a) > -\infty$, $\underline{s}(b) < \infty$, then

$$\frac{\overline{s}(b) - \overline{s}(\frac{|x-z|^2}{2})}{\overline{s}(b) - \overline{s}(a)} \leq Q_y\{I_{T_a \wedge T_b} < \infty\} \leq \frac{\underline{s}(b) - \underline{s}(\frac{|x-z|^2}{2})}{\underline{s}(b) - \underline{s}(a)}$$

b) $\overline{s}(b) = \infty$, then $Q_y\{I_{T_a \wedge T_b} < \infty\} = 1$

c) $\overline{s}(b) = \infty$, $\overline{\delta}(b) = \infty$, then $Q_y I_{T_a \wedge T_b} = \infty$.

3) If $\overline{k}(a) = \infty$, $\overline{k}(b) < \infty$, then $\{I_{T_a \wedge T_b} < \infty\} = \{T_b < T_a\}$ Q_y a.s.

If additionally :

a) $\underline{s}(a) > -\infty$, $\underline{s}(b) < \infty$, then

$$\frac{\underline{s}(\frac{|x-z|^2}{2}) - \underline{s}(a)}{\underline{s}(b) - \underline{s}(a)} \leq Q_y\{I_{T_a \wedge T_b} < \infty\} \leq \frac{\overline{s}(\frac{|x-z|^2}{2}) - \overline{s}(a)}{\overline{s}(b) - \overline{s}(a)}$$

b) $\overline{s}(a) = -\infty$, then $Q_y\{I_{T_a \wedge T_b} < \infty\} = 1$

c) $\overline{s}(a) = -\infty$, $\overline{\delta}(a) = -\infty$, then $Q_y I_{T_a \wedge T_b} = \infty$.

Using the function $\underline{k}$ we prove

Theorem 3.4. : Suppose (A) holds and $y = (t,x) \in \mathbb{R}_+ \times (B(z,b) - \overline{B(z,a)})$. Assume that one of the following conditions is satisfied :

a) $\underline{k}(a) < \infty$, $\underline{k}(b) < \infty$,

b) $\underline{k}(a) < \infty$, $\underline{k}(b) = \infty$, $\overline{s}(b) = \infty$,

c) $\underline{k}(a) = \infty$, $\overline{s}(a) = -\infty$, $\underline{k}(b) < \infty$.

Then $Q_y\{I_{T_a \wedge T_b} < \infty\} = 1$. If additionally $\underline{\delta}(b) < \infty$ in b) and $\underline{\delta}(a) > -\infty$ in c) then $Q_y I_{T_a \wedge T_b} < \infty$.

From the above results, putting $f \equiv 1$, one can obtain a criteria for transience, recurrence, positive recurrence and null recurrence of degenerated diffusions satisfying (A) (cf. Bhattacharya [2,3] for nondegenerated case). If $\underline{s}(\infty) < \infty$, then the diffusion is transient. If $\overline{s}(\infty)=\infty$, then it is recurrent. If $\overline{s}(\infty)=\infty$, $\underline{\delta}_1(\infty) < \infty$ and $\overline{s}(\infty)=\infty$, $\overline{\delta}_1(\infty)=\infty$, then it is, respectively, positive recurrent and null recurrent ($\underline{\delta}_1=\underline{\delta}$ with $f\equiv 1$).

Now using Theorem 2.3 we can prove

Theorem 3.6. : Assume (A) holds. If one of the following conditions holds :

a) $\underline{k}(a) < \infty$, $\underline{k}(b) < \infty$ and $\int_{t_0}^{b}\int_{t_0}^{t} \underline{\delta}(du)\overline{s}(dt) < 1$, where t_0 is given by

$$\underline{\delta}(t_0) \int_a^b \overline{s}(du) = \int_a^b \underline{\delta}(u)\, \overline{s}(du) ,$$

b) $\underline{k}(a) < \infty$, $\underline{k}(b) = \infty$, $\overline{s}(b) = \infty$, $\underline{\delta}(b) < \infty$ and $\int_a^b \int_t^b \underline{\delta}(du)\overline{s}(dt) < 1$,

c) $\underline{k}(a) = \infty$, $\overline{s}(a) = -\infty$, $\underline{\delta}(a) > -\infty$, $\underline{k}(b) < \infty$ and $\int_a^b \int_a^t \underline{\delta}(du)\overline{s}(dt) < 1$,

then there exists a constant c such that for $y \in \mathbb{R}_+ \times (B(z,b) - \overline{B(z,a)})$ we have

$$Q_y \exp(I_{T_a \wedge T_b}) \leq c < \infty .$$

4. EXAMPLE

We suppose now that $b(t,x) = 0$ and $a(t,x) = I$. In this case under $P_x=Q_{0,x}$ $x \in \mathbb{R}^d$, (X_t) is a d-dimensional standard Brownian motion ($d \geq 2$). Consider the integral

$$I_t = \int_0^t \gamma\left(\frac{|X_s|^2}{2}\right) ds , \quad 0 \leq t < \infty ,$$

where the function $\gamma : [0,\infty[\to \mathbb{R}_+$ is continuous. Note that with $z=0$ and $r > 0$ we have

$$\overline{e}(t) = \underline{e}(t) = \left(\frac{r}{t}\right)^{\frac{d}{2}}$$

$$\delta(t) = \underline{\delta}(t) = \underline{m}(t) = \overline{\delta}(t) = \overline{m}(t) = r^{-\frac{d}{2}} \int_r^t s^{\frac{d}{2}-1} \gamma(s)\, ds$$

$$k(t) = \underline{k}(t) = \overline{k}(t) = \int_r^t s^{-\frac{d}{2}} \int_r^s u^{\frac{d}{2}-1} \gamma(u)\, du\, ds .$$

Denote, as before, $T_a = \inf\{t \geq 0 ;\ |X_t| = (2a)^{1/2}\}$ and $B(a) = \{ x \in \mathbb{R}^d ;\ |x| < (2a)^{1/2}\}$. The above results yield the following statements.

Let $d = 2$, $0 < a < \infty$ and $x \in \mathbb{R}^2 - \overline{B(a)}$. Then

a) $P_x\{ I_{T_a} < \infty\} = 1$,

b) if $\int_a^\infty \gamma(t)\,dt = \infty$, then $P_x I_{T_a} = \infty$,

c) if $\int_a^\infty \gamma(t)dt < \infty$, then $P_x I_{T_a} = \infty$,

d) if $\int_a^\infty \frac{1}{t} \int_t^\infty \gamma(s)dsdt < 1$, then $P_x \exp(I_{T_a}) \le c < \infty$.

Let $d \ge 3$, $0 < a < \infty$ and $x \in \mathbb{R}^d - \overline{B(a)}$.Then

a) if $k(\infty) = \infty$,then $P_x\{ I_{T_a} < \infty \} = (2a|x|^{-2})^{\frac{d}{2}-1}$

b) if $k(\infty) < \infty$,then there exists $\lambda > 0$ such that

$$P_x \exp(\lambda I_{T_a}) \le c < \infty,$$

c) if $k(\infty) < \infty$ and $\int_{t_0}^\infty t^{-d/2} \int_{t_0}^t s^{d/2-1}\gamma(s)dsdt < 1$,where t_0 is given by $\delta(t_0) = (\frac{d}{2}-1)\, a^{d/2-1} \int_a^\infty \delta(t)\, t^{-d/2}\, dt$,then

$$P_x \exp(I_{T_a}) \le c < \infty.$$

Let $d \ge 3$, $0 < a < b < \infty$ and $x \in B(b) - \overline{B(a)}$. Then

$$P_x \exp\left(\frac{d/2-1}{b-a}\, T_a \wedge T_b\right) \le c < \infty .$$

ACKNOWLEDGEMENTS

The author would like to thank Professors A.Le Breton and B.Maisonneuve for comments and suggestions throughout the course of this research.

REFERENCES

[1] R.AZENCOTT .Behavior of diffusion semi-groups at infinity,Bull.Soc. Math.France, 102, 193 - 240 (1974).

[2] R.N.BHATTACHARYA.Criteria for recurrence and existence of invariant measures for multidimensional diffusions,Ann.Probab.,6,541 - 553 (1980).Correction note,Ibid.,8,1194 - 95 (1980).

[3] R.N.BHATTACHARYA,S.RAMASUBRAMANIAN.Recurrence and ergodicity of diffusions,J.of Multivariate Analysis 12,95 - 122 (1982).

[4] E.B.DYNKIN.Markov Processes,Sringer-Verlag,Berlin,1965.

[5] N.IKEDA,S.WATANABE.Stochasic Differential Equations and Diffusion Processes,North-Holland,Amsterdam;Kodansha LTD Tokyo,1981.

[6] Ju.M.KABANOV,R.S.LIPTSER,A.N.SHIRYAEV.Absolute continuity and singularity of locally absolutely continuous probability distributions. I;II,Math.Sbornik 107(149),n°3(11)364-415(1978);108(150),n°1,(1979).

[7] R.Z.KHAS'MINSKI.On positive solutions of the equation $\Lambda u + Vu = 0$. Theor.Probability Appl.,vol.4,n°3,309-318 (1959).

[8] R.Z.KHAS'MINSKI.Ergodic properties of recurrent diffusion processes and stabilization of the solution of the Cauchy problem for parabolic equation,Theor.Probability Appl.,vol.5,n°2,179-196 (1960).

[9] R.Z.KHAS'MINSKI.Stochastic stability of differential equations , Sijthoff and Noordhoff,Rockville,1980.

[10]H.P.Mc KAEN Jr.Stochastic Integrals.Academic Press,New-York, London 1969.

[11]K.NARITA.On explosion and growth order of inhomogeneous diffusion processes,Yokohama Math.J.vol.28,45-57 (1980).

[12]K.NARITA.Remarks on non explosion theorem for stochastic differential equations,Kodai Math.J.,5,395-401,(1982).

[13]K.NARITA.No explosion criteria for stochastic differential equations, J.Math.Soc.Japan,vol.34,n°2,191-203 (1982).

[14]D.W.STROOCK,S.R.S.VARADHAN.Multidimensional Diffusion Processes, Springer-Verlag,Berlin,Heidelberg,New-York,1979.

ON FIRST EXIT TIMES OF DIFFUSIONS

E. Platen
Akademie der Wissenschaften der DDR
Institut für Mathematik
DDR-1086 Berlin, Mohrenstraße 39

1. Introduction

The paper presents an assertion which was useful for the construction of approximate first exit times of diffusions (see [1]). But this result might be also of interest for other investigations in connection with first exit times. It provides an estimation of the expectation of the first exit time in dependence on the distance of the diffusion to the boundary.

2. Diffusion

Let $(\Omega, \underline{F}, \underline{\mathbb{F}}, P)$ be a filtered probability space, where $\underline{\mathbb{F}} = (\underline{F}_t)_{t\in[0,T]}$ is an increasing right-continuous family of complete sub-σ-fields of $\underline{F}$ and

$$w = \{w_t\}_{t\in[0,T]} = \{(w_t^j)_{j=1}^n\}_{t\in[0,T]}$$

an n-dimensional $\underline{\mathbb{F}}$-adapted Wiener process on this space.
We consider the m-dimensional diffusion process $x = \{x_t\}_{t\in[s,T]}$ which fulfils the Itô equation:

$$(1) \qquad x_t = x_s + \int_s^t b(r,x_r)\,dr + \int_s^t \sigma(r,x_r)\,dw_r,$$

where x_s is the fixed initial value at time $s \in (0,T)$ and b and σ are continuous.

3. Assumptions

Let Q denote a connected bounded open domain in $(0,T) \times R^m$ with $\bar{Q} \cap \{T\} \times R^m \neq \emptyset$.
For a given $\varepsilon > 0$ we define an inner neighbourhood

$$(2) \qquad Q_\varepsilon := \{(r,x) \in Q : 0 < \rho(r,x) < \varepsilon\}$$

of the part

$$\partial' Q := \overline{\partial Q \cap (0,T) \times R^m}$$

of the boundary ∂Q of Q, where

(3) $$\varrho(r,x) := \inf_{(r,y)\in \partial' Q} |y - x|$$

denotes the distance of the point $(r,x) \in (0,T) \times R^m$ to the part $\partial' Q$ of the boundary ∂Q.

$C^{1,2}(G)$ denotes the class of functions on a domain G which are continuous differentiable with respect to its first variable and twice continuous differentiable with respect to the remaining variables.

We introduce for functions $U \in C^{1,2}(\overline{Q_\varepsilon})$ the operator :

(4) $$L\,U(t,x) := \sum_{i,j=1}^{m} a^{i,j}(t,x) \frac{\partial^2}{\partial x^i \partial x^j} U(t,x) +$$

$$\sum_{i=1}^{m} b^i(t,x) \frac{\partial}{\partial x^i} U(t,x) + \frac{\partial}{\partial t} U(t,x),$$

$(t,x) \in \overline{Q_\varepsilon}$ with

$$a(t,x) := \frac{1}{2}\, \sigma(t,x)\, \sigma(t,x)^*.$$

σ^* denotes the transposed of the matrix σ.

We assume that the diffusion starts in Q_ε, that means:

(5) $$(s,x_s) \in Q_\varepsilon \ .$$

Now let us formulate the crucial assumptions: There exists a continuous function R on $\overline{Q}_\varepsilon$ such that

(6) $R \in C^{1,2}(\overline{Q_\varepsilon})$,

for all $(t,x) \in Q_\varepsilon$:

(7) $L\,R(t,x) \leq -1$,

for all $(t,x) \in Q_\varepsilon$:

(8) $R(t,x) \leq K\varrho(t,x)$

for all $(t,x) \in \partial Q_\varepsilon \setminus \partial Q$:

(9) $0 < c \leq R(t,x)$.

We remark, that $R(t,x) + t$ can be interpreted as a Ljapunov function on Q_ε.

For instance, (6), (8) and (9) are fulfilled if we set for $(t,x) \in Q_\varepsilon$:

(10) $$R(t,x) = \mu\,(1 - \exp(-\lambda\varrho(t,x)))$$

with $\lambda, \mu > 0$, $\varrho \in C^{1,2}(\overline{Q}_\varepsilon)$.

To illustrate the condition (7) we consider the case m = n = 1 and denote by d(t), $t \in (0,T)$, the upper part of the boundary of Q. Further, let uns consider a point $(t,x) \in Q_\varepsilon$ with $0 < d(t)-x < \varepsilon$. Then it follows from (4) and (10):

$$L\,R(t,x) \leq -\mu\lambda \exp(-\lambda\varepsilon)\left\{ \lambda a(t,x) + b(t,x) - \frac{\partial}{\partial t} d(t) \right\}.$$

Obviously, if the diffusion is nondegenerate, then we have $a(t,x) \geq a > 0$, and λ together with μ can be chosen large enough such that (7) is fulfilled. But we are not restricted to the nondegenerate case only. For instance, if in the degenerate case (e.g. $a(t,x) = 0$) it is:

$$b(t,x) - d'(t) \geq d > 0,$$

that means the drift is strictly outward directed, then μ can be chosen large enough such that (7) holds true.

One can say that (7) is a condition on the smoothness of the part $\partial'Q$ of the boundary and the behaviour of the drift and diffusion coefficients in its neighbourhood.

4. Main Theorem

We denote by

(11) $$\tau := \inf\{ t \in [s,T] : (t,x_t) \notin Q \}$$

the first exit time of the diffusion from Q.

Theorem:

Under the assumptions (5) - (9) it holds

(12) $$E(\tau - s) \leq K\,\varrho(s,x_s),$$

where the constant K does not depend on s and x_s.

The theorem gives an estimation of the expectation of the first exit time τ in dependence on the distance of the diffusion at s to the part of the boundary $\partial'Q$.

5. Proof

Let

(13) $\tau_\varepsilon := \inf\{ t \in [s,T] : (t,x_t) \notin Q_\varepsilon \}$

denote the first exit time of the diffusion from Q_ε .
It follows with the help of the Itô formula, from (6) and (7):

(14) $$R(s,x_s) = E\,R(\tau_\varepsilon, x_{\tau_\varepsilon}) - E\int_s^{\tau_\varepsilon} L\,R(r,x_r)\,dr$$

$$\geq E\,R(\tau_\varepsilon, x_{\tau_\varepsilon}) + E(\tau_\varepsilon - s).$$

Because of (9) and (14) we get:

(15) $$E\,\mathbb{1}_{\{\tau>\tau_\varepsilon\}} \leq E\,\mathbb{1}_{\{\tau>\tau_\varepsilon\}} R(\tau_\varepsilon, x_{\tau_\varepsilon})c^{-1}$$

$$\leq c^{-1} E\,R(\tau_\varepsilon, x_{\tau_\varepsilon})$$

$$\leq c^{-1} R(s,x_s).$$

Finally it follows from (14) and (8) with (15):

$$E(\tau - s) \leq R(s,x_s) + E(\tau - \tau_\varepsilon)$$

$$\leq R(s,x_s) + T\,E\,\mathbb{1}_{\{\tau>\tau_\varepsilon\}}$$

$$\leq (1 + T\,c^{-1})\,R(s,x_s)$$

$$\leq K\;E\,\varrho(s,x_s).$$

6. Reference:

[1] Platen, E.: Approximation of first exit times of diffusions and approximate solution of parabolic equations. Math. Nachr. 111 (1983), 127-164.

4. FILTERING

SMOOTHING FOR A FINITE STATE MARKOV PROCESS

ROBERT J. ELLIOTT

UNIVERSITY OF HULL, ENGLAND

1. INTRODUCTION

This paper is inspired by the results of Kunita [2], [3], and motivated by the work of Pardoux [4], [5] and Anderson and Rhodes [1]. In [2], Kunita discusses backward and forward stochastic differential equations, and in [3] uses these techniques to construct solutions of stochastic partial differential equations. In [1], Anderson and Rhodes discuss smoothing formulae for various kinds of signal and observation processes, and obtain, (in the diffusion case), similar stochastic partial differential equations.

Their methods involve the reverse time differentiation of various stochastic differential fomulae, and as Pardoux points out in [7], their arguments are sometimes a little formal. Using existence theorems from partial differential equations Pardoux obtains similar results for smoothing and prediction formulae in [5], [6], when the signal and observation process are diffusions. As Pardoux observes, the beautiful theory of Kunita can be used to write down such solutions.

In this paper we discuss a similar formula related, as in [1], to filtering and smoothing when the signal process is a finite state Markov chain and the observation process is a diffusion. The principal innovation is the derivation of reverse-time formulae. In the first section of the paper we derive some reverse - time formulae associated with Markov chains; however, we cannot apply these directly in the second part because of the mixture there of Markov chains and diffusions. Reverse-time formulae for diffusions with jumps will be discussed elsewhere.

2. MARKOV PROCESS

In this section we describe the Markov process and mention some related forward and backward equations. Write $e_i = (0, 0, \ldots 1, \ldots 0)$ for the ith unit vector in R^N. Consider a Markov process $\{X_t\}, t \geq 0$, defined on a probability space (Ω, G, P) whose state space is $\{e_1, \ldots, e_N\}$. Write $p_t^i = P\{X_t = e_i\}$. We shall suppose that $p_t = (p_t^1, \ldots, p_t^N)'$ satisfies the forward equation

$$\frac{dp_t}{dt} = A_t\, p_t$$

where the entries of the matrix A are uniformly bounded. The transition matrix associated with A will be denoted by $\Phi(t,s)$, so

$$\frac{d\Phi(t,s)}{dt} = A_t\, \Phi(t,s),\ \Phi(s,s) = I$$

and $$\frac{d\Phi(t,s)}{ds} = -\,\Phi(t,s)A_s,\ \Phi(t,t) = I$$

where I is the $N \times N$ identity matrix.

Consider the process in state $x \in \{ei\}$ at time s, and write $X_{s,t}(x)$ for its state at the later time t. Then $E[X_{s,t}(x)] = \Phi\,(t,s)x$.

Write $G_t^s(x) = \sigma\{X_{s,u}(x) : s \leq u \leq t\}$

and define a process $N_{s,t}(x)$ by

$$N_{s,t}(x) = X_{s,t}(x) - x - \int_s^t A_u X_{s,u}(x)\, du.$$

Then we have the following 'forward' equation for $X_{s,t}(x)$, (in t):

LEMMA 2.1. For $t \geq s$ the process $\{N_{s,t}(x)\}$ is a $G_t^s(x)$ martingale.

PROOF Note that if $s \leq t \leq r$

$$X_{s,r}(x) = X_{t,r}(X_{s,t}(x)) = \sum_{i=1}^{N} \delta_{e_i}(X_{s,t}(x))X_{t,r}(x).$$

For $h \geq 0$

$$N_{s,t+h}(x) - N_{s,t}(x) = X_{s,t+h}(x) - X_{s,t}(x) - \int_t^{t+h} A_u X_{s,u}(x)\, du.$$

$$= X_{t,t+h}(X_{s,t}(x)) - X_{s,t}(x) - \int_t^{t+h} A_u X_{t,u}(X_{s,t}(x))du.$$

Now $X_{s,t}(x)$ is $G_t^s(x)$ measurable, so using the Markov property:

$$E[N_{s,t+h}(x) - N_{s,t}(x) \,|\, G_t^s\,(x)]$$

$$= \Phi(t+h,t)X_{s,t}(x) - X_{s,t}(x) - \int_t^{t+h} A_u \Phi(u,t)X_{s,t}(x)du = 0.$$

REMARKS 2.2 We now consider a σ - field B_t^s which, for each $\sigma \in [s,t]$, contains the history of each $X_{\sigma,t}(e_i)$, $1 \le i \le N$ but does not specify any particular $X_{\sigma,t}(e_i)$.

Roughly, if $G_t^\sigma := \bigvee_{i=1}^N G_t^\sigma(e_i)$ then $B_t^s := \bigvee_{s \le \sigma \le t} G_t^\sigma$.

However, these σ-fields live on the much larger space of maps from $[o,\infty)^2$ into the set of functions from $\{e_i\}$ to itself.

For a vector $(\alpha_1, \ldots, \alpha_N) \in R^N$ we define the G_t^s measurable process

$$X_{s,t}(\alpha_1 e_1 + \ldots + \alpha_N e_N) := \alpha_1 X_{s,t}(e_1) + \ldots + \alpha_N X_{s,t}(e_N).$$

Then we have the following 'backward' equation for $X_{s,t}(x)$, (in s):

LEMMA 2.3 For $s \le t$ the process $\{\hat{N}_{s,t}(x)\}$ is a backward B_t^s martingale, where

$$\hat{N}_{s,t}(x) = X_{s,t}(x) - x - \int_s^t X_{u,t}(A_u x)\, du.$$

PROOF Write

$$X_{s,t}(e_i) = X^i \in R^N$$

and X for the matrix with rows X^i.

Then $X_{s,t}(x) = X(x)$.

Suppose $h > o$, so

$$\hat{N}_{s-h,t}(x) - \hat{N}_{s,t}(x) = X_{s-h,t}(x) - X_{s,t}(x) - \int_{s-h}^{s} X_{u,t}(A_u x)\, du$$

$$= X_{s,t}(X_{s-h,s}(x)) - X_{s,t}(x) - \int_{s-h}^{s} X_{s,t}(X_{u,s}(A_u x))du .$$

In terms of matrix X this is:

$$= X(X_{s-h,s}(x)) - X(x) - \int_{s-h}^{s} X(X_{u,s}(A_u x))du.$$

Now X is B_t^s measurable, and future behaviour of the collection of all sample paths $X_{s,u}(e_i)$, ... $X_{s,u}(e_N)$ is independent of the behaviour of $X_{s-h,v}(x)$, $s-h \le v \le s$.

Therefore

$$E[\hat{N}_{s-h,t}(x) - \hat{N}_{s,t}(x) | B_t^s]$$
$$= X.E[X_{s-h,s}(x) - x - \int_{s-h}^{s} X_{u,s}(A_u x)\, du | B_t^s]$$

$$= X.\Phi(s,s-h)x - X(x) - \int_{s-h}^{s} X\, \Phi(s,u)\, A_u x\, du$$

$$= X\,\Phi(s,\,s-h)x - X(x) + \int_{s-h}^{s} X\,\frac{d\Phi(s,u)}{du}\,x\,du = 0\,.$$

REMARKS 2.4 A function of $X_{s,t}(x)$ can be represented by a vector

$$f(t) = (f_1(t),\ldots,\,f_N(t)$$

so that

$$f(t,X_{s,t}(x) = \langle f(t),\,X_{s,t}(x)\rangle$$

where $\langle\,\rangle$ denotes the scalar product. Similar calculations to the above establish the following:

LEMMA 2.5 Suppose (the components of) $f(t)$ are differentiable in t. Then

$$f(t,\,X_{s,t}(x)) - f(s,x) = \int_s^t \langle f'(u),\,X_{s,u}(x)\rangle\,du + \int_s^t \langle f(u),\,A_u X_{s,u}(x)\rangle\,du$$

$$+ M_{s,t}(f),$$

where $M_{s,t}(f)$ is a forward $G_t^s(x)$, martingale (in t).

Also,

$$f(s,\,X_{s,t}(x)) - f(t,x) = -\int_s^t \langle f'(u),\,X_{u,t}(x)\rangle\,du$$

$$+ \int_s^t \langle f(u),\,X_{u,t}(A_u x)\rangle\,du + \hat{M}_{s,t}(f),$$

where $\hat{M}_{s,t}(f)$ is a backward B_t^s martingale (in s).

LEMMA 2.6 If $g(t)$ is a differentiable function similar to $f(t)$ the product function $fg\,(t, X_{s,t}(x))$ can be represented as the inner product of the vector with components $fg(t, X_{s,t}(e_i))$ and x and we have the following formulae:

$$fg(t,X_{s,t}(x)) - fg(s,x) =$$

$$\int_s^t \{fg'(u,X_{s,u}(x)) + f'g(u,X_{s,u}(x)) + fg(u,A_u X_{s,u}(x))\}du + M_{s,t}(fg),$$

where $M_{s,t}(fg)$ is a forward $G_t^s(x)$ martingale, in(t).

$$fg(s, X_{s,t}(x)) - fg(t,x) =$$

$$\int_s^t \{-fg'(u,X_{u,t}(x)) - f'g\,(u, X_{u,t}(x)) + fg(u,X_{u,t}(A_u x))\}du + \hat{M}_{s,t}(fg),$$

where $\hat{M}_{s,t}(fg)$ is a backward B_t^s martingale, (in s).

3. REVERSE TIME DIFFERENTIATION OF AN EXPONENTIAL.

Consider, as above, a finite state Markov process defined on a probability space (Ω, G, P) and with state space $e_1, \ldots, e_N$. Suppose the process is in state x at time $s \geq o$, and again write $X_{s,t}(x)$ for its state at time $t \geqq s$. Recall

$$G_t^s(x) = \sigma\,\{X_{s,u}(x) \colon s \leq u \leq t\}.$$

Suppose $w_t = (w_t^1, \ldots, w_t^m)$, $t \geq o$, is an m- dimensional Brownian motion, defined on a probability space (Ω', F, Q), which is independent of the Markov process X. Write, for $s \leq t$,

$$F_t^s = \sigma\,\{w_v - w_u \colon s \leq u \leq v \leq t\}.$$

For $o \leq j \leq m$ suppose that h^j is a bounded, real valued function on $\{e_i\} \times [o,\infty)$. Then h^j is represented as a vector of functions:

$h^j(u) = (h^j(e_1, u), h^j(e_2,u), \ldots, h^j(e_N,u))$, and we further suppose that the derivative $\frac{dh^j}{du}$ exists and is a bounded function of the same form. Then h^j evaluated at state x and time $u \in [o,\infty)$ can be represented as $< h^j(u), x > = h^j(x,u)$.

Write $w_t^o = t$. Then the theory of filtering and smoothing, when the signal process is the above Markov process and the observation process is a diffusion, leads one to consider an exponential expression of the form

$$\Lambda_{s,t}(x) = exp\{\sum_{j=o}^m \int_s^t h^j(X_{s,u}(x),u)dw_u^j\}.$$

Because x and w are independent it is immaterial whether the integrals are interpreted as Ito or Stratonovich integrals.

Write

$$\hat{\Lambda}_{s,t}(x) = E_P[\Lambda_{s,t}(x) \mid F_t^s].$$

The theory of smoothing requires one to investigate the derivative in s of $\tilde{\Lambda}_{s,t}(x)$. Following the method of Kunita, [2], this is the problem studied in the present section. We first state as Lemmas two simple inequalities.

LEMMA 3.1 For $n = 1,2,4$ etc.

$$E[\Lambda_{s,t}^{2n}(x)] \le exp\{C_n(t-s)\}$$

for constants C_n.

LEMMA 3.2. For any power M

$$E[|X_{s,t}(x) - x|^M] \le Const. |t-s|.$$

NOTATION 3.3. Recall that if $f(u)$, $o \le u \le t$, is an F_t^u predictable process such that $\int_s^t E[f(u)^2]\, du < \infty$ then the backward Ito integral, (see [2]), is defined, if $f(u)$ is continuous in probability, as:

$$\int_s^t f(u)\, \hat{d}w_u^j := \lim_{|\Delta| \to o} \sum_{k=0}^{n-1} f(t_{k+1})(w_{t_{k+1}}^j - w_{t_k}^j).$$

Here $\Delta = \{s = t_o < t_1 < \dots < t_n = t\}$ and

$$|\Delta| = \max_k |t_{k+1} - t_k|.$$

The backward Stratonovich integral is defined as

$$\int_s^t f(u) \circ \hat{d}\, w_u^j := \lim_{|\Delta| \to 0} \sum_{k=0}^{n-1} \tfrac{1}{2} (f(t_{k+1}) + f(t_k))(w_{t_{k+1}}^j - w_{t_k}^j).$$

For $1 \le i \le N$ we shall write

$$H^i(s,t) = exp\, \{\sum_{j=0}^{m} \int_s^t h^j(X_{s,u}(e_i),u) dw_u^j\} = \Lambda_{s,t}(e_i),$$

and H for the vector $(H^1, H^2, \dots, H^N)$.

Then $\Lambda_{s,t}(x) = \langle H(s,t), x \rangle$,

and with $\hat{\Lambda}_{s,t}(x) = E_P[\Lambda_{s,t}(x) | F_t^s]$

and $\hat{H}_{s,t} = E_P\, [H(s,t)\ | F_t^s]$

we have $\hat{\Lambda}_{s,t}(x) = \langle \hat{H}\, (s,t), x \rangle$,

Our main result states the following:

THEOREM 3.4 $\hat{\Lambda}_{s,t}(x)$ satisfies the following reverse time stochastic differential equation:

$$\hat{\Lambda}_{s,t}(x) - 1 = \int_s^t \hat{\Lambda}_{u,t}(A_u x)du + \sum_{j=0}^{m} \int_s^t \hat{\Lambda}_{u,t}(x)\, h^j(x,u) \circ d\hat{w}_u^j$$

SKETCH OF PROOF. The method used is an adaptation of that in Kunita's paper [2]. Consider a partition as above:

$$\Delta = \{s = t_0 < t_1 < \dots < t_n = t\}$$

with $|\Delta| = \max_k |t_{k+1} - t_k|$. Then

$$\Lambda_{s,t}(x) - 1 = \sum_{k=0}^{n-1} (\Lambda_{t_k,t}(x) - \Lambda_{t_{k+1},t}(x)).$$

However,

$$\Lambda_{t_k,t}(x) = \Lambda_{t_k,t_{k+1}}(x)).\ \Lambda_{t_{k+1},t}(X_{t_k,t_{k+1}}(x)).$$

So the kth term in the above sum is:

$$\Lambda_{t_k,t}(x) - \Lambda_{t_{k+1},t}(x) = \Lambda_{t_k,t_{k+1}}(x)\ (\Lambda_{t_{k+1},t}(X_{t_k,t_{k+1}}(x)) - \Lambda_{t_{k+1},t}(x))$$

$$+ \Lambda_{t_{k+1},t}(x)(\Lambda_{t_k,t_{k+1}}(x) - 1)$$

$$= J_k + K_k, \text{ say.}$$

Write $\hat{J}_k = E_P\,[J_k | F_t^s\,]$

$\hat{K}_k = E_P\,[K_k\,|\,F_t^s\,].$

Then by repeated applications of Schwarz's inequality and Lemmas 3.1 and 3.2 we show that

$$\sum_{k=0}^{n-1} \hat{J}_k \xrightarrow[|\Delta|\to 0]{} \int_s^t \hat{\Lambda}_{u,t}\,(A_u x)du$$

and

$$\sum_{k=0}^{n-1} \hat{K}_k \xrightarrow[|\Delta|\to 0]{} \sum_{j=0}^{m} \int_s^t \hat{\Lambda}_{u,t}(x) h^j(x,u) \circ d\hat{w}_u^j$$

and the result follows.

REFERENCES

[1] B.D.O. ANDERSON and I.B. RHODES. Smoothing algorithms for nonlinear finite dimensional systems, Stochastics, 9 (1983), 139-165.

[2] H. KUNITA, On backward stochastic differential equations, Stochastics, 6 (1982), 293-313.

[3] H. KUNITA, Stochastic partial differential equations connected with nonlinear filtering, Lecture Notes in Mathematics, Vol. 972. Springer-Verlag 1983.

[4] E. PARDOUX, Stochastic partial differential equations and filtering of diffusion processes. Stochastics 3(1979), 127-167.

[5] E. PARDOUX, Equations du filtrage non linéaire, de la prediction et du lissage, Stochastics, 6 (1982), 193-231.

[6] E. PARDOUX, Equations of non-linear filtering, and applications to stochastic control with partial observation, in Non Linear Filtering and Stochastic Control, Lecture Notes in Mathematics Vol. 972, Springer-Verlag, 1983.

[7] E. PARDOUX, Equations du lissage non linéaire, in Filtering and Control of Random Processes, Lecture Notes in Control and Information Sciences, to appear.

SOME REMARKS ON GAUSSIAN SOLUTIONS AND EXPLICIT FILTERING FORMULAE

Dimitar I. Hadjiev

Center of Mathematics and Mechanics
P.O.Box 373 1090-Sofia
Bulgaria

Introduction. This paper deals with filtering and extrapolation problems for Gaussian processes equivalent to Gaussian martingales. It turns out that explicit formulae can be obtained for the conditional means of such a process when observing a Gaussian process of similar type. As far as our processes are functionals corresponding to Gaussian martingales and a collection of functional parameters, the conditional means are expressed in terms of both these parameters and the observation process (Section 2). In Section 3 we characterize the class of processes under consideration by means of the equivalence of the corresponding Gaussian measures. Section 4 is devoted to a Gaussian solution problem that leads to the same class.
The results are partially published and partially reported at the ICM-82 (Warsaw, August 1983).

1. Notations and preliminaries. Let $(\Omega, \mathcal{F}, P)$ be a complete probability space and D_0 be the space of starting from zero ($f(0)=0$) right continuous and left-hand limited functions $f=f(t)$, $t \in R_+ = [0,+\infty)$. We consider a Gaussian martingale M (with trajectories in D_0) with respect to some filtration $F=(F_t \subseteq \mathcal{F},\ t \in R_+)$. Suppose that F is generated by a separable (may be vector-valued) Gaussian process which forms a Gaussian system with all the processes under consideration. In this case the square characteristic $\langle M \rangle$ of M is deterministic,

$\langle M\rangle_t = E(M_t^2)$, $t\in R_+$, and

(1) $$M_t = M_t^c + \sum_{s_n\le t} \Delta M_{s_n},$$

where M^c is a continuous Gaussian martingale, $(s_n, n\le 1)=(s:\Delta\langle M\rangle_s > 0)$ is the set (obviously countable) of jump times for $\langle M\rangle$ and $\Delta f(t) = f(t) - f(t-)$ (see [2]). Due to the independent increments property the distribution μ of M in D_0 is completely determined by $\langle M\rangle$.

We introduce the class C(M) of Gaussian processes of the type

(2) $$X_t = X_0 + \int_{(0,t]} (\alpha(v) - \int_{(0,v]} k(v,u)dM_u)\, d\langle M\rangle_v + M_t,$$

with X_0 independent of M, and (measurable) functions α and k satisfying the conditions

(3) $$\|\alpha\|^2 = \int_{(0,\infty)} \alpha^2(v)d\langle M\rangle_v < \infty ,$$

(4) $$k(v,u) = 0 \text{ for } v\le u, \text{ and}$$

(5) $$\|k\|^2 = \int_{(0,\infty)}\int_{(0,\infty)} k^2(v,u)d\langle M\rangle_u d\langle M\rangle_v < \infty .$$

The function k(v,u) (with properties (4) and (5)) will be called a strong Volterra kernel, $k\in L^2(R_+^2,\langle M\rangle)$. We do not distinguish two functions that coincide in the sense of the corresponding norms: $\alpha' = \alpha''$ means $\|\alpha'-\alpha''\| = 0$ and $k' = k''$ means $\|k'-k''\| = 0$. Further, α and k will be interpreted as parameters of X in C(M).

For every strong Volterra kernel $k\in L^2(R_+^2,\langle M\rangle)$ there exist a unique resolvent strong Volterra kernel $k^*\in L^2(R_+^2,\langle M\rangle)$ satisfying the identities

$$\left|\begin{array}{l} k(v,u) + k^*(v,u) - \int_{(v,u]} k(v,s)k^*(s,u)d\langle M\rangle_s = 0 \\ k(v,u) + k^*(v,u) - \int_{(v,u]} k^*(v,s)k(s,u)d\langle M\rangle_s = 0. \end{array}\right.$$

We emphasize the fact that the process X defined by (2) is the so-tion of the following stochastic equation (see Theorem 3 below)

$$X_t = X_o + \int_{(0,t]} (\alpha^*(v) + \int_{(0,v]} k^*(v,u)dX_u)d\langle M\rangle_v + M_t \tag{6}$$

with

$$\alpha^*(v) = \alpha(v) - \int_{(0,v]} k^*(v,u)\alpha(u)d\langle M\rangle_u. \tag{7}$$

The problem of calculating k^* when knowing k is not simple but it is well studied in many analysis books. Up to calculating difficulties one observes that knowing (α,k) and $\langle M\rangle$ the process X is a functional of (X_o,M) as well as its martingale part M is a functional of X (see (6)).

2. Filtering and extrapolation results. Let (M,N) be a two-dimensional Gaussian martingale independent of the Gaussian vector (X_o,Y_o) and let (X,Y) be a Gaussian process such that $X\in C(M)$, $Y\in C(N)$. More precisely, we consider the system

$$\left|\begin{array}{l} X_t = X_o + \int_{(0,t]} (\alpha(v) - \int_{(0,v]} k(v,u)dM_u)d\langle M\rangle_v + M_t \\ Y_t = Y_o + \int_{(0,t]} (\beta(v) - \int_{(0,v]} l(v,u)dN_u)d\langle N\rangle_v + N_t \end{array}\right.$$

corresponding to the collection of parameters (α,k) for X and (β,l) for Y. Suppose X is the observable component of the system and Y is the non-observable one. As usual, the (completed by P-null sets of $\mathcal{F}$) σ-algebra $F_t^X = \sigma(X_s,\ s\le t)$ is interpreted as the collection of observable events up to the time moment $t\in R_+$.

We introduce the cadlag modification $\pi_t(Y)$ of the conditional means $E(Y_t/\ F_t^X)$, $t\in R_+$, and the cadlag modification $\pi_{s,t}(Y)$ of $E(Y_t/\ F_s^X)$, $0\le s\le t$ (for fixed $t\in R_+$). The process $\pi_t(Y)$ is called the optimal filter and $\pi_{s,t}(Y)$ is called the optimal extrapolation estimator of Y.

It turns out that (see [4]) these two estimators can be described explicitly in our situation. Denote $\langle M,N\rangle_t = E(M_t N_t)$, $t\in R_+$, and let

(9) $$\bar{N}_t = \int_{(0,t]} \frac{d\langle M,N\rangle_u}{d\langle M\rangle_u}\, dM_u,\quad t\in R_+,$$

be the projection of N onto M. We take for granted the fact that $\langle M,N\rangle$ is absolutely continuous with respect to $\langle M\rangle$ (see [8]). It is clear that $\langle M,N\rangle$ determines the statistical connection between M and N (or between the components X and Y of our system (8)).

<u>Theorem 1</u>. The next explicit formulae are valid:

(10) $$\pi_t(Y) = \pi_0 + \int_{(0,t]} (\beta(v) - \int_{(0,v]} l(v,u)\,d\bar{N}_u)\,d\langle N\rangle_v + \bar{N}_t$$

where

(11) $$\pi_0 = E(Y_0) + \mathrm{cov}(X_0,Y_0).(\mathrm{cov}(X_0,X_0))^{\oplus}(X_0 - E(X_0))$$

($b^{\oplus}$ means b^{-1} if $b\neq 0$, and 0 if $b=0$), and

(12) $$\pi_{s,t}(Y) = \pi_s(Y) + \int_{(s,t]} (\beta(v) - \int_{(0,s]} l(v,u)\,d\bar{N}_u)\,d\langle N\rangle_v.$$

The proof is based on Fubini-type theorem for stochastic integrals (see [6] and [4]).

<u>Remark</u>. The identity (11) which is usually written for finite-dimensional vectors X_0 and Y_0, forms an essential part of the so called normal correlation theorem ([9],Ch.13). Our formulae (10) and (12) may be considered as extension of this theorem to an infinite-dimensional case. Certainly, the result can be extended to the case of a process $Y_t = Y_0 + \sum_n Y_t^n$, where $Y^n\in C(N^n)$ and (N^n,M) form Gaussian martingales for every n as well. Due to (9) and (6) the estimators $\pi_t(Y)$ and $\pi_{s,t}(Y)$ can be also expressed as explicit fuctionals of the observation process X trajectories and the parameters involved.

3. <u>An equivalence problem for Gaussian measures.</u> In this section we clarify the origin of the class C(N) generated by a Gaussian martingale N. To this end we introduce the measure μ_N generated by N in the space D_o. Let X be a Gaussian process (with trajectories in D_o) that generates a measure μ_X in D_o. It is well known that the following dichotomy is true: either μ_X and μ_N are equivalent ($\mu_X \sim \mu_N$) or they are singular ($\mu_X \perp \mu_N$). The first case is much more interesting for statistical problems. Our next theorem says in this case one reaches one of the classes C(M).

<u>Theorem 2</u>. The measure μ_X is equivalent to the measure μ_N in D_o corresponding to a Gaussian martingale N if and only if X is a semi-martingale with Gaussian martingale part M such that

(i) μ_M and μ_N are equivalent ($\mu_M \sim \mu_N$), and

(ii) $X \in C(M)$ for some (α, k) satisfying (3), (4) and (5).

Moreover, in this case the parameters (α, k) are unique and the measures μ_X and μ_M are equivalent ($\mu_X \sim \mu_M$) too.

For more details as well as for the proof we refer the reader to [10] and [11].

It should be noted that, in general, the equivalence of μ_M and μ_N does not imply the coincidence of these distributions. In fact, the relation $\mu_M \sim \mu_N$ means that $\langle M^c \rangle = \langle N^c \rangle$, $(s : \Delta\langle N \rangle_s > 0) = (s : \Delta\langle M \rangle_s > 0) = (s_n,\ n \geq 1)$ and also $\sum_{n \geq 1} \left(1 - \frac{\Delta\langle M \rangle_{s_n}}{\Delta\langle N \rangle_{s_n}}\right)^2 < \infty$ (see [10], or [5] and the references there).

4. <u>A problem of Gaussian solutions.</u> Let us consider the stochastic

equation

$$Z_t = \int_{(0,t]} \gamma(s,Z)\,d\langle M\rangle_s + M_t \tag{13}$$

with $\gamma = \gamma(t,x)$, $t \in R_+$, $x \in D_o$, being a predictable fuctional with respect to the natural filtration in D_o.

We say the equation (13) has a weak Gaussian solution corresponding to both the predictable functional γ and a fixed non-decreasing function $\langle M\rangle$ (certainly $\langle M\rangle$ must belong to D_o) if there exist

1) a complete probability space $(\Omega, \mathcal{F}, P)$ with a filtration $F=(F_t, t \in R_+)$ satisfying the usual conditions [1], and

2) F-adapted random processes Z and M such that

a) (M,F,P) is a Gaussian martingale with square characteristic $E(M_t^2) = \langle M\rangle_t$, $t \in R_+$;

b) $\int_{(0,\infty)} \gamma^2(s,Z)\,d\langle M\rangle_s < \infty$ P-a.s.;

c) the equality (13) holds P-a.s. for every $t \in R_+$, and

d) the random process (Z,P) (resp. the probability measure μ_Z generated by (Z,P) in D_o) is Gaussian.

For given predictable functional γ, probability space $(\Omega, \mathcal{F}, P)$ with a filtration F and a Gaussian martingale M with square characteristic $\langle M\rangle_t = E(M_t^2)$, $t \in R_+$, the F-adapted random process Z satisfying 2b), 2c) and 2d) is called a strong Gaussian solution of the stochastic equation (13).

In this section we give necessary and sufficient conditions under which (weak or strong) Gaussian solutions of (13) exist. Naturally, these conditions are to determine the type of drift coefficient as a functional of the solution Z trajectories (see [7] in the case of M being a Wiener process, that is $\langle M\rangle_t = t$, and [3] in the case of zero mean Gaussian solutions Z).

It should be expected that $\gamma(.,Z)=\bar{\gamma}(.,M)$, where

(14) $$\bar{\gamma}(s,M) = \alpha(s) - \int_{(0,s]} k(s,u)dM_u, \quad s\in R_+.$$

So one has to give reasons for this equality and then deduce the explicit form of the functional γ. Denote by μ the measure in D_o generated by any Gaussian martingale with square characteristic $\langle M\rangle$, a fixed function.

<u>Theorem 3</u>. Let the predictable functional $\gamma = \gamma(t,x)$ satisfy the condition

(15) $$\mu(x\in D_o: \int_{(0,\infty)} \gamma^2(s,x)d\langle M\rangle_s < \infty) = 1.$$

Then the equation (13) has a weak Gaussian solution if and only if we have

(16) $$\gamma(t,x)=\int_{(0,t]} k^*(t,s)dx_s + \alpha^*(t) \qquad (\nu\times\mu)\text{-a.e.}$$

where $\alpha^*(t)$ is defined in (7), ν is the measure on R_+ with $\nu(dt) = d\langle M\rangle_t$, and (α,k^*) are parameters satisfying (3), (4) and (5) with k^* instead of k. The parameters (α,k^*) are unique.

<u>Proof</u>. Let γ satisfy (16) for given (α,k^*) and $\langle M\rangle$. We consider an arbitrary probability space $(\Omega,\mathcal{F},P)$ in which one can define a Gaussian martingale M with $E(M_t^2) = \langle M\rangle_t$, $t\in R_+$ (for instance, the space $\Omega = D_o$ with natural filtration F and the measure $P = \mu$). In this probability space, the random process $Z_t = X_t$ defined by (2) with $X_o = 0$ and k the resolvent strong kernel for k^*, is Gaussian and it has the property (see [10])

(17) $$\begin{aligned}\bar{Z}_t &= Z_t - \int_{(0,t]} \alpha(s)d\langle M\rangle_s \\ &= -\int_{(0,t]}\int_{(0,s]} k(s,u)dM_u d\langle M\rangle_s + M_t \\ &= \int_{(0,t]}\int_{(0,s]} k^*(s,u)d\bar{Z}_u d\langle M\rangle_s + M_t.\end{aligned}$$

Cosequently, Z satisfies (13).

On the other hand, by Theorem 2 we have $\mu_Z \sim \mu$. Because of the properties of both the parameters (α, k^*) and the stochastic integrals we obtain (15) which immediately implies 2b).

The sufficience of (16) is proved.

In order to prove the necessity of (16) let us suppose the Gaussian process Z (defined in some probability space $(\Omega, \mathcal{F}, P)$ with a Gaussian martingale M) is a solution of (13) corresponding to a given functional γ and to the fixed non-decreasing function $\langle M \rangle$. As a consequence of 2b) and Theorem 17 in [5] we get $\mu_Z \ll \mu$. As far as both measures are Gaussian the latter means $\mu_Z \sim \mu$. Then by our Theorem 2 the representation

$$Z_t = \int_{(0,t]} \bar{\gamma}(s,M)\,d\langle M\rangle_s + M_t \tag{18}$$

holds with $\bar{\gamma}$ defined by (14) for some (α, k) satisfying (3), (4) and (5). Comparing (13) and (18) we get that the random processes $\int_{(0,t]} \gamma(s,Z)\,d\langle M\rangle_s$, $t \in R_+$, and $\int_{(0,t]} \bar{\gamma}(s,M)\,d\langle M\rangle_s$, $t \in R_+$, are indistinguishable [1].

On the other hand (see (17)) one easily comes to

$$\begin{aligned}\int_{(0,t]} \bar{\gamma}(s,M)\,d\langle M\rangle_s &= \int_{(0,t]} \Big(\alpha(s) - \int_{(0,s]} k(s,u)\,dM_u\Big)\,d\langle M\rangle_s \\ &= \int_{(0,t]} \Big(\alpha(s) + \int_{(0,s]} k^*(s,u)\,d\bar{Z}_u\Big)\,d\langle M\rangle_s \\ &= \int_{(0,t]} \Big(\int_{(0,s]} k^*(s,u)\,dZ_u + \alpha^*(s)\Big)\,d\langle M\rangle_s\end{aligned}$$

where k^* is the resolvent strong kernel for k and α^* is defined in (7). So the set

$$S = \Big\{(s,x) \in R_+ \times D_0 : \gamma(s,x) \neq \int_{(0,s]} k^*(s,u)\,dx_u + \alpha^*(s)\Big\}$$

satisfies the condition $(\nu \times \mu_Z)(S) = 0$ and, consequently $(\nu \times \mu)(S) = 0$.

We have proved (16) and the proof of Theorem 3 is completed.

Corollary. If Z is a Gaussian solution of the equation (13) then the completed by P-null sets filtrations, corresponding to Z and M, coincide: $F_t^Z = \sigma(Z_s, s \leq t) = \sigma(M_s, s \leq t) = F_t^M$ (mod P).

This result follows immediately from the representations (13) and (18). The process Z itself has no independent increments property, but the filtration $F^Z = (F_t^Z, t \in R_+)$ is generated by an (innovation) process M with independent increments. The advantage of this fact in random processes statistics problems is well known (in particular, the martingales with respect to F^Z can be represented as stochastic integrals, [11]).

Our next statement claims that the existence of a weak Gaussian solution guarantees the existence and uniqueness of a strong Gaussian solution of (13) in probability spaces that are rich enough.

Theorem 4. Let the predictable functional $\gamma = \gamma(t,x)$ satisfy the condition (15) with a given function $\langle M \rangle$ and the equation (13) have a weak Gaussian solution. Let $(\Omega, \mathcal{F}, P)$ be any probability space in which a Gaussian martingale M **with** $E(M_t^2) = \langle M \rangle_t$, $t \in R_+$, can be defined. Then, there exists a (unuque up to indistinguishability) strong Gaussian solution Z of the equation (13) in this probability space.

The proof of this theorem follows straightforward. The solution Z is defined by (18) and (14) for (α, k^*) determined by (16) and k being the resolvent strong kernel for k^*.

Remark. According to the results in [6] the representation (18) can be rewritten in the form

$$Z_t = \int_{(0,t]} \alpha(s) d\langle M \rangle_s + \int_{(0,t]} (1 - \int_{(u,t]} k(s,u) d\langle M \rangle_s) dM_u .$$

The latter indicates that the centred process $\overline{Z} = Z - E(Z)$ is of multiplicity one:

$$Z_t = \int_{(0,t]} F(t,u)dM_u, \quad t \in R_+,$$

with the kernel

$$F(t,u) = 1 - \int_{(u,t]} k(s,u)d\langle M\rangle_s, \quad 0 \leq u \leq t \in R_+.$$

References.

1. C.Dellacherie, Capacités et Processus Stochastiques, Springer-Verlag, Berlin Heidelberg New York, 1972.

2. D.I.Hadjiev, On the structure of Gaussian martingales, Serdika 4, 1978, pp.224-231. (in Russian)

3. D.I.Hadjiev, Gaussian solutions of some stochastic equations, C.R.Bulgarian Acad.Sci.34, 1981, pp.1647-1649.

4. D.I.Hadjiev, An example of effective filtering and extrapolation, C.R.Bulgarian Acad.Sci.36, 1983, pp.1379-1382. (in Russian)

5. Ju.M.Kabanov, R.S.Liptser and A.N.Shiryaev, Absolute continuity and singularity of locally absolutely continuous probability distributions II, Math.Sbornik 108 (150), 1979, pp.32-61.

6. T.Kailath, A.Segal and M.Zakai, Fubini-type theorems for stochastic integrals, Sankhya 40, Series A, 1978, pp.138-143.

7. G.Kallianpur, Stochastic filtering theory, Applications of Mathematics Vol.13, Springer-Verlag, New York Heidelberg Berlin, 1980.

8. R.S.Liptser, Gaussian martingales and generalization of Kalman-Bucy filter, Theory Proba.and Appl.20, 1975, pp.292-308.

9. R.S.Liptser and A.N.Shiryaev, Statistics of Random Processes, Nauka, Moscow, 1974. (Engl.transl.: Applications of Mathematics Vol.5 and 6, Springer-Verlag, 1977/78.)

10. L.D.Minkova and D.I.Hadjiev, Representation of Gaussian processes equivalent to a Gaussian martingale, Stochastics 3, 1980, pp.251-266.

11. L.D.Minkova and D.I.Hadjiev, Equivalence and singularity of some Gaussian measures, Pliska 7, 1984, pp.163-169. (in Russian)

WHITE NOISE THEORY OF FILTERING-SOME ROBUSTNESS AND CONSISTENCY RESULTS

G. Kallianpur

University of North Carolina at Chapel Hill

1. <u>Introduction</u>. In earlier papers R.L. Karandikar and I have developed a theory of nonlinear filtering based on finitely additive white noise [5,6,7]. Since the talk on which the material of this article is based was given, the extension of the theory to general state space valued signal processes and Hilbert space valued observation processes has appeared in print [8]. We refer the reader to it for the proofs of the existence of unique solutions to the measure-valued equations for the optimal filter.

Further progress in the white noise approach to the subject would require a generalization of some basic notions given in our earlier work, in particular, the concepts of conditional expectation and absolute continuity in the finitely additive setup. Among the problems to which my colleagues and I hope to apply the extended theory may be mentioned the following:

(i) Prediction and smoothing.

(ii) Filtering theory of random fields (regarded as processes of many parameters).

(iii) Filtering when signal and noise are not assumed to be independent.

(iv) Discrete approximations in filtering, prediction and smoothing.

(v) Robustness questions in the infinite dimensional case.

(vi) Filtering when the signal process is given on a finitely additive probability space,

We do not as yet have the most general theoretical framework to solve all these problems though we have succeeded in finding more inclusive definitions of the basic concepts mentioned above. As a result, we study the robustness properties of the white noise theory more thoroughly than we were able to do in [5] and to give more satisfactory versions of some of the results stated in Section 8 of [6]. The present report is largely devoted to these new results of which only the more important ones will be mentioned for lack of space. The investigation of the optimal filter as a process on a finitely additive probability space and a proof of its Markov property will appear in [9]. As is the case with the previous work, the results presented here have been obtained in collaboration with R.L. Karandikar.

2. Abstract statistical model and Bayes formula.

Let H be a separable Hilbert space, C-the field of finite dimensional cylinder sets and P = class of $\perp$ projections on H with finite dimensional range. For $P \in P$, let $C_P = \{P^{-1}B$, B - a Borel set in Range P$\}$. Suppose that (Ω, A) is an arbitrary measurable space, (A= a σ-field). Let (E, E) be a quasi-cylindrical measurable space, i.e., $E = \Omega \times H$, $E = \bigsqcup_{P \in P} E_P$, $E_P = A \otimes C_P$.

Let m be the canonical Gauss (cylinder) measure on H, i.e. such that $\int_H \exp\{i(h,h_1)\}dm(h) = \exp\{-\frac{1}{2}|h_1|^2\}$. Let α be the finitely additive probability measure determined by the family $\{\alpha_P\}$ where $\alpha_P = \Pi \otimes m_P$ on E_P $(P \in P)$, m_P being the restriction of m to C_P. Let ξ $(\Omega, A, \Pi) \to (H, B(H))$ be a measurable map. The identity map e on H considered as a mapping from (H, C, m) into (H, C) is called Gaussian white noise.

Let y: E $\to$ H be given by $y = \xi + e$, i.e.,

$y(\omega,h) = \xi(\omega) + e(h) = \xi(\omega) + h.$

We call (1) the abstract statistical filtering model.

An important tool in nonlinear filtering and other statistical problems is the Bayes formula (See [4] for the usual version).

Theorem 1. (finitely additive Bayes Formula).

Let Q be a $\perp$ projection on H, $H' = QH$.

(a) Let g be an integrable function on (Ω, A, Π). Then $E_\alpha(g|Qy)$ exists.

(b) $E_\alpha(g|Qy) = \dfrac{\sigma(g, Qy)}{\sigma(1, Qy)}$

where for $h' \in H'$ and

$$\sigma(g,h') = \int_\Omega g(\omega)\exp[(h',Q\xi(\omega)) - \tfrac{1}{2}|Q\xi(\omega)|^2]d\Pi(\omega).$$

3. The nonlinear filtering problem for a finite dimensional signal process.

$X = (X_s : 0 \le s \le T)$ is a Markov process taking values in (S, S) and defined on (Ω, A, P). Let $h: [0,T] \times S \to \mathbb{R}^m$ be Borel measurable and such that

$\int_0^T |h_s(X_s)|^2 ds < \infty$ a.s.

Let

$H = \{\eta: [0,T] \to \mathbb{R}^m : |\eta| \in L^2[0,T]\}$,

$\xi_s(\omega) = h_s(X_s(\omega))$, $0 \le s \le T$. Then $\xi_\cdot(\omega) \in H$

(Conditions on X and h are imposed to ensure the measurability of the map $\omega \to \xi_{\cdot}(\omega)$).

The model $y = \xi + e$ on $(E, \mathcal{E}, \alpha)$ now takes the form

$$y_s = h_s(X_s) + e_s, \quad 0 \le s \le T, \tag{2}$$

where $e = (e_s)$ is the white noise $\perp\!\!\!\perp$ X. y_s is the observation "process."

Let Q_t be the $\perp$ projection onto $H_t = \{\eta \in H: \int_t^T |\eta_s|^2 ds = 0\}$. The Bayes formula then gives, for $f \circ X_t \in L'(\Omega, A, \Pi)$, $f: S \to \mathbb{R}$,

$$E_\alpha[f(X_t)|Q_t y] = \frac{\sigma_t(f, y)}{\sigma_t(1, y)}$$

where $(Q_t y)(s) = y(s)$, $0 \le s \le t$, $= 0$ for $t < s \le T$ and

$$(4) \quad \sigma_t(f,\eta) = \int_\Omega f(X_t(\omega)\exp\left\{\sum_{j=1}^m \int_0^t \eta_s^j h_s^j(X_s(\omega))ds - \tfrac{1}{2}\int |h_s(X_s(\omega))|^2 ds\right\} d\Pi(\omega).$$

If $\sigma_t(f,\eta) = \int_{\mathbb{R}^m} f(x)p_t(x,\eta)dx$, then $p_t(x,\eta)$ is called the unnormalized conditional density. Now assume that X is an $\mathbb{R}^d$-valued diffusion with $L_t = \sum_{i,j} a_{ij}(t,x) \frac{\partial^2}{\partial x^i \partial x^j} + \sum_{i=1}^d b_i(t,x) \frac{\partial}{\partial x^i}$ and $C_0^{1,2}([0,\infty) \times \mathbb{R}^d) \subseteq \mathcal{D}$. Let L_t^* denote the formal adjoint of L_t.

<u>Theorem 2</u>.

Make the following assumptions:

$h: [0,T] \times \mathbb{R}^d \to \mathbb{R}^m$ is locally Hölder continuous. The drift and diffusion coefficients b,a of X satisfy

(i) $\sum_{i,j=1}^d a_{ij}(t,x)z_i z_j \ge K_1 |z|^2$ $(K_1 < \infty)$ for all t,x) and $z = (z_1,\dots,z_d) \in \mathbb{R}^d$;

(ii) a_{ij}, $\frac{\partial a_{ij}}{\partial x^i}$, $\frac{\partial^2 a_{ij}}{\partial x^i \partial x^j}$, b_i, $\frac{\partial b_i}{\partial x^i}$ are locally Hölder continuous

functions satisfying the growth condition

$$|g(t,x)| \le K_2(1 + |x|^2)^{\frac{1}{2}}.$$

Suppose that the distribution of X_0 has a continuous density ϕ satisfying

$$|\phi(x)| \le \exp\{K_3(1 + |x|^2)^{\frac{1}{2}-\varepsilon} \quad \text{for some } \varepsilon > 0,\ K_2 < \infty.$$

Let $H_0 = \{y \in H: \ y_t$ is Hölder continuous$\}$. Then for all $y \in H_0$, the PDE

$$\frac{\partial p_t(x,y)}{\partial t} = L_t^* p_t(x,y) + \left\{ \sum_{j=1}^{m} h_t^j(x) y_t^j - \tfrac{1}{2}|h_t(x)|^2 \right\} p_t(x,y) \qquad (5)$$

with the initial condition

$$p_o(x,y) = \phi(x)$$

has a unique solution in the class

$$G = \left\{ g \in C^{1,2}([0,T] \times \mathbb{R}^d): \ |g(t,x)| \le \exp K_4(1 + |x|^2)^{\frac{1}{2}} \right\}$$

for some $K_4 < \infty$.

Furthermore the unique solution $p_t(x,y)$ $(y \in H_0)$ is the unnormalized conditional density of X_t given $Q_t y$ i.e.

$$E_\alpha[f(X_t)|Q_t y] = \frac{\int f(x) p_t(x,y) dx}{\int p_t(x,y) dx} .$$

Note: The function $h: [0,T] \times \mathbb{R}^m$ is not assumed bounded.

We now obtain a characterization of $p_t(x,y)$ as the unique solution of (5) for <u>all</u> $y \in H$:

<u>Theorem 3</u>.

Assume conditions of Theorem 2. Further, suppose that $E \exp \alpha |X_0|^2 < \infty$ for some $\alpha > 0$, and for all i.j.k

(i) a_{ij} is bounded,

(ii) $h^k, \dfrac{\partial h^k}{\partial x^i}, \dfrac{\partial^2 h^k}{\partial x^i \partial x^j}, \dfrac{\partial h^k}{\partial t}$ are locally Hölder continuous in (t,x).

(iii) $\dfrac{\partial h^k}{\partial t}, \sum_j a_{ij} \dfrac{\partial h^k}{\partial x^i}, \sum_{i,j} a_{ij} \left(\dfrac{\partial^2 h^k}{\partial x^i \partial x^j} + \dfrac{\partial h^k}{\partial x^i} \cdot \dfrac{\partial h^k}{\partial x^j} \right)$ and $\sum_i b_i \dfrac{\partial h^k}{\partial x^i}$ satisfy the growth condition of Theorem 2. Then,

(a). for all $y \in H$, the PDE (5) admits a unique solution $p_t(x,y)$ in the class G.

(b). The map $y \to p_t(x,y)$ is continuous in the following sense: If $y_n \to y$ in H, then $p_.(\cdot, y_n) \to p_.(\cdot,y)$ uniformly on compact subsets of $[0,T] \times \mathbb{R}^d$.

(c). For all $y \in H$, the unique solution $p_t(x,y)$ of (5) is the unnormalized conditional density of the filtering problem.

4. Consistency of the white noise theory with the stochastic calculus approach.

Recent work in the stochastic calculus (or conventional) approach to nonlinear filtering has centered around creating a pathwise or robust theory [2,3,11]. The impetus for it has naturally come from the need to bring the subject closer to applications. The latter has, indeed, been the motivation for developing the white noise theory.

In this section we shall present some of the main results we have obtained very recently, concerning the relationship between the two theories. The two-fold purpose of this section is to show that the white noise theory is naturally robust and, furthermore, that it is consistent with the conventional theory in the sense that the robustness of the latter theory can be recovered from the white noise results.

Let $(\tilde{\Omega},\tilde{A},\tilde{\Pi}) = (\Omega,A,\Pi) \otimes (\Omega_0,A_0,\Pi_0)$ where $\Omega_0=C_0([0,T],\mathbb{R}^m)$ is the space of continuous functions $\omega_0: [0,T] \to \mathbb{R}^m$ with $\omega_0(0)=0$ with the topology of uniform convergence and Π_0 is standard Wiener measure. Let $\{Z_t\}$ be the co-ordinate maps on Ω_0: $Z_t(\omega_0)=\omega_0(t)$, so that $\{Z_t\}$ is a Brownian motion under Π_0. The stochastic calculus filtering model may be assumed to be given by

$$Y_t = \int_0^t h_s(X_s)ds + Z_t$$

where $Y_t(\tilde{\omega})$ $(0\le t\le T)$ is the sample path of the observation process (Here $\tilde{\omega}= (\omega,\omega_0)$). Also let $\Omega^* = \{\omega_0 \in \Omega_0: t \to \omega_0(t)$ is Hölder continuous$\}$ and H be the reproducing kernel Hilbert space of m-dimensional Wiener process.

Theorem 4.

Suppose that the conditions of Theorem 3 are satisfied. Let $p_t(x,\eta)$ be the unnormalized conditional density of X_t given by Theorem 3. Then

(i) There exists a continuous function
$\hat{p}(\cdot,Z)$: $\Omega^* \to C([0,T] \times \mathbb{R}^d)$
such that for all $\eta \in H$,
$\hat{p}_t(x,Z) = p_t(x,\eta)$, $0\le t\le T$, $x \in \mathbb{R}^d$
where $Z \in H$ is given by

$$Z_t = \int_0^t \eta_s ds;$$

(ii) $\hat{p}_t(x,Y)$ is a version of the unnormalized conditional density of X_t given F_t^Y.

If the conditions of the above theorem are strengthened by replacing linear growth by boundedness then the conclusion remains true with Ω^* replaced by Ω_0.

Theorem 5.

Let $\{Y_k\}$ be the polygonal approximation to Y ($Y \in \Omega_0$) defined by

$$Y_t(\tilde{\omega})(s) = Y(\tilde{\omega})(t^k_{j-1}) + \frac{Y(\tilde{\omega})(t^k_j) - Y(\tilde{\omega})(t^k_{j-1})}{t^k_j - t^k_{j-1}}(s-t^k_{j-1})$$

for $t^k_{j-1} \leq s \leq t^k_j$, $1 \leq j \leq m_k$, where $\tilde{\omega} \in \tilde{\Omega}$ and $\{t^k_j\}$ $(j=0,\ldots,m_k)$ is a sequence of partitions of $[0,T]$ $(k=1,2,\ldots)$ such that

$$\lim_{k\to\infty} \sup_{1\leq j\leq m_k} |t^k_j - t^k_{j-1}| = 0.$$

Then we have

(a) $p_{.}(\cdot,\dot{Y}_k) \to \hat{p}_{.}(\cdot,Y)$ $\tilde{\Pi}$ - a.e.

Furthermore, for all real valued continuous functions f on $\mathbb{R}^d$ with compact support,

(b) $\sigma_t(f,\dot{Y}_k) \to \hat{\sigma}_t(f,Y)$ $\tilde{\Pi}$ - a.e.,

and

(c) $\Pi_t(f,\dot{Y}_k) \to \tilde{\Pi}_t(f,Y)$ $\tilde{\Pi}$ - a.e., where

Π and $\tilde{\Pi}$ represent the conditional expectation in the white noise and stochastic calculus theories.

In (b) and (c) $\dot{Y}_k(\tilde{\omega})$ is the element in $L^2[0,T]$ such that

$$Y_k(\tilde{\omega})(t) = \int_0^t \dot{Y}_k(\tilde{\omega})(s)ds \quad \text{for } 0\leq t\leq T.$$

References

[1] Balakrishnan, A.V. (1980). Non-linear white noise theory, Multivariate Analysis V, P.R. Krishnaiah, Ed., North Holland, Amsterdam.

[2] Clark, J.M.C. (1978). The design of robust approximations to the stochastic differential equations of nonlinear filtering in Communications Systems and Random Process Theory, ed., J.K. Skwirzynski, NATO Advanced Study Institute Series. Alphen aan den Rijn: Sijthoff and Noordhoff.

[3] Davis, M.H.A. (1979). Pathwise solutions and multiplicative functionals in nonlinear filtering, 18th IEEE conference on decision and control, Fort Lauderdale, Florida.

[4] Kallianpur, G. (1980). Stochastic filtering theory, Springer-Verlag, New York.

[5] Kallianpur, G. and Karandikar, R.L., (1983a). A finitely additive white noise approach to nonlinear filtering, J. Appl. Math. Optim. 10, 159-185.

[6] Kallianpur, G. and Karandikar, R.L., (1983b). Some recent developments in nonlinear filtering theory, Acta Applicandae Mathematicae 1, 399-434.

[7] Kallianpur, G. and Karandikar, R.L. (1984a). The nonlinear filtering problem for the unbounded case. To appear in Journal of Stochastic Processes and Their Applications.

[8] Kallianpur, G. and Karandikar, R.L. (1984b). Measure valued equations for the optimum filter in finitely additive nonlinear filtering theory, Z. Wahrsch. Verw. Geb. 66, 1-17.

[9] Kallianpur, G. and Karandikar, R.L. (1984c). Markov property of the filter in the finitely additive white noise approach to nonlinear filtering, To appear in Stochastics.

[10] Krylov, N.V. and Rozovskii, B.L. (1981). Stochastic evolution equations, J. Soviet Math. 16, 1233-1276.

[11] Pardoux, E. (1979). Stochastic partial differential equations and filtering of diffusion processes, Stochastics, 3, 127-168.

A MARTINGALE PROBLEM FOR CONDITIONAL DISTRIBUTIONS AND UNIQUENESS FOR THE NONLINEAR FILTERING EQUATIONS

Thomas G. Kurtz
Department of Mathematics
University of Wisconsin-Madison
Madison WI 53706

and

Daniel Ocone
Department of Mathematics
Rutgers University
New Brunswick NJ 08903

1. Martingale problems. A martingale problem, as developed by Stroock and Varadhan (1979), provides a method of characterizing a Markov process in terms of its generator A, by the requirement that certain functionals of the process be martingales. Specifically let E be a complete, separable metric space, and let A be a linear operator with domain and range in $B(E)$, the bounded measurable functions on E. Let ν be in $\mathcal{P}(E)$, the probability measures on E. A measurable process Z is a solution of the martingale problem for A (for (A,ν)) if there exists a filtration $\{\mathcal{F}_t\}$ such that

$$f(Z(t)) - \int_0^t Af(Z(s))ds \tag{1.1}$$

is a $\{\mathcal{F}_t\}$-martingale for every $f \in \mathcal{D}(A)$ (and $P\{Z(0) \in \Gamma\} = \nu(\Gamma)$).

The solution of the martingale problem for (A,ν) is unique (in distribution), if any two solutions have the same finite dimensional distributions. If a solution for (A,ν) exists and is unique for each $\nu \in \mathcal{P}(E)$, then the martingale problem is said to be well-posed.

If E is compact and $\mathcal{D}(A)$ is dense in $C(E)$ (in the sup norm), then any solution of the martingale problem for A has a modification with sample paths in $D_E[0,\infty)$. Similarly, if E is locally compact and separable and $\mathcal{D}(A)$ is dense in $\hat{C}(E)$ (the space of continuous functions that vanish at infinity), then any solution has a modification with sample paths in $D_{E^\Delta}[0,\infty)$, where $E^\Delta = E \cup \{\Delta\}$ denotes the one point compactification of E. (In most cases of interest it can be shown that the process never hits Δ.) In fact, if $\mathcal{D}(A)$ is dense in $\hat{C}(E)$, then, extending $f \in \hat{C}(E)$ to $f \in C(E^\Delta)$ by defining $f(\Delta) = 0$, we can extend A to an operator with $\mathcal{D}(A)$ dense in $\hat{C}(E^\Delta)$ by defining $A1 = 0$. Of course any solution of the original martingale problem is a solution of the new martingale problem, and any solution of the new martingale problem satisfying $P\{Z(t) = \Delta\} = 0$, $t \geq 0$, has a modification which is a solution of the original martingale problem. With this observation in mind, we will assume that all solutions considered have sample paths in $D_E[0,\infty)$, the space of right continuous E-valued functions having left limits at each $t \geq 0$. Under the usual Skorohod topology, $D_E[0,\infty)$ is a complete separable metric space if E is.

We now assume $E = E_1 \times E_2$, where E_1 and E_2 are locally compact and separable, and $Z = (X,Y)$. For the filtering problem, Y will denote the observed component and X the unobserved. Assume that the underlying probability space is complete and set $\mathcal{F}_t^Y = \sigma(Y(s):s\leq t) \vee \mathcal{N}$ where $\mathcal{N}$ denotes the collection of sets of probability zero. Let π_t^0 denote the conditional distribution of $X(t)$ given $\mathcal{F}_t^Y$, that is π_t^0 is a $\mathcal{P}(E_1)$-valued random variable satisfying

$$E[f(X(t))\,|\,\mathcal{F}_t^Y] = \int f(x)\,\pi_t^0(dx) \tag{1.2}$$

for every $f \in B(E)$. For technical reasons it is convenient to consider the conditional distribution of $X(t)$ given $\mathcal{F}_{t+}^Y$ which we denote π_t. The following lemma shows that this makes little difference. Note $\{\pi_t\}$ and $\{\pi_t^0\}$ are $\mathcal{P}(E_1)$-valued processes. We take the weak topology on $\mathcal{P}(E_1)$.

<u>1.1 Lemma</u> Let $E = E_1 \times E_2$ be locally compact and separable, and $\mathcal{D}(A)$ be a dense subspace of $\hat{C}(E)$. If $Z = (X,Y)$ is a solution of the martingale problem for A and π^0 and π are defined as above, then π has a modification with sample paths in $D_{\mathcal{P}(E_1)}[0,\infty)$, and $\pi_t = \pi_t^0$ a.s. for all but countably many $t \geq 0$.

<u>Proof</u> See Yor (1977). ∎

For any measure μ, we will let μf denote $\int f d\mu$. Under the assumptions of Lemma 1.1

$$\pi_t f(\cdot,Y(t)) - \int_0^t \pi_s Af(\cdot,Y(s))ds \tag{1.3}$$

is a $\{\mathcal{F}_{t+}^Y\}$-martingale for all $f \in \mathcal{D}(A)$. Our goal is to exploit this observation in order to characterize the distribution of $\{(\pi_t, Y(t))\}$ considered as a $\mathcal{P}(E_1) \times E_2$-valued process. In Section 3, we will see how this characterization can be used to prove uniqueness for the stochastic partial differential equations arising in filtering theory.

Just requiring (1.3) to be a martingale with respect to some filtration for each $f \in \mathcal{D}(A)$ is clearly not sufficient. In particular $\{\delta_{X(t)}, Y(t)\}$ would meet this requirement. Consequently, we introduce the notion of the filtered martingale problem for A. A $\mathcal{P}(E_1) \times E_2$-valued process $\{(\mu_t, U(t))\}$ is a solution of the filtered martingale problem for A if μ is $\{F^U_{t+}\}$-adapted and for each $f \in \mathcal{D}(A)$

$$\mu_t f(\cdot, U(t)) - \int_0^t \mu_s Af(\cdot, U(s))ds \tag{1.4}$$

is a $\{F^U_{t+}\}$-martingale.

In the next section we will show, in considerable generality, that uniqueness for the martingale problem for A implies uniqueness for the filtered martingale problem for A.

2. <u>Uniqueness for the filtered martingale problem</u>. Let (μ, U) be a process with sample paths in $D_{\mathcal{P}(E_1) \times E_2}[0,\infty)$, and suppose (1.4) is a martingale for each $f \in \mathcal{D}(A)$. Define $\nu_t \in \mathcal{P}(E)$ by

$$\nu_t f = E[\mu_t f(\cdot, Y(t))], \quad f \in B(E). \tag{2.1}$$

Then

$$\nu_t f = \nu_0 f + \int_0^t \nu_s Afds \tag{2.2}$$

for every $f \in \mathcal{D}(A)$, and $\{\nu_t\}$ is in some sense a weak solution of

$$\frac{d}{dt}\nu_t = A^* \nu_t . \tag{2.3}$$

There are two sets of conditions under which (2.2) uniquely determine ν_t (for given ν_0). If the closure of A generates a strongly continuous contraction semigroup $\{T(t)\}$, then (2.2) implies

$$\frac{d}{ds}\nu_s T(t-s) f = 0 \tag{2.4}$$

and hence

$$\nu_t f = \nu_0 T(t) f, \quad f \in \mathcal{D}(A). \tag{2.5}$$

Consequently if $\mathcal{D}(A)$ is separating (i.e. $\eta f = \varphi f$ for all $f \in \mathcal{D}(A)$ implies $\eta = \varphi$), $\{\nu_t\}$ is uniquely determined. The second set of conditions is from Echeverria (1982), who treated the special case $\nu_0 Af = 0$ for $f \in \mathcal{D}(A)$. (In this case uniqueness implies $\nu_t = \nu_0$, $t \geq 0$, i.e. ν_0 is a stationary distribution.)

<u>2.1 Lemma</u> Let E be locally compact and separable. Suppose A is a linear operator on $\hat{C}(E)$ and that $\mathcal{D}(A)$ is an algebra and dense in $\hat{C}(E)$. Suppose $\nu: [0,\infty) \to \mathcal{P}(E)$ satisfies (2.2) for each

$f \in \mathcal{D}(A)$. If the martingale problem for A is well-posed, then

$$\nu_t f = E[f(Z(t))] \tag{2.6}$$

where Z is the solution of the martingale problem for (A,ν_0).

Proof See Ethier and Kurtz (1985). ∎

If (μ,U) is a solution of the filtered martingale problem, A satisfies the conditions of Lemma 2.1, and ν_t is defined by $\nu_t f = E[\mu_t f(\cdot.U(t))]$, then ν_t satisfies (2.2) and Lemma 2.1 gives

$$E[\mu_t f(\cdot,U(t))] = E[f(X(t),Y(t))] \tag{2.7}$$

for each $f \in \mathcal{D}(A)$, which is enough to conclude that $U(t)$ and $Y(t)$ have the same distribution for each $t \geq 0$. This, of course, is not enough for our purposes, but it indicates how we will proceed. Let $g_i \in \hat{C}(E_2)$, $i = 1,\ldots,m$, and define $V_i(t) = \int_0^t g_i(U(s))ds$, and $Z_i(t) = \int_0^t g_i(Y(s))ds$. If for each m and each choice of $g_1,\ldots,g_m$, we can show that

$$\begin{aligned} &E[\mu_t f(\cdot,U(t),\, V_1(t),\ldots,V_m(t))] \\ &\qquad = E[f(X(t),\, Y(t),\, Z_1(t),\ldots,Z_m(t))], \quad t \geq 0, \end{aligned} \tag{2.8}$$

for a collection of functions f that is separating in $E_1 \times E_2 \times \mathbb{R}^m$ then U and Y have the same finite dimensional distributions (knowing the joint distributions of $Y(t)$, $Z_1(t),\ldots,Z_m(t)$ for all $t \geq 0$ and all choices of the g_i determines the finite dimensional distributions of Y). Furthermore, since μ has sample paths in $D_{P(E_1)}[0,\infty)$, for all but countably many t, $\mu_t = \mu_{t-}$ and hence μ_t is F_t^u measurable. For each such t there exists a Borel measurable function $G_t: D_{E_2}[0,\infty) \to P(E_1)$ such that $\mu_t = G_t(U_t)$ a.s. where $U_t = U(\cdot \wedge t)$. Consequently

$$\begin{aligned} &E[G_t(Y_t) f(\cdot,Y(t),Z_1(t),\ldots,Z_m(t))] \\ &\quad = E[G_t(U_t) f(\cdot,U(t),V_1(t),\ldots,V_m(t)] \\ &\quad = E[f(X(t),Y(t),Z_1(t),\ldots,Z_m(t))], \end{aligned} \tag{2.9}$$

and it follows that $\pi_t^0 = G_t(Y_t)$.

Observe that (2.8) has the same form as (2.7) with U replaced by $(U,V_1,\ldots,V_m)$. With this in mind, for $f_0 \in \mathcal{D}(A)$ and $f_i \in C_0^\infty(\mathbb{R})$, $i = 1,\ldots,m$, define

$$f(x,y,z_1,\ldots,z_m) = f_0(x,y)\prod_{i=1}^{m} f_i(z_i) \tag{2.10}$$

and

$$\begin{aligned} Bf(x,y,z_1,\ldots,z_m) &= \prod_{i=1}^{m} f_i(z_i)Af_0(x,y) \\ &\quad + \sum_{i=1}^{m} g(z_i)f_i'(z_i)f_0(x,y)\prod_{j\neq i} f_j(z_j). \end{aligned} \tag{2.11}$$

Extend B linearly, and note that $\mathcal{D}(B)$ is an algebra that is dense in $\hat{C}(E \times \mathbb{R}^m)$. We claim that the martingale problem for B is well-posed and hence that B satisfies Echeverria's conditions, i.e. the conditions of Lemma 2.1.

Existence of solutions of the martingale problem for B follows immediately from existence for A, since if (X,Y) is a solution for A and

$$Z_i(t) = \int_0^t g_i(Y(s))ds, \tag{2.12}$$

then $(X,Y,Z_1,\ldots,Z_m)$ is a solution for B. Conversely, if $(X,Y,Z_1,\ldots,Z_m)$ is a solution for B, it follows easily that (X,Y) is a solution for A, and uniqueness for B is a consequence of the following lemma. We leave the proof to the reader.

<u>2.2 Lemma</u> Let $\{\mathcal{F}_t\}$ be a filtration, and U and V be right continuous real-valued processes adapted to $\{\mathcal{F}_t\}$. If

$$f(U(t)) - \int_0^t V(s)f'(U(s))ds \tag{2.13}$$

is a $\{\mathcal{F}_t\}$-martingale for each $f \in C_0^\infty(\mathbb{R})$, then

$$U(t) = U(0) + \int_0^t V(s)ds. \tag{2.14}$$

We now have the tools to prove the main uniqueness theorem.

<u>2.3 Theorem</u> Suppose A satisfies the conditions of Lemma 2.1. Let (X,Y) be a solution of the martingale problem for A with sample paths in $D_E[0,\infty)$, and let π_t be the conditional distribution of $X(t)$ given $\mathcal{F}_{t+}^Y$. For each $t \geq 0$, let $H_t : D_{E_2}[0,\infty) \to \mathcal{P}(E_1)$ be Borel measurable and satisfy $\pi_t = H_t(Y)$ a.s. (Such a H always exists.) If (μ,U) is a solution of the filtered martingale problem for A with sample paths in $D_{\mathcal{P}(E_1)\times E_2}[0,\infty)$, and

$$E[f(X(0),Y(0))] = E[\mu f(\cdot,U(0))] \tag{2.15}$$

for each $f \in B(E)$, then for each $t \geq 0$, $\mu_t = H_t(U)$ a.s.

Proof For $g_i \in C(E_2)$, $i = 1,\dots,m$, define $V_i(t) = \int_0^t g_i(U(s))ds$.

Then $(\mu,U,V_1,\dots,V_m)$ is a solution of the filtered martingale problem for B, and since B satisfies Echeverria's conditions, Lemma 2.1 gives (2.8). For all but countably many $t, \pi_t = \pi_{t-}$ and $\mu_t = \mu_{t-}$. For each such t, let G_t be as in (2.9). Then $G_t(Y_t) = \pi_t^0 = \pi_t = H_t(Y)$ a.s., and since U and Y have the same distribution, $\mu_t = G_t(U_t) = H_t(U)$ a.s. The conclusion extends to all t by right continuity. ■

2.4 Corollary Suppose that (μ,U) is a process with sample paths in $D_{P(E_1)\times E_2}[0,\infty)$ satisfying (2.15), that μ is $\{F_{t+}^U\}$-adapted, that τ is a $\{F_t^U\}$-stopping time, and that for each $f \in \mathcal{D}(A)$

$$\mu_{t\wedge\tau} f(\cdot,U(t)) - \int_0^{t\wedge\tau} \mu_s Af(\cdot,U(s))ds \tag{2.16}$$

is a $\{F_{t+}^U\}$-martingale. Then for each $t \geq 0$

$$\mu_t \chi_{\{t<\tau\}} = H_t(U)\chi_{\{t<\tau\}} \text{ a.s.} \tag{2.17}$$

where H_t is as in Theorem 2.3.

Proof A process $(\tilde{\mu},\tilde{U})$ can be constructed on an enlarged sample space such that $(\tilde{\mu}_{t\wedge\tau},\tilde{U}(t\wedge\tau)) = (\mu_{t\wedge\tau},U(t\wedge\tau))$, $t \geq 0$, and $(\tilde{\mu},\tilde{U})$ is a solution of the filtered martingale for A_0. Then (2.17) follows by uniqueness. ■

3. Uniqueness of solutions to the filtering equations. In situations where the observation is obtained by adding white noise to the signal, π_t satisfies the Kushner-Stratonovich equations of nonlinear filtering (see 3.3). In addition there exists another measure-valued process σ_t such that $\pi_t = \sigma_t/\sigma_t 1$ and σ_t satisfies the Zakai equations (see 3.13)). These are derived by explicitly representing the martingale $\pi_t f - \int_0^t \pi_s Af\, ds$ as a stochastic integral with respect to Y for each $f \in \mathcal{D}(A)$ such that $f(x,y) = f(x)$ depends only on the x-variable (see Fujisaki, Kallianpur, and Kunita The resulting equations define, in essence, stochastic p.d.e.'s in weak form for π_t and σ_t. we will be concerned with the question of the uniqueness of solutions of these nonlinear filtering equations. That is, if μ_t

is another solution of, say the Kushner-Stratonovich equations, under what conditions can we conclude that $\mu_t = \pi_t$? We will use the martingale problem characterization of π_t in Theorem 2.4 to show that, in fact, the solutions of both the Zakai and Kushner-Stratonovich equations are unique under rather weak hypotheses.

We consider the following model. All processes are defined on the probability space (Ω, F, P). The signal process X is an E-valued solution of the martingale problem for (A_0, π_0), where E is a complete, separable, locally compact metric space, A_0 maps $\mathcal{D}(A_0) \subset \hat{C}(E)$ into $\hat{C}(E)$, and $\pi_0 \in P(E)$. The observation process is defined to be

$$Y(t) = \int_0^t h(x(s))ds + B(t)$$

where $h: E \to \mathbb{R}^p$, and B is a Brownian motion independent of X. We shall impose the following assumptions

$$\begin{array}{ll} \text{a)} & A_0 \text{ satisfies Echeverria's conditions} \\ \text{b)} & f(x)h(x) \text{ is bounded for every } f \in \mathcal{D}(A_0) \\ \text{c)} & h(x) \text{ is continuous} \\ \text{d)} & \text{For some } T > 0, \quad E\int_0^T |h(x(t))|^2 dt < 0. \end{array} \tag{3.1}$$

Note that by (3.1.d) and independence of B and X, the distribution of $\{Y(t) | t \leq T\}$ is mutually absolutely continuous with respect to that of Brownian motion, and, hence, $\{F_t^Y\}$ is right-continuous. By Lemma 1.1, it is possible to choose a right-continuous, $\{F_t^Y\}$-adapted version of $\pi_t = \pi_t^0$. π_t satisfies the Kushner-Stratonovich equation

$$\begin{aligned} \pi_t f = \pi_0 f &+ \int_0^t \pi_s A_0 f\, ds \\ &+ \int_0^t [\pi_s hf - (\pi_s h)(\pi_s f)] \cdot [dY(s) - \pi_s h\, ds] \end{aligned} \tag{3.2}$$

for all $f \in \mathcal{D}(A)$.

Of course, (3.2) is valid under much more general hypotheses than those posited in (3.1) (see, for example, Lipster and Shiryayev (1977)). However, these assumptions will be needed for the uniqueness result. The process $\nu(t) = Y(t) - \int_0^t \pi_s h\, ds$, which appears as a stochastic differential in (3.2), is called the <u>innovations process</u> and is a Brownian motion. The main theorem follows.

Theorem 3.1 Assume the conditions in (3.1). Let $\{\mu_t\}$ be a $\{F_t^Y\}$-adapted, right continuous, $P(E)$-valued process such that for every $f \in \mathcal{D}(A_0)$ and $t \leq T$

$$\mu_t f = \pi_0 f + \int_0^t \mu_s A_0 f\, ds + \int_0^t [\mu_s hf - (\mu_s h)(\mu_s f)] \cdot [dY - \mu_s h\, ds]\,. \tag{3.3}$$

Then $\mu_t = \pi_t$ a.s. for all $t \leq T$.

Remarks 1. Since h may be unbounded, $\mu_s h$ does not automatically make sense. Therefore we regard as implicit in the assumption that μ_s satisfy (3.3), the conditions that h be a.s. μ_s-integrable and $\int_0^T (\mu_s h)^2\, ds < \infty$ a.s.

2. If μ_t solves (3.3), then μ_t is actually a.s. continuous. In particular, π_t is a.s. continuous.

We will sketch the main ingredients of the proof of Theorem 3.1. The first step is to define a generator A on $\hat{C}(E \times \mathbb{R}^p)$ for the pair (X,Y), since the filtered martingale problem characterizes (π,Y) in termes of A and not just A_0. Therefore, let

$$\mathcal{D}(A) = \text{span}\{\, f(x)\psi(y) \mid f \in \mathcal{D}(A_0),\ \psi \in C_0^\infty(\mathbb{R}^p)\} \tag{3.4}$$

and define

$$Af(x)\psi(y) = (A_0 f(x))\psi(y) + f(x)[\tfrac{1}{2}\Delta\psi(y) + h(x)\cdot\nabla\psi(y)] \tag{3.5}$$

The following lemma may be proved without too much difficulty. Note that assumption (3.1.b) insures that A maps $\mathcal{D}(A)$ into $\hat{C}(E \times \mathbb{R}^p)$.

Lemma 3.2 a) (X,Y) solves the martingale problem for $(A, \pi_0 \times \delta_0)$.

b) A satisfies Echevarria's conditions. In particular, the martingale problem for A is well-posed.

Proof of Theorem 3.1: Let $\psi \in C_0^\infty(\mathbb{R}^p)$. By Ito's rule

$$\psi(Y(t)) = \psi(0) + \int_0^t [\tfrac{1}{2}\Delta\psi(Y(s)) + \mu_s h\cdot\nabla\psi(Y(s))]\, ds + \int_0^t \nabla\psi(Y(s))\cdot[dY(s) - \mu_s h\, ds]$$

where we have added and subtracted a $(\mu_s h)\, ds$ integral to make the stochastic integral term conform to that in (3.3). Again applying Ito's rule, if $f \in \mathcal{D}(A_0)$,

$$(\mu_t f)\psi(Y(t)) = (\mu_0 f)\psi(0) + \int_0^t \mu_s[Af\psi(\cdot,Y(s))]ds + \int_0^t [\psi(Y(s))[\mu_s hf - (\mu_s h)(\mu_s f)] + (\mu_s f)\nabla\psi(Y(s))]\cdot d\nu_\mu(s) \tag{3.7}$$

where

$$\nu_\mu(t) = Y(t) - \int_0^t \mu_s h\, ds\ . \tag{3.8}$$

If we knew a priori that $\nu_\mu(t)$ were a martingale, we would be done since then (3.7) would imply that $\mu_t\varphi(\cdot,Y(t)) - \int_0^t \mu_s A\varphi(\cdot,Y(s))\, ds$ is a local martingale for every $\varphi \in \mathcal{D}(A)$. To make $\nu_\mu(t)$ locally a martingale, let $\xi(s) = \mu_s h - \pi_s h$ and for $T > 0$ define the stopping time

$$\tau_N(\omega) = \inf\{t: \int_0^t |\xi(s)|^2 ds \geq N\} \cup \{T\}\quad . \tag{3.9}$$

Then by Girsanov's theorem, $\nu_\mu(t)$ is a Brownian motion up to time τ_N on (Ω,F,Q_N), where

$$\frac{dQ_N}{dP} = \exp[\int_0^{\tau_N} \xi(s)\cdot d\nu(s) - \frac{1}{2}\int_0^{\tau_N} |\xi(s)|^2 ds]\ . \tag{3.10}$$

Thus on (Ω,F,Q_N)

$$\mu_{t\wedge\tau_N}\varphi(\cdot,Y(t\wedge\tau_N)) - \int_0^{t\wedge\tau_N} \mu_s A\varphi(\cdot,Y(s))\, ds$$

is a bounded, local martingale, and hence a martingale for every $\varphi \in \mathcal{D}(A)$. By Lemma 3.2 and Corollary 2.4

$$\mu_t\chi_{\{t<\tau_N\}} = H_t(Y(\cdot\wedge\tau_N))\chi_{\{t<\tau_N\}} \tag{3.11}$$

a.s. Q_N and therefore a.s. P, since $P << Q_N$. Now τ_N eventually reaches T for each ω as $N \to \infty$. Thus (3.11) implies

$$\mu_t = H_t(Y\cdot\wedge T)) = \pi_t$$

a.s. for $t < T$. Since μ_t and π_t are a.s. continuous (see Remark 3) $\mu_T = \pi_T$ a.s. also.

The Zakai equation is a filtering equation for an unnormalized conditional distribution. Let P_0 be a new probability measure on Ω defined by $dP_0/dP = L_T^{-1}$ where

$$L_t = \exp[\int_0^t h(x(s))\cdot dY(s) - \frac{1}{2}\int_0^t |h(x(s))|^2 ds] \tag{3.12}$$

(P_0 is a probability measure because of assumption 3.1.d and the independence of X and B.). On (Ω, F, P_0), Y is a Brownian motion independent of X. Define the $M^+(E)$-valued $\{F_t^Y\}$- progressively measurable process $\sigma_t = E_0[L_t | F_t^Y]\pi_t$. ($M^+(E)$ = {positive bounded Borel measures on E}.) Then $\sigma_t f = E_0[f(x(t))L_t | F_t^Y]$ a.s. for all $t \leq T$ and all $f \in B(E)$.

<u>Theorem 3.3</u> Assume the conditions in (3.1). Let $\{\eta_t\}$ be a right-continuous, $\{F_t^Y\}$-adapted, $M^+(E)$-valued process such that

$$\eta_t f = \pi_0 f + \int_0^t \eta_s A_0 f\, ds + \int_0^t (\eta_s hf)\cdot dY(s) \tag{3.13}$$

for every $f \in \mathcal{D}(A_0)$ and $t \leq T$. Then $\eta_t = \sigma_t$ a.s. for all $t \leq T$.

<u>Remarks</u> 1. As with $\mu_s h$ we are implicitly assuming h is a.s. σ_s-integrable and $\int_0^T (\sigma_s h)^2 ds < \infty$ a.s.

2. (3.13 is called the Duncan-Mortensen-Zakai equation.

3. This theorem improves over previous uniqueness results for Zakai's equation given by Szpirglas (1978), Pardoux (1979), Pardoux (1982) by allowing h to be unbounded (as long as (3.1.b) and (3.1.d) hold). When X is the solution of a stochastic differential equation on $\mathbb{R}^n$ with uniformly Lipshitz coefficients and Brownian white noise input, one may take $\mathcal{D}(A_0) = C_0^\infty(\mathbb{R}^n)$, and (3.1.b) is satisfied automatically.

<u>Proof</u> Ito's rule shows that $\rho_t = \eta_t/\eta_t 1$ satisfies (3.3). Thus by Theorem 3.1 $\eta_t/\eta_t 1 = \pi_t = \sigma_t/\sigma_t 1$ a.s. Another application of Ito's rule shows that $d[\eta_t 1/\sigma_t 1] = 0$. Thus $\eta_t 1 = \sigma_t 1$ a.s., and hence, $\eta_t = \sigma_t$ a.s. ∎

Framing uniqueness theorems for filtering equations in other situations is simple in principle. Besides assuming that A satisfies

Echeverria's condition, one must verify that if μ_t is a solution of the filtering equations, $\mu_t\varphi(\cdot,Y(s)) - \int_0^t \mu_s A\varphi(\cdot,Y(s))\,ds$ is a $\{F^Y_{t+}\}$-martingale on (Ω,F,Q) where Q is mutually absolutely continuous with respect to the original measure P. (A stopping time argument as in Theorem 3.1 might also be necessary). In this way, for example, it is easy to formulate a uniqueness theorem for Kushner-Stratonovich and Zakai equations that allows correlation between B and X. The reader is invited to provide the details.

References

1. Echeverria, P. E. (1982). A criterion for invariant measures of Markov processes. Z. Wahrsch. verw. Gebiete 61, 1-16.
2. Ethier, S. N. and Kurtz, T. G. (1985). Markov Processes: Characterization and Convergance (to appear).
3. Fujisaki, M., Kallianpur, G. and Kunita, H. (1972). Stochastic differential equations for the nonlinear filtering problem. Osaka J. Math. 9, 19-40.
4. Lipster, R. S. and Shiryayev, A. N. (1977). The Statistics of Random Processes I: General Theory. Springer-Verlag, Berlin.
5. Pardoux, E. (1979). Stochastic partial differential equations and filtering of diffusion processes. Stochastics 3, 127-167.
6. Pardoux, E. (1982). Equations du filtrage nonlinéaire, de la prediction, et du lissage. Stochastics, 193-232.
7. Stroock, D. W. and Varadhan, S. R. S. (1979). Multidimensional Diffusion Processes. Springer-Verlag, Berlin.
8. Szpirglas, J. (1978). Sur l'equivalence d'equations différentielles stochastique á valeurs mesures intervenant dans le filtrage markovien nonlinéaire. Ann. Inst. Henri Poincaré, XIV, 33-59.
9. Yor, M. (1977). Sur les théories du filtrage et de la prediction. Séminaire de Probabilités XI, Lect Notes in Math 581, Springer-Verlag, Berlin.

CONTINUOUS VERSIONS OF THE CONDITIONAL STATISTICS OF NONLINEAR FILTERING

H.J. Sussmann
Mathematics Department
Rutgers University
New Brunswick, NJ 08903 U.S.A.

In this paper we consider nonlinear filtering problems that arise from the specification of the following data:

(i) a probability space (Ω, A, P),

(ii) three filtrations $\{A_t : t \geq 0\}$, $\{F_t : t \geq 0\}$, $\{V_t : t \geq 0\}$, such that $F_t \cup V_t \subseteq A_t \subseteq A$, and F_t is independent from V_t for every $t \geq 0$,

(iii) a "signal process" $S = \{S(t) : t \geq 0\}$, progressively measurable with respect to $\{F_t\}$, and with values in $\mathbb{R}^m$, for some m,

(iv) a standard $\mathbb{R}^m$-valued Wiener process $V = \{V(t) : t \geq 0\}$, progressively measurable with respect to $\{V_t\}$, such that $\{(V(t), V_t)\}$ is a martingale, and all the sample paths of V are continuous.

We assume that the signal process satisfies

$$\int_0^T \|S(t)\|^2 dt < \infty \quad \text{a.s.} \tag{1}$$

for every $T > 0$. (Here $\|\ldots\|$ denotes Euclidean norm.)

The observation process $Y = \{Y(t)\}$ is defined by

$$Y(t) = \int_0^t S(\tau)d\tau + V(t), \tag{2}$$

and the observation filtration $\{Y_t\}$ by

$$Y_t = \sigma(\{Y(\tau) : 0 \leq \tau \leq t\}). \tag{3}$$

One should think of F_t as the σ-algebra that contains all the information about the state up to time t of some process, of $S(t)$ as some function of the state which is being observed, and of the time-derivative of $V(t)$ as "observation noise". The nonlinear filtering problem considered here is that of determining the conditional probability measure of the state given the observations. That is, if $A \in F_t$ and $\eta \in C_o^o([0,t], \mathbb{R}^m)$ (the space of continuous functions $[0,t] \to \mathbb{R}^m$ that vanish at 0), then we want to study $\pi_t(A, \eta)$, the conditional probability of the event A given that the path up to time t of the observation process has turned out to be η. Of course, the usual definition of a conditional expectation as a Radon-Nikodym derivative does not make it possible to define $\pi_t(A, \eta)$ uniquely for every A and η. However, one can define, if Φ is an arbitrary F_t-measurable bounded real-valued function, the conditional expectation $\mathbb{E}(\Phi / Y_t)$, which is a Y_t-measurable real-valued function, and is unique modulo P-almost sure equality. Let Y^t denote the "path up to time t" map of the observation process, i.e. let $Y^t(\omega)$ be the map $\tau \to Y(\tau)(\omega)$, $0 \leq \tau \leq t$. Then $Y^t(\omega) \in C_o^o([0,t], \mathbb{R}^m)$ for almost all ω. So Y^t is a map from

Ω to $C_o^o([0,t],\mathbb{R}^m)$. We will always assume that the latter space is equipped with the σ-algebra $\mathcal{B}_{o,m,t}$ generated by the evaluation process $E_{o,m,t}$. (Precisely, $E_{o,m,t}$ is the family of maps $\{E_{o,m,t}(\tau): 0 \leq \tau \leq t\}$ from $C_o^o([0,t],\mathbb{R}^m)$ to $\mathbb{R}^m$ given by $E_{o,m,t}(\tau)(\eta) = \eta(\tau)$, and $\mathcal{B}_{o,m,t}$ is the σ-algebra generated by the $E_{o,m,t}(\tau)$.) Then Y^t is a $(\mathcal{Y}_t,\mathcal{B}_{o,m,t})$-measurable map. If $Y_*^t(P)$ denotes the pushforward of the probability measure P by Y^t, then $\mathbb{E}(\Phi/\mathcal{Y}_t)$ has a factorization

$$\mathbb{E}(\Phi/\mathcal{Y}_t) = \hat{\Phi}_t \circ Y^t \tag{4}$$

where $\hat{\Phi}_t: C_o^o([0,t],\mathbb{R}^m) \to \mathbb{R}$ is measurable, and is unique modulo $Y_*^t(P)$-almost sure equality.

Let $\mathcal{P}_t^m$ denote standard Wiener measure on $C_o^o([0,t],\mathbb{R}^m)$. Then it is well known that $\mathcal{P}_t^m$ and $Y_*^t(P)$ are mutually absolutely continuous, and so $\hat{\Phi}_t$ is unique modulo $\mathcal{P}_t^m$-almost sure equality. For each $A \in \mathcal{F}_t$, we can then define $\pi_t(A,\eta) = \hat{\chi}_{A,t}(\eta)$, where χ_A is the indicator function of A. Then $\pi_t(A,\cdot)$ is well defined, modulo $\mathcal{P}_t^m$-almost sure equality. Under suitable technical conditions, one can choose a regular version, that is, one can choose versions of the $\hat{\chi}_{A,t}$ such that the resulting function $(A,\eta) \to \pi_t(A,\eta)$ has the property that, for $\mathcal{P}_t^m$-almost all η, the function $A \to \pi_t(A,\eta)$ is a probability measure.

The purpose of this paper is to discuss some recent results on the <u>existence of continuous versions</u> of the functions $\hat{\Phi}_t$ and of the π_t. Precisely, let us say that a particular nonlinear filtering problem has <u>property</u> (C_t^o) if the following holds:

(C_t^o) <u>for every bounded $\mathcal{F}_t$-measurable $\Phi: \Omega \to \mathbb{R}$, the function $\hat{\Phi}_t$ has a version $\hat{\hat{\Phi}}_t$ which is a continuous function from $C_o^o([0,t],\mathbb{R}^m)$ to $\mathbb{R}$.</u>

(Here $C_o^o([0,t],\mathbb{R}^m)$ is given the topology of uniform convergence.)

More generally, we can consider, for $0 < \alpha$, the property (C_t^α) which is defined in the same way as (C_t^o), except that $\hat{\hat{\Phi}}_t$ is only required to be continuous on $C_o^\alpha([0,t],\mathbb{R}^m)$, the space of Hölder continuous functions $f: [0,t] \to \mathbb{R}^m$ that satisfy $f(0) = 0$, with the norm

$$\|f\|_{\alpha,t} = \sup\left\{\frac{|f(\tau_1)-f(\tau_2)|}{(\tau_1-\tau_2)^\alpha}: 0 \leq \tau_2 < \tau_1 \leq t\right\}. \tag{5}$$

We say that <u>property</u> (C^α) holds if (C_t^α) holds for every $t > 0$.

If property (C_t^α) holds for some $t > 0$ and some $\alpha \in [0,\frac{1}{2}[$, then the versions $\hat{\hat{\Phi}}_t$ are unique on $C_o^\alpha([0,t],\mathbb{R}^m)$, because two continuous functions which agree $\mathcal{P}_t^m$-almost surely on $C_o^\alpha([0,t],\mathbb{R}^m)$ necessarily coincide. Hence one can define a "continuous version" π_t of the conditional probabilities by

$$\pi_t(A,\eta) = \hat{\hat{\chi}}_{A,t}(\eta),\ A \in \mathcal{F}_t,\ \eta \in C_o^\alpha([0,t],\mathbb{R}^m), \tag{6}$$

and there is no need to worry about choosing regular versions. The functions $\pi_t(\cdot,\eta)$ are probability measures, and an unambiguous sense has been given to $\pi_t(A,\eta)$ for each $A \in F_t$ and each individual sample path $\eta \in C_o^\alpha([0,t],\mathbb{R}^m)$.

The well-known Kallianpur-Striebel formula says that

$$\hat{\Phi}_t = \frac{\Phi_t^*}{1_t^*}, \tag{7}$$

where

$$\Phi_t^*(\eta) = \int_\Omega \Phi(\omega)e^{Q_t(\eta)(\omega)}\,dP(\omega) \tag{8}$$

and

$$Q_t(\eta)(\omega) = \int_0^t <S(\tau)(\omega),\ d\eta(\tau)> - \frac{1}{2}\int_0^t \|S(\tau)(\omega)\|^2 d\tau. \tag{9}$$

In the first integral in (9), $<\ldots,\ldots>$ denotes Euclidean inner product in $\mathbb{R}^m$. The precise interpretation of this integral is as a <u>stochastic integral</u>: both the signal process S and the evaluation process $E_{o,m,t}$ are well defined on the product space $(\Omega\times C_o^o([0,t],\mathbb{R}^m),\ A\times B_{o,m,t},\ P\times P_t^m)$, and $E_{o,m,t}$ is a standard Wiener process. So the stochastic integral

$$I_1 = \int_0^t <S(\tau),\ dE_{o,m,t}(\tau)> \tag{10}$$

is well defined, modulo $P\times P_t^m$-almost sure equality, as a function of ω and η. The first integral in (9) is then interpreted to be $I_1(\omega,\eta)$. Clearly, $(\omega,\eta) \to Q_t(\eta)(\omega)$ is well defined modulo $P\times P_t^m$-almost sure equality, and then Φ_t^* is well defined modulo P_t^m-almost sure equality. Naturally, 1_t^* means Φ_t^* for the particular function Φ given by $\Phi \equiv 1$. From now on, we will write $Z_t^*(\eta)$ instead of $1_t^*(\eta)$.

The functions Φ_t^* are called the <u>unnormalized conditional statistics</u>. In particular, if we define π_t^* by

$$\pi_t^*(A,\eta) = \chi_{A,t}^*(\eta), \tag{11}$$

then $A \to \pi_t^*(A,\eta)$ is the <u>unnormalized conditional probability measure</u> given the observation path η. The true (i.e. normalized) conditional probability is obtained by dividing by the normalization constant $\pi_t^*(\Omega,\eta)$, i.e. $Z_t^*(\eta)$.

The functions Φ_t^* have obvious "pathwise" counterparts $\Phi_t^{*,P}$ which are defined for paths η that are of class C^1 or, slightly more generally, for paths in H_1 (the space of absolutely continuous functions with a square-integrable derivative). Indeed, if η is such a path, then the first integral in (9) is well defined for that individual path, and the exponential that appears in (8) satisfies

$$e^{Q_t(\eta)(\omega)} \leq e^{\frac{1}{2}\int_0^t \|\dot{\eta}(\tau)\|^2 d\tau} . \tag{12}$$

Therefore $\Phi_t^*(\eta)$ is well defined, for bounded Φ and $\eta \in H_1$. This then defines a functional $\Phi_t^{*,p}$ on $H_1 \cap C_o^o([0,t],\mathbb{R}^m)$, and it is not hard to prove that this functional is continuous with respect to the norm $(\int_0^t \|\dot{\eta}(\tau)\|^2 d\tau)^{\frac{1}{2}}$.

One can similarly define $\hat{\Phi}_t^p$, $\pi_t^{*,p}$, π_t^p. The question then arises of how the "pathwise" objects $\Phi_t^{*,p}$, $\hat{\Phi}_t^p$, $\pi_t^{*,p}$, π_t^p are related to the "statistical" objects Φ_t^*, $\hat{\Phi}_t$, π_t^*, π_t. The "obvious" answer, i.e. that the former are the restrictions of the latter to $H_1 \cap C_o^o([0,t],\mathbb{R}^m)$, does not work, because the restriction referred to here is to a set of measure zero, and the "statistical" objects are only defined modulo P_t^m-almost sure equality.

It was proved by Kallianpur and Striebel (and brought to our attention by S. Varadhan) that a result analogous to the Wong-Zakai approximation theorem of stochastic differential equations holds here as well. We shall refer to this result as the Wong-Zakai-Kallianpur-Striebel Approximation Lemma or, simply, the WZKS Lemma. To state the WZKS Lemma, we let $E_{o,m,t}^n$ be the process $\{E_{o,m,t}^n(\tau): 0 \leq \tau \leq t\}$ such that $E_{o,m,t}^n(\tau)(\eta) = \eta(\tau)$ for $\tau = \frac{kt}{n}$, $k = 0,\dots,n$, and $\tau \to E_{o,m,t}^n(\eta)(\tau)$ is linear affine on each interval $[\frac{(k-1)t}{n}, \frac{kt}{n}]$, $k = 1,\dots,n$. Then the sample paths of $E_{o,m,t}^n$ are in $H_1 \cap C_o^o([0,t],\mathbb{R}^m)$. If we define Q_t^n, $\Phi_t^{*,n}$ exactly like Q_t, Φ_t^*, but with $E_{o,m,t}$ replaced by $E_{o,m,t}^n$, then $\Phi_t^{*,n}(\eta) = \Phi_t^{*,p}(E_{o,m,t}^n(\eta))$. The WZKS Lemma then says that $\Phi_t^{*,n} \to \Phi_t^*$ a.s. This implies, in particular, that the statistical objects listed above are completely determined by their pathwise counterparts.

For $\alpha \geq 0$, $t > 0$, we say that property (E_t^α) holds if

(E_t^α) <u>for every bounded F_t-measurable function</u> $\Phi: \Omega \to \mathbb{R}$, <u>the map</u> $\Phi_t^{*,p}: H_1 \cap C_o^\alpha([0,t],\mathbb{R}^m) \to \mathbb{R}$ <u>has a continuous extension to</u> $C_o^\alpha([0,t],\mathbb{R}^m)$.

Then it follows from the WZKS Lemma that (E_t^α) implies (C_t^α) if $0 \leq \alpha < \frac{1}{2}$.

To obtain sufficient conditions for (E_t^α) to hold, we use a trick originally due to J.M. Clark. We assume that the signal process is differentiable, in the sense that there exist, for some integer $\mu \geq 0$, processes W, A, B, such that

(DIFF) W, A, B <u>are progressively measurable with respect to the</u> F_t; W <u>is</u> $\mathbb{R}^\mu$<u>-valued</u>, A <u>is</u> $\mathbb{R}^m$<u>-valued</u>, B <u>is</u> $\mathbb{R}^{m\times\mu}$<u>-valued</u>; W <u>is a standard Wiener process and a</u> F_t<u>-martingale</u>; $\int_0^t \|A(s)\| ds < \infty$ <u>a.s. for all</u> $t > 0$, $\int_0^t \|B(s)\|^2 ds < \infty$ <u>a.s. for all</u> $t > 0$, <u>and</u>

$$S(t) = S(0) + \int_0^t A(s)ds + \int_0^t B(s)dW(s) \tag{13}$$

<u>for all</u> $t > 0$.

In this case, the stochastic integral $I_1 = \int_0^t \langle S(\tau), d\eta(\tau)\rangle$ can be rewritten as

$$I_1(\omega,\eta) = [\langle\eta(t),S(t)\rangle - \int_0^t \langle\eta(\tau),dS(\tau)\rangle](\omega), \tag{14}$$

and therefore the integral is well defined (as a function of ω, modulo P-almost sure equality) for each individual path η. So we can construct a "canonical version" Φ_t^c of Φ_t^* by letting

$$\Phi_t^c(\eta) = \int_\Omega \Phi(\omega)e^{Q_t^c(\eta)(\omega)}dP(\omega) \tag{15}$$

where

$$Q_t^c(\eta)(\omega) = \langle\eta(t),S(t)(\omega)\rangle - \int_0^t \langle\eta(\tau),dS(\tau)(\omega)\rangle - \frac{1}{2}\int_0^t \|S(\tau)(\omega)\|^2 d\tau. \tag{16}$$

Clearly, Φ_t^c agrees with $\Phi_t^{*,p}$ on H_1, and so a sufficient condition for (E_t^α) to hold is that, for all Φ, the function Φ_t^c be itself continuous on $C_o^\alpha([0,t],\mathbb{R}^m)$.

The following elementary lemma shows that the continuity of the functions Φ_t^c follows if a certain estimate can be proved. We let

$$Z_t^c(\eta) = \int_\Omega e^{Q_t^c(\eta)(\omega)}dP(\omega) \tag{17}$$

(i.e. $Z_t^c = 1_t^c$).

LEMMA 1. <u>Suppose that, for some</u> $\alpha \in [0,1]$, $t > 0$, <u>the quantity</u>

$$J(\alpha,t,D) = \sup\{Z_t^c(\eta): \eta \in C_o^\alpha([0,t];\mathbb{R}^m),\ \|\eta\|_{\alpha,t} \le D\} \tag{18}$$

<u>is finite for every</u> $D > 0$. <u>Then</u> (E_t^α) <u>holds. Moreover, for every bounded</u> F_t<u>-measurable</u> $\Phi: \Omega \to \mathbb{R}$ <u>the function</u> Φ_t^c <u>is locally Lipschitzian on</u> $C_t^\alpha([0,t],\mathbb{R}^m)$.

To prove this, fix a $D > 0$ and let η, ζ satisfy $\|\eta\|_{\alpha,t} \le D$, $\|\zeta\|_{\alpha,t} \le D$. Then

$$\Phi_t^c(\eta)-\Phi_t^c(\zeta) = \int_0^1\int_\Omega R_t(\eta-\zeta,\omega)\Phi(\omega)e^{Q_t^c(\lambda\eta+(1-\lambda)\zeta)(\omega)}dP(\omega)d\lambda, \tag{19}$$

where

$$R_t(\eta-\zeta,\omega) = \int_0^t \langle S(\tau)(\omega), d(\eta-\zeta)(\tau)\rangle. \tag{20}$$

Using the inequality $|x| \le \frac{\gamma}{2}\operatorname{Cosh}(\frac{x}{\gamma})$, with $\gamma = \|\eta-\zeta\|_{\alpha,t}$, we get

$$|R_t(\eta-\zeta,\omega)e^{Q_t^c(\lambda\eta+(1-\lambda)\zeta)(\omega)}| \leq \frac{1}{2}\|\eta-\zeta\|_{\alpha,t}(e^{Q_t^c(\theta_{+,\lambda})(\omega)} + e^{Q_t^c(\theta_{-,\lambda})(\omega)}) \tag{21}$$

where

$$\theta_{\pm,\lambda} = \lambda\eta+(1-\lambda)\zeta \pm \frac{\eta-\zeta}{\|\eta-\zeta\|_{\alpha,t}} . \tag{22}$$

Therefore

$$\|\Phi_t^c(\eta)-\Phi_t^c(\zeta)\| \leq J(\alpha,t,1+D)\|\Phi\|_{L^\infty}\|\eta-\zeta\|_{\alpha,t}, \tag{23}$$

and the conclusion follows.

To prove the required estimates we use the following:

THEOREM I. <u>Assume that the signal process</u> S <u>is differentiable, and</u> A, B, W, μ <u>are such that</u> <u>Condition</u> (DIFF) <u>holds.</u> <u>Let</u> $\beta \geq 0$, $T > 0$, <u>and assume that there exist nonnegative real-valued processes</u> $\{F_i(t): 0 \leq t \leq T\}$, $i = 1,2$, <u>adapted to the</u> F_t, <u>such that</u> $F_i(t) \in L^p(\Omega, A, P)$ <u>for</u> $0 \leq t \leq T$, $i = 1,2$, $1 \leq p < \infty$, <u>and there is a constant</u> C <u>such that the inequalities</u>

$$\|A(t)\| \leq C\|S(t)\|^2 + F_1(t) \quad \text{a.s.}, \tag{24}$$

$$\|B(t)\| \leq C\|S(t)\| + F_2(t) \quad \text{a.s.} \tag{25}$$

(<u>where the norm</u> $\|M\|$ <u>of a matrix</u> $M = (M_{ij})$ <u>is defined by</u> $(\sum_{ij} M_{ij}^2)^{1/2}$),

$$\|F_1(t)\|_{L^p} \leq Cp^{\beta+1/2}, \tag{26}$$

$$\|F_2(t)\|_{L^p} \leq Cp^{\beta}, \tag{27}$$

<u>hold for</u> $0 \leq t \leq T$, $1 < p < \infty$.

<u>Then the quantity</u>

$$J^r(\alpha,t,D) = \sup\{\int_\Omega e^{rQ_t^c(\eta)(\omega)} dP(\omega): \eta \in C_o^\alpha([0,t],\mathbb{R}^m),\ \|\eta\|_{\alpha,t} \leq D\} \tag{28}$$

is finite for every $t \in [0,T]$, $D > 0$, $r \in [1,\infty[$, provided that

$$1 > \alpha > \max(0, 1 - \frac{1}{2\beta}). \tag{29}$$

<u>Proof</u>. We begin by decomposing η, for each $\rho > 0$, as a sum $\eta = \zeta_\rho+\xi_\rho$, where

$$\zeta_\rho(s) = \frac{1}{\rho}\int_{s-\rho}^{s} \eta(\tau)d\tau,$$

$$\xi_\rho(s) = \frac{1}{\rho}\int_{s-\rho}^{s} [\eta(s)-\eta(\tau)]d\tau.$$

Assume that $\eta \in C_o^\alpha([0,t],\mathbb{R}^m)$, and let $K = \|\eta\|_{\alpha,t}$. Then the function ζ_ρ is of class C^1, and satisfies $\|\dot{\zeta}_\rho(s)\| \leq K\rho^{\alpha-1}$, while ξ_ρ is continuous and satisfies $\|\xi_\rho(s)\| \leq (1+\alpha)^{-1}K\rho^\alpha$.

If $\phi: [0,t] \to \mathbb{R}^m$ is an arbitrary continuous function such that $\phi(0) = 0$, we can define

$$I_1(\omega,\phi) = \int_0^t \langle S(\tau),d\phi(\tau)\rangle(\omega), \tag{30}$$

where the integral is, by definition, what results after integrating by parts. Also, we can define

$$I_2(\omega,\phi) = \int_0^t \langle\phi(\tau),dS(\tau)(\omega)\rangle, \tag{31}$$

$$I_3(\omega,\phi) = \int_0^t \langle\phi(\tau),S(\tau)(\omega)\rangle d\tau, \tag{32}$$

and

$$I_4(\omega) = \int_0^t \|S(\tau)(\omega)\|^2 d\tau. \tag{33}$$

Then $Q_t^c(\eta)(\omega) = I_1(\omega,\eta) - \frac{1}{2}I_4(\omega)$. We decompose $I_1(\omega,\eta)$ into the sum of $I_1(\omega,\zeta_\rho)$ and $I_1(\omega,\xi_\rho)$. Since ζ_ρ is of class C^1, $I_1(\omega,\zeta_\rho)$ equals $I_3(\omega,\dot{\zeta}_\rho)$. We rewrite $I_1(\omega,\xi_\rho)$ using integration by parts, and get

$$I_1(\omega,\xi_\rho) = \langle\xi_\rho(t),S(t)(\omega)\rangle - I_2(\omega,\xi_\rho). \tag{34}$$

The term $\langle\xi_\rho(t),S(t)(\omega)\rangle$ satisfies

$$\langle\xi_\rho(t),S(t)\rangle = \int_0^t \langle\frac{\tau\xi_\rho(t)}{t}, dS(\tau)\rangle - \int_0^t \langle\frac{\xi_\rho(t)}{t}, S(\tau)\rangle d\tau. \tag{35}$$

Therefore, if we let (for a fixed t)

$$\theta_\rho(s) = \dot{\zeta}_\rho(s) - \frac{\xi_\rho(t)}{t},\ \psi_\rho(s) = \frac{\tau\xi_\rho(t)}{t} - \xi_\rho(s) \tag{36}$$

we get

$$Q_t^c(\eta)(\omega) = I_2(\omega,\psi_\rho) + I_3(\omega,\theta_\rho) - \frac{1}{2}I_4(\omega). \tag{37}$$

The function θ_ρ satisfies $\|\theta_\rho(s)\| \leq 2K\rho^{\alpha-1}$ if $0 < \rho \leq t$. From this it follows easily that

$$I_3(\omega,\theta_\rho) - \frac{1}{4}I_4(\omega) \leq 4K^2\rho^{2\alpha-2} \quad \text{if} \quad \rho \leq t. \tag{38}$$

To estimate $I_2(\omega,\psi_\rho) - \frac{1}{4}I_4(\omega)$ we use $dS = Ad\tau + BdW$. Then $I_2(\omega,\psi_\rho) = I_{2,1}(\omega,\psi_\rho) + I_{2,2}(\omega,\psi_\rho)$, where

$$I_{2,1}(\omega,\psi_\rho) = \int_0^t \langle\psi_\rho(\tau), A(\tau)(\omega)\rangle d\tau \tag{39}$$

$$I_{2,2}(\omega,\psi_\rho) = \int_0^t \langle\psi_\rho(\tau), B(\tau)dW(\tau)\rangle(\omega). \tag{40}$$

In view of (24), we can decompose A as $A^*+A^\#$, where, for $0 \leq \tau \leq t$

$$\|A^*(\tau)\| \leq C\|S(\tau)\|^2, \quad \|A^\#(\tau)\| \leq F_1(\tau) \quad \text{a.s.} \tag{41}$$

This gives a decomposition $I_{2,1} = I^*_{2,1} + I^\#_{2,1}$, where $I^*_{2,1}$, $I^\#_{2,1}$ are obtained from $I_{2,1}$ by substituting A^*, $A^\#$ for A. Then it is easy to verify (using the bound $\|\psi_\rho(\tau)\| \leq 2K(1+\alpha)^{-1}\rho^\alpha$) that

$$I^*_{2,1}(\omega,\psi_\rho) - \frac{1}{8}I_4(\omega) \leq \left(\frac{2KC\rho^\alpha}{1+\alpha} - \frac{1}{8}\right)\int_0^t \|S(\tau)(\omega)\|^2 d\tau. \tag{42}$$

In particular, we have

$$I^*_{2,1}(\omega,\psi_\rho) - \frac{1}{8}I_4(\omega) \leq 0 \quad \text{if} \quad \rho^\alpha \leq \frac{1+\alpha}{16KC}. \tag{43}$$

A similar decomposition $B = B^* + B^\#$, with $\|B^*(\tau)\| \leq C\|S(\tau)\|$ a.s. and $\|B^\#(\tau)\| \leq F_2(\tau)$ a.s., can be obtained for B. This gives rise to a decomposition $I_{2,2} = I^*_{2,2} + I^\#_{2,2}$. Let $H_\rho(\omega) = I^*_{2,2}(\omega,\psi_\rho) - \frac{1}{8}I_4(\omega)$. It is not possible to obtain a bound for H_ρ similar to those in Formulas (42), (43). However, one can prove that $e^{\nu H_\rho}$ is integrable for any $\nu > 0$, as long as ρ is smaller than some $\bar{\rho}(\nu)$. To see this, write $H_\rho = H^1_{\rho_1} + H^2_{\rho_1}$, where

$$H^1_{\rho,\nu} = \int_0^t \langle\psi_\rho(\tau), B^*(\tau)dW(\tau)\rangle - \frac{\nu}{2}\int_0^t \sum_{j=1}^{\mu}\left(\sum_{i=1}^{m} \psi_{\rho,i}(\tau)B^*_{i,j}(\tau)\right)^2 d\tau, \tag{44}$$

and $\{\psi_{\rho,i}: i = 1,\dots,m\}$, $\{B^*_{i,j}: i = 1,\dots,m,\ j = 1,\dots,\mu\}$ are, respectively, the components of the vector ψ_ρ and the matrix B^*.

It is then easy to see that $\mathbb{E}(e^{\nu H^1_{\rho,\nu}}) \leq 1$. On the other hand, $H^2_{\rho,\nu}$ is pointwise bounded above by $\int_0^t \left[\frac{2\nu K^2\rho^{2\alpha}}{(1+\alpha)^2}\|B^*(\tau)\|^2 - \frac{1}{8}\|S(\tau)\|^2\right]d\tau$. Therefore

$$H^2_{\rho,\nu} \leq 0 \quad \text{a.s.} \quad \text{if} \quad \rho^\alpha \leq \frac{1+\alpha}{4\nu^{\frac{1}{2}}kC}. \tag{45}$$

It then follows that

$$E(e^{\nu H_\rho}) \leq 1 \quad \text{if} \quad \rho^\alpha \leq \frac{1+\alpha}{4\nu^{\frac{1}{2}}KC}. \tag{46}$$

In particular, if we apply Chebischev's inequality, we get

$$P(e^{\nu H_\rho} \geq e^{\frac{\nu\lambda}{3}}) \leq e^{-\frac{\nu\lambda}{3}}, \tag{47}$$

i.e.

$$P(H_\rho \geq \frac{\lambda}{3}) \leq e^{-\frac{\nu\lambda}{3}}, \tag{48}$$

provided that

$$\rho^\alpha \leq \frac{1+\alpha}{4\nu^{\frac{1}{2}}KC}. \tag{49}$$

Clearly,

$$Q_t^c(\eta) = (I_3(\cdot,\theta_s) - \frac{1}{4}I_4) + (I_{2,1}^*(\cdot,\psi_3) - \frac{1}{8}I_4) + H_\rho + \tilde{H}_\rho, \tag{50}$$

where

$$\tilde{H}_\rho = I_{2,1}^{\#}(\cdot,\psi_\rho) + I_{2,2}^{\#}(\cdot,\psi_\rho). \tag{51}$$

The first three terms have already been controlled by inequalities (38), (43) and (48). We now control $\tilde{H}_\rho$ by estimating its distribution function. First, it follows from (26) that

$$\|I_{2,1}^{\#}(\cdot,\psi_\rho)\|_{L^p} \leq \frac{2KC\rho^\alpha t p^{\beta+\frac{1}{2}}}{1+\alpha}. \tag{52}$$

To estimate the L^p norm of $I_{2,2}^{\#}(\cdot,\psi_\rho)$ we use (27) and the Burkholder-Hunt-Gundy inequality. We get, for $p \geq 2$:

$$\|I_{2,2}^{\#}(\cdot,\psi_\rho)\|_{L^p} \leq \frac{2KC\, m\mu\rho^\alpha t^{\frac{1}{2}} p^{\beta+\frac{1}{2}}}{1+\alpha}, \tag{53}$$

so that

$$\|\tilde{H}_\rho\|_{L^p} \leq K_1 \rho^\alpha p^{\beta+\frac{1}{2}} \quad \text{if} \quad p \geq 2, \tag{54}$$

where $K_1 = \frac{2K}{1+\alpha}(Ct + C\, t^{\frac{1}{2}})$.

We pick a $\lambda > 0$, a $\rho > 0$, and then we let

$$p = \lambda^\gamma K_1^{-\gamma} \rho^{-\alpha\gamma} e^{-1} \tag{55}$$

where $\gamma = (\frac{1}{2}+\beta)^{-1}$. We then apply Chebischev's inequality, and conclude that

$$P(\tilde{H}_\rho \geq \frac{\lambda}{3}) \leq e^{-K_2\lambda^\gamma\rho^{-\alpha\gamma}}, \tag{56}$$

provided that (54) can be applied, i.e. provided that the p given by (55) is ≥ 2. (The constant K_2 is equal to $3^{-\gamma}(\beta+\frac{1}{2})K_1^{-\gamma}e^{-1}$.)

We now choose ρ as a function of λ. Specifically, we let

$$\rho = \left(\frac{\lambda}{12K^2}\right)^{-\delta}, \quad \delta = (2-2\alpha)^{-1}. \tag{57}$$

Then (38) becomes

$$I_3(\omega,\theta_\rho) - \frac{1}{4}I_4(\omega) \leq \frac{\lambda}{3}, \tag{58}$$

while (56) gives

$$P(\tilde{H}_\rho \geq \frac{\lambda}{3}) \leq e^{-K_3\lambda^\sigma}, \tag{59}$$

where

$$\sigma = \frac{2-\alpha}{(1-\alpha)(2\beta+1)}, \quad K_3 = K_2(12K^2)^{-\alpha\gamma\delta}. \tag{60}$$

Next we apply (48) with ν chosen so that (49) is an equality. We get

$$P(H_\rho \geq \frac{\lambda}{3}) \leq e^{-K_4\lambda^{1+2\alpha\delta}}, \tag{61}$$

where $K_4 = (1+\alpha)^2(4KC)^{-2}(12K^2)^{-2\alpha\delta}$.

If we use (50) together with (58), (59), (61) and (43), we get

$$P(Q_t^c(\eta) \geq \lambda) \leq e^{-K_3\lambda^\sigma} + e^{-K_4\lambda^{1+2\alpha\delta}}. \tag{62}$$

Naturally, the conditions for the validity of (62) are those required for (38), (56) and (43) to be applicable. These conditions turn out to amount to the requirement that $\lambda \geq \bar{\lambda}$, where

$$\bar{\lambda} = \max([2eK_1^\gamma(12K^2)^{\alpha\delta\gamma}]^{1/\sigma}, 12K^2t^{-1/\delta}, 12K^2[\frac{16KC}{1+\alpha}]^{\frac{2}{\alpha}-2}). \tag{63}$$

Hypothesis (29) implies that $\sigma > 1$. This fact, together with (62), imply that $e^{Q_t^c(\eta)}$ is integrable, and that its integral over $Q_t^c(\eta) \geq \bar{\lambda}$ is bounded by a constant K_5 which remains bounded as long as K_3 and K_4 are bounded below. The formulas obtained above for K_1, K_2, K_3, K_4 and $\bar{\lambda}$ show that K_3 and K_4 are bounded below as long as K is bounded, and $\bar{\lambda}$ is bounded if K is bounded. Since

$$\mathbb{E}(e^{Q_t^c(\eta)}) \leq e^{\bar{\lambda}} + K_5, \tag{64}$$

the desired conclusion follows for $r = 1$. The proof for $r > 1$ is similar.

Q.E.D.

With slightly stronger hypotheses, it is possible to prove a theorem for $\alpha = 0$ as well, but we shall not do it here.

We now illustrate the use of Theorem I by considering the case when the signal is a scalar polynomial function of some Wiener process, i.e.

$$S(t) = h(X_1(t),\dots,X_n(t)) \tag{65}$$

where $h: \mathbb{R}^n \to \mathbb{R}$ is a polynomial function and $\{(X_1(t),\dots,X_n(t)): t \geq 0\}$ is a standard Wiener process. In this case, A and B are also polynomials in $X_1(t),\dots,X_n(t)$, given by

$$A(t) = \frac{1}{2}(\Delta h)(X_1(t),\dots,X_n(t)) \tag{66}$$

$$B(t) = (\nabla h)(X_1(t),\dots,X_n(t)). \tag{67}$$

If $\|\nabla h\|$ is bounded by some constant times $1+|h|$, and $|\Delta h|$ by some constant times $1+h^2$, then the conditions of Theorem I hold with $\beta = 0$, and (E_t^α) holds for $0 < \alpha \leq 1$, $t > 0$. (Actually, it can be proved in this case that (E_t^o) holds as well.) This happens, in particular, for arbitrary h, if $n = 1$. If $n > 1$, then the conditions of the theorem hold if, for instance, h is of the form $x_1^{2m}+\dots+x_n^{2m}$ for some positive integer m.

An interesting example where Theorem I does not yield a satisfactory result is the "two dimensional cubic sensor". Here $S(t) = X_1(t)^3+X_2(t)^3$, where $X = (X_1,X_2)$ is a standard two-dimensional Wiener process. The conditions of Theorem I are satisfied with $\beta = 1$. This implies that (E_t^α) holds for $\alpha > \frac{1}{2}$, but then one is not yet in a position to construct continuous versions of the Φ_t^*.

Actually, it turns out that, for the two-dimensional cubic sensor, the situation is as bad as the preceding considerations suggest. Indeed, we have

THEOREM II. _For the two-dimensional cubic sensor,_ (E_t^α) _does not hold if_ $t > 0$, $0 \leq \alpha < \frac{1}{2}$.

The proof of Theorem II will be omitted, since it is quite long. The idea is to construct for an arbitrary C^1 function η, a sequence $\{\eta^n\}$ of piecewise C^1 functions that converge to η in the C^α norm but not in the C^1 norm, and are such that $Z_t^c(\eta^n) \to +\infty$ as $n \to \infty$. To make $Z_t^c(\eta^n)$ go to ∞, one makes sure that the maximum value of the exponential whose integral is $Z_t^c(\eta^n)$ is quite large, and that the exponential remains large on a Wiener tube about the maximizing path whose measure is not too small.

S. Varadhan (personal communication) has recently suggested an argument which proves, among other things, that if Φ_t^* has a continuous version, then $\Phi_t^{*,p}$ is the restriction of this version to the set of H_1 paths, and so $\Phi_t^{*,p}$ has a continuous extension. Using this, Theorem II yields the stronger assertion that, under the same conditions, (C_t^α) also fails to hold.

Properties (C_t^α), (E_t^α) have obvious analogues $(\hat{C}_t^\alpha)$, $(\hat{E}_t^\alpha)$ for the normalized statistics. It is clear that, whenever (C_t^α) (or (E_t^α)) holds, then $(\hat{C}_t^\alpha)$ (or $(\hat{E}_t^\alpha)$) must hold as well. But it is conceivable that (E_t^α) might fail to hold for some problem because the normalization constant $Z_t^c(\eta)$ is badly behaved, even though $(\hat{E}_t^\alpha)$ holds. It would be nice to find examples where $(\hat{E}_t^\alpha)$ fails to hold. We suspect this happens for the two-dimensional cubic sensor, but we have not yet been able to prove it.

5. CONTROL THEORY

HOMOGENIZATION OF BELLMAN EQUATIONS

A. Bensoussan (*) L. Boccardo (**) F. Murat (***)

INTRODUCTION

Homogenization concerns the behaviour of the solution of partial differential equations containing highly oscillatory coefficients (see A. Bensoussan - J.L. Lions G. Papanicolaou [3] for a presentation of the theory). This aspect has a probabilistic counterpart, in the case of elliptic or parabolic 2nd order equations. Indeed the solution of the P.D.E. is then interpreted as the value of a functional computed along the trajectory of a stochastic differential equation. This equation will have highly oscillatory drift and variance terms. We consider here a similar situation with control and we are interested in the infimum of a set of functionals. From the P.D.E. point of view, this amounts to studying the homogenization of non linear elleptic or parabolic equations of Bellman type. The non linearity occurs only on the gradient or the function itself. Situations of this sort have been considered already in B.L.P., loc. cit., A. Bensoussan [1] and L. Boccardo - F. Murat [4]. The study is limited to the case of operators written in divergence form. In the first two references only non linearities with growth are considered. In the third reference, general quadratic growth is permitted.

We consider here the situation of operators not written in divergence form, with quadratic growth non linearities. This situation is quite natural in the context of stochastic control problems where the divergence form operators do not come in naturally.

The extension is not evident and not quite intuitive, although the result that we obtain is natural. The case of linear operators not written in divergence form has already been considered in B.L.P. and in A. Bensoussan, loc. cit. There, the importance of the role played by the formal adjoint operator has been stressed (the divergence form operators appear as self adjoint operators and thus things simplify a lot).

It is thus natural to expect similar considerations for the non linear operators. Our limit results are obtained by P.D.E. methods, the treatment of the non linearities being delicate require some sophisticated tools of analysis (namely MEYERS estimate, cf. [7]). However, the limit problem can also been interpreted as a Bellman equation. Therefore the limit function is also the infimum of a set of functionals computed along the trajectory of a process, which is the solution of a stochastic differential equation. This provides a useful approximated device for the control of diffusion processes with highly oscillatory drift and variance terms.

1 - STATEMENT OF THE PROBLEM

1.1 - Assumptions-Notation

Consider function $a_{ij}(x,y)$, $b_i(x,y)$ in $(R^n)^2$ such that

(1.1) a_{ij}, b_i are of class C^3 with bounded derivates of all orders

(1.2) $a_{ij} = a_{ji}$, $a_{ij}\ \xi_i\ \xi_j \geq \alpha|\xi|^2 \quad \alpha > 0, \forall \xi \in R^n$

(using the summation convention)

(1.3) a_{ij}, b_i are periodic in y with period 1 in all components.

The regularity assumptions made above are more than necessary. Only some third derivatives will be necessary in the sequel.

We shall consider the 2nd order differential operator

(1.4) $$A^\varepsilon = -a_{ij}(x,\frac{x}{\varepsilon})\frac{\partial^2}{\partial x_i \partial x_j} - \frac{1}{\varepsilon} b_i(x,\frac{x}{\varepsilon})\frac{\partial}{\partial x_i}$$

which is not given in divergence form. As a consequence the symmetry of the matrix a_{ij} can always be assumed, without loss of generality. This operator is associated to a diffusion. Let us denote by $\sigma(x,t)$ the square root of the matrix a_{ij}. Consider a probability space $(\Omega,\mathcal{A},P)$, a filtration F^t and a standard F^t Wiener process with values in R^n. We solve the equation

(1.5) $$dy^\varepsilon = \frac{1}{\varepsilon} b(y^\varepsilon, \frac{y^\varepsilon}{\varepsilon})dt + \sigma(y^\varepsilon, \frac{y^\varepsilon}{\varepsilon})\,dw$$

$$y^\varepsilon(0) = x.$$

This can be done in a strong sense by virtue of the regularity assumptions which have been made. The drift and variance terms are highly oscillatory because of the term $\frac{y^\varepsilon}{\varepsilon}$ (non uniformly since the functions depend also on y^ε). Besides the drift is very large.

Consider next :

(1.6) U_{ad} non empty subset of a metric space U

(1.7) $g(x,y,v) : R^n \times R^n \times U_{ad} \to R$, periodic in y, continuous

$$|g| \leq \bar{g}(1 + |v|)$$

We then introduce a control to modify the trajectory (1.5). An admissible control v(.) is a process adapted to F^t, with values in U_{ad}, with a.s. bounded values. Note that the bound on the values of the control may depend on the particular control considered.

For a particular control v(.) define the process

(1.8) $$\alpha^\varepsilon(t) = \sigma^{-1}(y^\varepsilon(t), \frac{y^\varepsilon(t)}{\varepsilon})\, g(y^\varepsilon(t), \frac{y^\varepsilon(t)}{\varepsilon}, b(t)).$$

because of the assumption (1.1),(1.2) and (1.7) the process $\alpha^\varepsilon(t)$ is a.s. bounded by a bound depending on the control). Therefore we can define the Girsanov transformation.

(1.9) $$\left.\frac{d\,P^\varepsilon_{v(.)}}{dP}\right|_{F^t} = \exp\left\{\int_0^t \alpha^\varepsilon(s)\, dw(s) - \frac{1}{2}\int_0^t |\alpha^\varepsilon(s)|^2 ds\right\}$$

considering the process

(1.10) $$w^\varepsilon(t) = w(t) - \int_0^t \alpha^\varepsilon(s)\, ds$$

then on the probability space $(\Omega, \mathcal{A}, P^\varepsilon_{v(.)})$ the process w becomes a standard F^t Wiener and the process $y^\varepsilon(t)$ appears as the solution of the equation

(1.11) $$dy^\varepsilon = [g(y^\varepsilon, \frac{y^\varepsilon}{\varepsilon}, v) + \frac{1}{\varepsilon}\, b\,(y^\varepsilon, \frac{y^\varepsilon}{\varepsilon})]dt + \sigma(y^\varepsilon, \frac{y^\varepsilon}{\varepsilon})\, dw$$

$$y^\varepsilon(0) = x.$$

Let $\mathcal{O}$ be a smooth domain of R^n and τ^ε denote the 1st exit time of $y^\varepsilon(t)$ from the domain $\mathcal{O}$

Let us consider functions

(1.12) $\ell(x,y,v) : R^n \times R^n \times U_{ad} \to R$, periodic in y, continuous

$$\ell \geq \ell_0\, |v|^2 - \ell_1 \qquad \ell_0 > 0 \quad , \quad \ell_1 \geq 0$$

$$\ell \leq \bar{\ell}_0\,(1+|v|^2)$$

(1.13) $c(x,y,v) = R^n \times R^n \times U_{ad} \to R$ periodic in y continuous and bounded

$$0 < \beta \leq c \leq \bar{c}$$

We can then define the pay off function

(1.14) $$J_x^\varepsilon(v(.)) = E_{v(.)}\left[\int_0^{\tau^\varepsilon} \ell(y^\varepsilon, \frac{y^\varepsilon}{\varepsilon}, v)\ (\exp - \int_0^t c(y^\varepsilon, \frac{y^\varepsilon}{\varepsilon}, v)ds)dt\right]$$

and set

(1.15) $$u^\varepsilon(x) = \operatorname*{Inf}_{v(.)} J_x^\varepsilon(v(.))$$

1.2 - The problem

It consists in studying the behaviour of $u^\varepsilon(x)$ as ε tends to 0. The function $u^\varepsilon(x)$ is solution of Bellman equation which we write now.

Define the Hamiltonian

(1.16) $$H(x,y,\lambda,p) = \operatorname*{Inf.}_{v} \{\ell(x,y,v) + p.g.(x,y,v) - \lambda c(x,y,v)\}$$

Then u^ε is the solution of

(1.17) $$A^\varepsilon u^\varepsilon = H(x, \frac{x}{\varepsilon}, u^\varepsilon, Du^\varepsilon)$$

$$u^\varepsilon|_\Gamma = 0$$

where Γ is the boundary of $\mathcal{O}$

From LADYZHENSKAJA-URAL'TSEVA [6], we can assert that for any $\varepsilon > 0$, there exists a unique solution of (1.17) which belongs to $W^{2,p}(\mathcal{O}), \forall\, 2\in]1,\infty[$.
We are going to obtain the limit of u^ε as ε tends to 0, by P.D.E. methods. The limit u will also be the solution of a Bellman equation and thus will correspond to the value function of a stochastic control problem. From indications due to VARADHAN, a probabilistic approach is possible in more simple cases, namely the case when U_{ad} is compact.

2 - PREMILINARIES - INVARIANT MEASURES

2.1 - Invariant measures

Let us consider the 2nd order differential operator (in y)

(2.1) $$A = A_x = -a_{ij}(x,y) \frac{\partial^2}{\partial y_i \partial y_j} - b_i(x,y) \frac{\partial}{\partial y_i}$$

and its formal adjoint

$$(2.2) \qquad A^* = -\frac{\partial}{\partial y_i \partial y_j}(a_{ij}(x,y).) + \frac{\partial}{\partial y_i}(b_i(x,y).)$$

$$= -\frac{\partial}{\partial y_i \partial y_j}(a_{ij}(x,y)\frac{\partial}{\partial y_j}\cdot) + \frac{\partial}{\partial y_i}((b_i - \frac{\partial a_{ij}}{\partial y_i}).)$$

For any fixed x, we solve the problem

$$(2.3) \qquad A^* m = 0\,,\ y \to m(x,y) \in W^{2,p}_{loc}\,,\ p \geq 2,\ p < \infty$$

$$\text{periodic, } m > 0, \int_Y m(x,y)\,dy = 1$$

By the regularity of the coefficients the function m is continuous in x,y. Since we consider only the values of x in $\bar{\mathcal{O}}$, we may assume without loss of generality that

$$(2.4) \qquad 0 < \bar{m} \leq m(x,y) \leq M.$$

The function m defines for any x, a probability density on Y called the invariant measure (cf. J.L. Doof [5] or BLP, loc. cit.)

Note that when $b_i = \frac{\partial a_{ij}}{\partial y_j}$ (the divergence case) then m = 1

2.2 - The main assumption

In the sequel, we shall need the fundamental assumption

$$(2.5) \qquad \int_Y m(x,y)\ b(x,y)\,dy = 0$$

Note that this assumption is automatically satisfied in the divergence case.

Let us set

$$(2.6) \qquad \begin{aligned} \tilde{b}(x,y) &= b(x,y)m(x,y) \\ \tilde{a}_{ij}(x,y) &= a_{ij}(x,y)\ m(x,y) \\ \tilde{\beta}_i(x,y) &= \tilde{b}_i(x,y) - \frac{\partial}{\partial y_j}(\tilde{a}_{ij}) \end{aligned}$$

It is easy to check that

$$(2.7) \qquad \frac{\partial}{\partial y_i} \tilde{\beta}_i(x,y) = 0$$

Now, by virtue of the assumption (2.5) we can assert that there exist functions (called "correctors" as usual in homogenization) defined by

$$(2.8) \qquad A\, X^{\ell} = - b_{\ell}\,(x,y)$$

$$y \to X^{\ell}(x,y) \text{ periodic}, \quad \int_Y X^{\ell}(x,y)\,dy = 0$$

The functions X satisfy the following regularity properties

$$(2.9) \qquad X^{\ell} \;,\; \frac{\partial}{\partial} \quad \text{are of class } C^2$$

3 - CONVERGENCE RESULTS

3.1 - Notation

Let us introduce the homogenized Hamiltonian

$$(3.1) \qquad \mathcal{H}(x,\lambda,p) = \int_Y m(x,y)\; H\;(x,y,\lambda(I-DX)p)\,dy$$

where DX denotes the matrix $\frac{\partial X^{\ell}}{\partial y_i}$ (i.e $D\,X.\,p = \frac{\partial X^{\ell}}{\partial y_j} p^{\ell}$)

Define also the homogenized coefficients

$$(3.2) \qquad a_{ij}(x) = \int_Y m(x,y)\; [a_{ij}(x,y) - a_{i\ell} \frac{\partial X^j}{\partial y_{\ell}} - a_{j\ell} \frac{\partial X^i}{\partial y_{\ell}} - \frac{1}{2}(b_i X^j + b_j X^i)]\; dy$$

$$(3.3) \qquad r_j(x) = - \frac{\partial}{\partial x_i} \int_Y m(x,y)\;(a_{ij}(x,y) - a_{i\ell} \frac{\partial X^{\ell}}{\partial y_{\ell}})\; dy$$

It can be checked easily (cf. B.L.P.) that

$$(3.4) \qquad q_{ij}(x) = \int m[a_{ij} - a_{i\ell} \frac{\partial X^j}{\partial y_{\ell}} - a_{j\ell} \frac{\partial X^i}{\partial y_{\ell}} + a_{k_{\ell}} \frac{\partial X^j}{\partial y_{\ell}} \frac{\partial X^i}{\partial y_{\ell}}]\; dy$$

and thus a_{ij} is uniformly positive definite.

Consider then the 2nd order elliptic operator

$$(3.5)\qquad \mathcal{A} = - a_{ij}(x) \frac{\partial^2}{\partial x_i \partial x_j} - r_j(x) \frac{\partial}{\partial x_j}$$

and the Dirichlet problem

$$(3.6)\qquad \mathcal{A}u = \mathcal{H}(x,y,Du)\ ,\quad u|\Gamma = 0$$

There exists one and only one solution of (3.6)

$$\text{in } W^{2,p}(\mathcal{O}) \quad \forall\, p \in [2,\infty[$$

3.2 - Statement of the result

The main result is the following

Theorem 3.1

Assume (1.1), (1.2), (1.3), (1.6), (1.7), (1.12), (1.13), (2.5).

Then

$$(3.7)\qquad u^\varepsilon \to u \ \text{ in } W^{1,p_0} \text{ weakly, for some } p_0 \in \,]2,\infty[\text{ and in } L^\infty \text{ weakly star}$$

$$(3.8)\qquad \frac{\partial u^\varepsilon}{\partial x_j} - \frac{\partial u}{\partial x_j} + \frac{\partial u}{\partial x_\ell} \frac{\partial X^\ell}{\partial y_j} \to 0 \ \text{ in } L^2 \text{ strongly}$$

3.3 - Interpretation of the limit problem

It is of course important from the point of view of stochastic control to interpret the limit function u as the value function of a stochastic control problem. This can be done easily, after recognizing that $\mathcal{H}$ can be written as an Hamiltonian.

From the definition (3.1) we have

$$(3.9)\qquad \mathcal{H}(x,\lambda,p) = \int_Y m(x,y) \operatorname*{Inf}_v [\ell(x,y,v) + (p_i - \frac{\partial X^\ell}{\partial y_i} p^\ell)\, g_i(x,y,v) - \lambda c(x,y,v)]$$

Introduce the set $\mathcal{V}$ of Borel function $v(y) : Y \to U$ and $\mathcal{V}_{ad}$ the subset of functions with values in $\mathcal{V}_{ad}$.

One can notice the following algebraic relation

$$(3.10)\qquad \inf_{v(.)\in\mathcal{V}_{ad}} \int_Y F(y,v(y))\,dy \qquad \int_Y \operatorname*{Inf}_{v\in U_{ad}} F(y,v)\,dy$$

which holds for any function F which is continuous. Making use of this remark and defining the functions

$$(3.11)\qquad \tilde{\ell}(x,v(.)) = \int_Y \ell(x,y,v(y))\,m(x,y)\,dy$$

$$(3.12)\qquad \tilde{g}_i(x,v(.)) = \int_Y [g_i(x,y,v)) - \frac{\partial x^i}{\partial y_\ell}\,g_\ell\,(x,y,v(y))]\,m(x,y)\,dy$$

$$(3.13)\qquad \tilde{c}\,(x,v(.)) = \int_Y c(x,y,v(y))\,m(x,y)\,dy$$

We can write $\mathcal{H}$ as follows

$$(3.14)\qquad \mathcal{H}(x,\lambda,p) = \operatorname*{Inf}_{v(.)} \{\tilde{\ell}(x,v(.))+p.\,\tilde{g}(x,v(.)) - \lambda\tilde{c}(x,v(.))\}$$

Therefore $\mathcal{H}$ can indeed be interpreted as a Hamiltonian. The main differences between H and $\mathcal{H}$ lies in the set of admissible controls. Instead of $v \in U_{ad}$, one needs to consider $\mathcal{V}_{ad}$, the set of Borel functions from Y into U_{ad}.

Then, $\tilde{\ell}$, $\tilde{c}$ are the average of the functions ℓ, c in y using the function m(x,y). The averaging of g is more delicate and involves the correctors.

The equation (3.6) is thus the Bellman equation of a stochastic control problem which we can easily describle.

Consider the diffusion process related to the operator $\mathcal{A}$,

$$(3.15)\qquad \begin{aligned} dx &= r(x)\,dt + \sqrt{2q(x)}\,dw \\ x(0) &= x \end{aligned}$$

constructed on a convenient probability space $(\Omega,\mathcal{A},P)$ equipped with a filtration F^t, in which a F^t Wiener process w(t) is given.

An admissible control is now a process adapted to F^t, with values in U_{ad}. Therefore it is in fact a stochastic process depending on y, v(t,y) taking values in U_{ad}. Given such a process v(t,.) define

$$(3.17) \qquad \frac{dP_{v(.)}}{dP}\Big|_{F^t} = \exp\left\{ \int_0^t \alpha(s).dw(s) - \frac{1}{2}\int |\alpha(s)|^2 ds\right\}$$

Considering the process $\tilde{w}(t)$ defined by

$$(3.18) \qquad \tilde{w}(t) = w(t) - \int_0^t \alpha(s)\,ds$$

it follows that on $(\Omega, \mathcal{A}, P_{v(.)})$, w is a F^t Wiener and x appears as the solutions of

$$(3.19) \qquad dx = (\tilde{g}\ (x(t)\ ;\ v(t,.)) + r(x,t))\,dt + \sqrt{2q(x(t))}\,d\tilde{w}$$

$$w(0) = x$$

Call the 1st exit time of x(t) from the domain $\mathcal{O}$ and define the pay off

$$(3.20) \qquad J_x(v(.) = E_{v(.)}\Big[\int_0^\tau \tilde{\ell}(x(t);v(t,.)) \left(\exp - \int_0^t \tilde{c}(x(s);v(s;.))\,ds\right)dt\Big]$$

The solution u(x) of (3.6) can be interpreted as

$$(3.21) \qquad u(x) = \operatorname*{Inf}_{v(.)}\ J_x(v(.))$$

<u>Remark 3.1</u>

Although theorem 3.1 gives only the convergence of the value function, it is important to give an interpretation of the optimal feedback for the limit problem.

Define in (1.16), $\hat{V}(x,y,\lambda,p)$ to be the optimal v realizing the infimum (in fact pick a Borel selection), then the optimal feedback for the ε problem is

$$\hat{V}(x,\frac{x}{\varepsilon},u^\varepsilon,Du^\varepsilon) = \hat{v}^\varepsilon(x)$$

On the other hand from (3.9) one deduces that

$$\hat{v}(x,y) = \hat{V}(x,y,y(x),(I-D_y X(x,y))^* Du(x))$$

is the optimal feedback for the limit problem (note it is necessarily a function of x,y).

4 - SOME IDEAS ON THE PROOF OF THEOREM 3.1.

We shall not present here the full proof for which we refer to A. Bensoussan - L. Boccardo - F. Murat [2]. We shall only present the main steps.

4.1 - A priori estimates

One starts with

Lemma 4.1 : There exists a constant M such that

$$\|u^\varepsilon\|_{L^\infty} \leq \frac{M}{\beta} \tag{4.1}$$

This follows from standard maximum principle arguments, or from the interpretation (1.15) of u^ε, using the assumptions on ℓ and c.

Lemma 4.2 : The following estimate holds

$$\|u^\varepsilon\|_{H^1_0} \leq C \tag{4.2}$$

Proof :

Multiply the equation (1.17) by m. After some rewriting and using the functions defined in (2.6), we obtain

$$- \frac{\partial}{x_i}(\tilde{a}^\varepsilon_{ij} \frac{\partial u^\varepsilon}{\partial x_j}) - \frac{1}{\varepsilon} \tilde{\beta}^\varepsilon_i \frac{\partial u^\varepsilon}{\partial x_i} + \frac{\partial u^\varepsilon}{\partial x_i} \frac{\partial}{\partial x_i} \tilde{a}_{ij} \tag{4.3}$$

$$= m\, H(x, \frac{x}{\varepsilon}, u^\varepsilon, Du^\varepsilon)$$

where the notation

$$\tilde{a}^\varepsilon_{ij}(x) = \tilde{a}_{ij}(x, \frac{x}{\varepsilon})$$

has been used. We then test (4.3) with the function $v = e^{\lambda(u^\varepsilon)^2} u^\varepsilon$, where λ is a convenient parameter. The L^2 norm of $|Du^\varepsilon|$ can then be estimated.

Lemma 4.3 : There exists $p_0 > 2$ such that

(4.4) $$\|u^\varepsilon\|_{W^{1,p_0}} \leq C$$

This requires more sophisticated techniques of Analysis. We refer to B.B.M. for details.

4.2 - Proof of Theorem 3.1

Pick a subsence of u^ε , still denoted u^ε such that

(4.5) $$u^\varepsilon \rightharpoonup u \quad \text{in } W^{1,p_0} \text{ weakly and } L^\infty \text{ weak star}$$

then one proves that

(4.6) $$\frac{\partial u^\varepsilon}{\partial x_j} - \frac{\partial u}{\partial x_j} + \frac{\partial u}{\partial x} \frac{\partial x^\ell}{\partial y_j} \to 0 \qquad \text{in } L^2 , \forall j$$

and

(4.7) $$m(x,\frac{x}{\varepsilon})\ H(x,\frac{x}{\varepsilon},u^\varepsilon,Du^\varepsilon) \to \mathscr{H}(x,u,Du)$$

in the sense of measures.

These results allow to pass to the limit in (1.7) and one can then see that u is a solution of (3.6).

By the uniqueness of the solution of (3.6), the convergence result follows.

Remark 4.1 :

It is possible also to pass to the limit without using the sharp estimate of Lemma 4.3.

(*) University Paris Dauphine and INRIA

(**) University of Rome II

(***) CNRS, Laboratory d'Analyse Numérique, University Pierre and Marie Curie

REFERENCES

[1] A. Bensoussan "Homogenization Theory, Atti del 3° seminario di Analisi funzionale ed applicazioni" Latezza, Bari, 1979.

[2] A. Bensoussan - L. Boccardo - F. Murat
To be published.

[3] A. Bensoussan - J.L. Lions - G. Papanicolaou "Asymptotic Methods in Periodic Structures" North Holland, 1978.

[4] L. Boccardo - F. Murat : Homogénisation de problèmes quasi-linéaires, Atti del Convegno "Studio di problemi limite della analisi funzionale" Bressonone 7.9. Sett. 1981, Pitagora Editrice, Bologna (1982) pp. 13-51.

[5] J.L. Doob Stochastic Processes
Wiley, 1953

[6] O.A. Ladyzhenskaya - N.N. Ural'tseva "Equations aux dérivées partielles de type elliptique" Dunod, Paris 1968, Traduit du russe.

[7] N.G. Meyers "An L^p estimate for the gradient of solutions of second order elliptic divergence equations" Ann. Scu. Normale Sup. Pisa (17) (1963), pp. 189-206.

PARTIALLY OBSERVED STOCHASTIC CONTROLS BASED ON A CUMULATIVE DIGITAL READ OUT OF THE OBSERVATIONS

Nigel J. Cutland,
Department of Pure Mathematics,
University of Hull,
England.

§ 1. INTRODUCTION

Consider a partially observed stochastic control system with past-dependent dynamics of the state x_t and observation y_t described as follows:

(1.1)

(a) $$dx_t = f(t,x,y,u(t,y))dt + g(t,x,y,u(t,y))\,db_t \qquad (x_0 = 0)$$

(b) $$dy_t = \bar{f}(t,x,y,u(t,y))dt + \bar{g}(t,y)d\bar{b}_t \qquad (y_0 = 0)$$

We discuss such systems for which the information available to the controller is a 'cumulative digital read-out' of the observation process y, as described below. Notice that the noise in the state equation is susceptible to control, whereas the noise in the observation is not.

We consider both ordinary and relaxed (or generalised) controls, and a very general cost function $J(u)$. We show that there is a natural topology on the controls for which the ordinary controls are dense, the relaxed controls form a compact set, and the cost and measure (on path space) are continuous as functions of the control. It follows that there is an optimal relaxed control, which under certain convexity conditions gives an optimal ordinary control.

The results described here extend those of the papers [4], [5], where only the drift was controlled, and the information pattern was more restricted. Proofs, which use techniques from nonstandard analysis (especially the Loeb measure construction) are sketched only. Full details will appear in a comprehensive paper on nonstandard methods in optimal control theory, in preparation. The simple idea used in establishing optimal controls is essentially as follows. Let J_0 be the minimum cost, and let (u_n) be a minimising sequence of controls, with $|J_0 - J(u_n)| < 1/n$. Nonstandard analysis provides immediately a nonstandard control u_N for infinite N, with $J(u_N) \approx J_0$. The hard work is then to show that from u_N we can construct a standard relaxed control v having $J(v) \approx J(u_N)$; hence v is optimal.

For a comprehensive introduction to nonstandard measure theory (especially the Loeb measure construction and its applications), as well as the basics of nonstandard analysis, see the survey paper [6].

§2. DESCRIPTION OF THE CONTROL SYSTEM

The evolution of $(x_t, y_t) \in \mathbb{R}^{d+m}$ for $0 \le t \le 1$ is governed by equations (1.1), in which b, $\bar{b}$ are independent Brownian motions of dimensions d, m respectively. Conditions on the coefficients are described at the end of this section.

Observations and Information Assume a fixed *information filtration*

$$\mathcal{G} = (\mathcal{G}_t)_{t \le 1} \text{ with } \mathcal{G}_t \subseteq \mathcal{F}^m_t = \sigma\{y_s : s \le t,\ y \in C^m[0,1]\},$$

such that $\mathcal{G}_1$ is generated by a countable number of atoms $(D_i)_{i \in \mathbb{N}}$. (Right continuity is not assumed) Writing $\mathcal{C} = C[0,1]$, we have:

Proposition 2.1 There are measurable functions $(r_t)_{t \le 1}$, $r_t : \mathcal{C}^m \to \mathbb{N}$ such that

(a) $\mathcal{G}_t = \sigma(\{r_t\})$;

(b) if $r_t(y) = r_t(y')$ then $r_s(y) = r_s(y')$ all $s \le t$;

(c) $r_t(y)$ increases with t.

Functions $(r_t)_{t \le 1}$ as in Proposition 2.1 constitute a *cumulative digital read-out;* the cumulation property is (b). This models the digitalisation of observations by a computer, for instance, prior to feedback into controls.

Controls. The class of (ordinary) *admissible controls* $\mathcal{U}$ is the class of measurable $\mathcal{G}$- adapted functions $u : [0,1] \times \mathcal{C}^m \to M$, where M is a fixed compact metric space. The *relaxed controls* $\mathcal{V}$ are those taking values in the space $\mathcal{M}(M)$ of Borel probability measures on M, regarded as an extension of M. A relaxed control v is interpreted in the drift f (or $\bar{f}$) by defining, for $\nu \in \mathcal{M}(M)$,

$$f(t, x, y, \nu) = \int_M f(t, x, y, a)d\nu(a);$$

and in the diffusion g by defining

$$g(t, x, y, \nu) = (\int_M g^2(t, x, y, a)d\nu(a))^{\frac{1}{2}}.$$

Here we assume that g is positive definite symmetric, and $(\ \ )^{\frac{1}{2}}$ denotes the unique positive definite symmetric square root matrix. This definition for g recognises that it is the covariance g^2 rather than g itself that influences the dynamics.

Cost With each control u (or v) is associated a cost $J(u)$ given by

$$J(u) = E\left(\int_0^1 h(t, x, y, u(t,y))dt + \bar{h}(x,y)\right)$$

where (x,y) is the trajectory for control u. It is assumed that $h, \bar{h} \geq 0$.

Conditions on coefficients The measurable coefficients $f, \bar{f}, g, \bar{g}, h, \bar{h}$ are assumed to satisfy the following conditions, which are fairly minimal conditions needed to make sense of the above and to provide solutions to (1.1) that are unique in distribution for each fixed control.

(2.2) (a) $f, \bar{f}, g, \bar{g}, h$ are adapted in (x,y) to $(\mathcal{F}_t^{d+m})$;

(b) $f, \bar{f}, g, h, \bar{h}$ are continuous in (x, u) for each fixed (t, y);

(c) g is uniformly Lipschitz in x ; $\bar{g}$ is uniformly Lipschitz in y;

(d) $g, \bar{g}$ are positive definite, and g is symmetric;

(e) $f, \bar{f}, g, \bar{g}, h, \bar{h}, g^{-1}f$ and $\bar{g}^{-1}\bar{f}$ satisfy linear growth conditions of the form

$$\|\theta(t, x, y, a)\| \leq \varkappa\,(1 + \|(x,y)\|),$$

where $\varkappa$ is constant.

§ 3. COMPACTNESS OF CONTROLS

A natural topology on controls is defined as follows. Let

$$\mathcal{K} = \{\theta\colon \theta\colon [0,1] \times M \to \mathbb{R};\ \theta \text{ is bounded, measurable; } \theta(t,\cdot) \text{ continuous}\}.$$

For each $v \in \mathcal{V}$ define a sequence v_i of deterministic controls

by $v_i(t) = v(t,y)$ for any $y \in D_i$.

Define

$$v(\theta,i) = \int_0^1 \theta(t, v_i(t))dt$$

for $\theta \in \mathcal{K}$; this gives a weak* topology on $\mathcal{V}$ having as subbase the sets

$$\{v : |v(\theta,i)| < \varepsilon\}_{\theta\in\mathcal{K},\ i\in\mathbb{N},\ \varepsilon>0}.$$

Theorem 3.1 (a) $\mathcal{V}$ is compact;

(b) $\mathcal{U}$ is dense in $\mathcal{V}$.

Proof (a) Let $V \in {}^*\mathcal{V}$; the nonstandard criterion for compactness requires that we construct $v \in \mathcal{V}$ with $V \approx v$. Assume first that V is deterministic. Define a nonstandard (i.e. internal) *Borel measure Q on $^*[0,1] \times {}^*M$ by

$$Q(C \times D) = \int_C V_\tau(D)d\tau.$$

A standard Borel measure q is defined on $[0,1] \times M$ by

$$q(X) = Q_L(st^{-1}(X))$$

where Q_L is the Loeb measure on $*[0,1] \times *M$ constructed from Q and

$$st: *[0,1] \times *M \to [0,1] \times M$$

is the standard part mapping. Now q disintegrates to give a deterministic $v \in \mathcal{V}$ such that

$$q(A \times B) = \int_A v_t(B)dt,$$

and it is routine to check that $v \approx V$.

For general $V \in *\mathcal{V}$, take $v^i \approx V_i$ as above for each $i \in \mathbb{N}$, and then $v \in \mathcal{V}$ is obtained by glueing the v^i together.

(b) is established along the same lines as in [3] and [5].

§ 4. CONTINUITY OF SOLUTION MEASURES

The conditions on the coefficients ensure that for each fixed control v there is a solution $z^v = (x^v, y^v)$ to the equations (1.1), determining a unique probability measure μ^v on $\mathcal{C}^n$ $(n = d + m)$. In the following we write $f_v(t,x,y)$ (etc.) for $f(t,x,y,v(t,y))$ (etc.). The main theorem of this section is the following.

4.1 Theorem The measure μ^v is continuous as a function of v with respect to the weak* topology on $\mathcal{M}(\mathcal{C}^n)$.

Proof Let $V \in *\mathcal{V}$ and $v \approx V$ as in Theorem 3.1 (a). The nonstandard criterion for continuity requires that we show that $\mu^v \approx \mu^V$, where μ^V is the internal measure on $*\mathcal{C}^n$ given by the internal control V. Loeb [9] and Anderson - Rashid [2] show that this is equivalent to showing that

(4.2) $\quad \mu^v(A) = \mu_L^V(st^{-1}(A))$ for all Borel sets A,

where μ_L^V is the Loeb measure on $*\mathcal{C}^n$ given by μ^V, and $st : *\mathcal{C}^n \to \mathcal{C}^n$ is the standard part map for the uniform norm topology.

By assumption, there is an internal solution $Z^V = (X^V, Y^V)$ to equations (1.1) for the control V, living on an internal space $(\Omega, \mathcal{A}, (\mathcal{A}_t)_{t \le 1}, \nu)$:

(4.3) (a) $dX_t^V = {}^*f_V(t,X^V,Y^V)dt + {}^*g_V(t,X^V,Y^V)dB_t$

(b) $dY_t^V = {}^*\overline{f}_V(t,X^V,Y^V)dt + {}^*\overline{g}(t,Y^V)d\overline{B}_t$

where $(B, \overline{B})$ is an internal Brownian motion. Keisler's $\mathcal{S}$- continuity theorem [8] shows that almost all paths of Z^V are $\mathcal{S}$- continuous (i.e. nearstandard in ${}^*\mathcal{C}^n$), so we can define a *standard* continuous process z^V by

$$z^V = {}^\circ(Z^V) \qquad ({}^\circ \text{ denotes standard part}),$$

a.s. with respect to the Loeb measure ν_L on Ω. From $(\mathcal{A}_t)$ we construct a standard filtration $(\mathcal{F}_t)$ on Ω, so writing $P = \nu_L$, the process z^V lives on the standard adapted space $(\Omega, \mathcal{F}, (\mathcal{F}_t)_{t \leq 1}, P)$ satisfying the usual conditions. The proof now consists in establishing the following claim.

4.4 Claim The process z^V is a solution to (1.1) for the control v.

To see that this is sufficient, we have

$$\mu^v(A) = P(z^V \in A) \quad \text{(by the claim)}$$
$$= P({}^\circ Z^V \in A) = P(Z^V \in st^{-1}(A)) = \nu_L(Z^V \in st^{-1}(A)) = \mu_L^V(st^{-1}(A)),$$

since $\nu(Z^V \in B) = \mu^V(B)$ for all internal sets B by definition of μ^V. Thus (4.2) is established and the theorem is proved, if we can justify Claim 4.4; this is sketched in a series of lemmas below.

First consider the equations

(4.5) (a) $dx_t = g_v(t,x,y)db_t$

(b) $dy_t = \overline{g}(t,y)d\overline{b}_t$.

Let λ^v be the measure induced on $\mathcal{C}^n$ by the solution to 4.5, and let $\lambda_1^{v,\eta}$ be the measure induced on $\mathcal{C}^d$ by the solution to 4.5(a) for $y = \eta \in \mathcal{C}^m$.

4.6 Lemma Let $K \subseteq \mathcal{C}^m$ be compact. The measures $(\lambda_1^{v,\eta})_{v \in \mathcal{V}, \eta \in K}$ are uniformly tight.

Proof By Prohorov's theorem it is sufficient to show that this set of measures is relatively compact, which is established by showing that $\lambda_1^{V,Y}$ is nearstandard for every $V \in {}^*\mathcal{V}$ and $Y \in {}^*K$. Taking a solution $\xi^{V,Y}$ to *4.5 (a) for fixed Y and V, we see that $\xi^{V,Y}$ is a.s. $\mathcal{S}$-continuous (by Keisler's Theorem) and so $st^{-1}(\mathcal{C}^d)$ has measure 1 under the Loeb measure $(\lambda_1^{V,Y})_L$. This (by [2], [9]) shows that $\lambda_1^{V,Y}$ is nearstandard.

4.7 Lemma For each of $\theta = f, g, \overline{f}, \overline{g}$: for almost all (τ, Z) with respect to $({}^*\text{Leb} \times \lambda^V)_L$

$${}^*\theta(\tau, Z, \alpha) \approx \theta({}^\circ\tau, {}^\circ Z, {}^\circ\alpha) \text{ all } \alpha \in {}^*M.$$

<u>Proof</u> Taking $\theta = f$, for example: for each compact $K \subseteq \mathcal{C}^m$ and compact set $K_\varepsilon \subseteq \mathcal{C}^d$ given by Lemma 4.6, apply Anderson's Lusin Theorem [1] to the function $\hat{f} = [0,1] \times K \to C(K_\varepsilon \times M, \mathbb{R}^d)$ defined by $\hat{f}(t,y)(x,\alpha) = f(t,x,y,\alpha)$. Now use the fact that the measure λ_2 induced on $\mathcal{C}^m$ by the solution to 4.5(b) is tight, and the disintegration of λ^v as

$$\lambda^v(A \times B) = \int_B \lambda_1^{v,\eta}(A)d\lambda_2(\eta)$$

to obtain the result.

We actually require Lemma 4.7 to hold for μ^V rather than λ^V, so the next stage is to analyse the relationship between these measures. Standard theory tells us that $\mu^v \ll \lambda^v$ with $d\mu^v/d\lambda^v = \rho^v$, say, given by the Girsanov formula. Careful analysis of the internal version of this for the internal control V gives:

4.8 <u>Lemma</u> ρ^V is $\mathcal{S}$-integrable (w.r.t. λ^V); hence $\mu_L^V \ll \lambda_L^V$ and $d\mu_L^V/d\lambda_L^V = {}^\circ\rho^V$.

4.9 <u>Corollary</u> Lemma 4.7 holds with μ^V in place of λ^V.

An immediate consequence of Corollary 4.9, together with the fact that $v \approx V$ is the following:

4.10 <u>Lemma</u> For a.a.ω, for all t:

$${}^\circ\!\int_0^t {}^*f_V(\tau, Z^V)d\tau = \int_0^t f_v(s,z^V)ds;$$

and similarly for $\overline{f}$.

Lemma 4.10 handles the drift terms in equations 4.3. For the diffusion in 4.3(b), note first that $\overline{B}$ has paths that are a.s. $\mathcal{S}$-continuous, and thus $\overline{b} = {}^\circ\overline{B}$ defines a Brownian motion. Moreover, using Corollary 4.9 for $\overline{g}$, and the nonstandard construction of the Itô integral, we have:

4.11 <u>Lemma</u> For a.a.ω, for all t,

$${}^\circ\!\int_0^t {}^*\overline{g}(\tau, Y^V)d\overline{B}_\tau = \int_0^t \overline{g}(s,y^V)d\overline{b}_s.$$

From this and Lemma 4.10 we see that

$$y_t^V = {}^\circ Y_t^V = \int_0^t \overline{f}_v(s,z^V)ds + \int_0^t \overline{g}(s,y^V)d\overline{b}_s \quad \text{a.s,}$$

as required for a solution. Thus the diffusion term in equation 4.3(a) is all that remains, and this is handled as follows.

4.12 <u>Lemma</u> There is a d-dimensional Brownian motion $\tilde{b}$ on Ω, independent of $\overline{b}$, such that a.s.

(4.13) $\int_0^t g_v(s,z^V)d\tilde{b}_s = {}^\circ\!\int_0^t {}^*g_V(\tau,Z^V)dB_\tau$ all t.

Proof Note first that $M_t = {}^\circ\!\int_0^t {}^*g_V(\tau,Z^V)dB_\tau$ is an L^2 - martingale, with

$$[M]_t = {}^\circ\!\int_0^t {}^*g_V^2(\tau,Z^V)d\tau.$$

(where we mean $(g_V)^2$, not$(g^2)_V$).

Using Corollary 4.9 for g^2 as it was used for f in Lemma 4.10, and remembering the interpretation of relaxed controls in the diffusion, we see that

$$[M]_t = \int_0^t g_v^2\ (s,z^V)ds.$$

From this, follwoing [7, p449], if we define

$$\tilde{b}_t = \int_0^t g_v^{-1}(s,z^V)dM_s$$

we see that $\tilde{b}$ is the required Brownian motion and (4.13) holds.

Combining Lemmas 4.12 and 4.10 we see that

$$x_t^V = {}^\circ X_t^V = \int_0^t f_v(s,z^V)ds + \int_0^t g_v(s,z^V)d\tilde{b}_s,$$

and so Claim 4.4. is justified. Theorem 4.1 is established.

§ 5. CONTINUITY OF COST AND OPTIMAL CONTROL

Routine extension of the proof of Theorem 4.1 proves the following:

Theorem 5.1 The cost function $J(v)$ is continuous.

Proof Using the notation and methods of the previous section, given

$V \in {}^*\mathcal{V}$ with $V \approx v \in \mathcal{V}$, we see that

a.s.

$$ {}^\circ\!\int_0^1 {}^*h_V(\tau,\ Z^V)d\tau \approx \int_0^1 h_v(t,z^V)dt$$

and

$$ {}^*\bar{h}(Z^V) \approx h(z^V).$$

This is sufficient (given the linear growth conditions) to show that $J(V) \approx J(v)$, as required.

An immediate consequence (in the light of the compactness of $\mathcal{V}$ and density of $\mathcal{U}$) is:

Theorem 5.2 (a) $\inf_{u\in\mathcal{U}} J(u) = \inf_{v\in\mathcal{V}} J(v)$;

(b) there is an optimal relaxed control.

It is routine to obtain optimal ordinary controls given certain convexity assumptions on the coefficients. This is well known, and uses standard measurable selection theorems.

§ 6. ANOTHER SYSTEM

Results similar to those described above can be established for systems of the form

(6.1) $dx_t = f(t,x_t, u(t,x))dt + g(t,x_t,u(t,x))db_t,$

with a modified information pattern as described below. We are here thinking of $x_t \in \mathbb{R}^n$ as combining state and observation; the noise in both components can be controlled. The drift and diffusion are however directly dependent on only the present x_t of x rather than the past; and the dependence of controls on observations is more restricted, as follows.

Observations and Information There are fixed times $0 = t_0 < t_1 <\ldots< t_p < t_{p+1} = 1$; the control $u(t,x)$ depends on a digital read-out made at times $(t_i)_{t_i \le t}$. In detail, there are instantaneous read-out functions $r_k : \mathcal{C}^n \to \mathbb{N}$ such that r_k is $\mathcal{F}^n_{t_k}$ - measurable. A cumulative digital read-out function $r(t,x)$ is defined by

$$r(t,x) = (r_1(x), r_2(x), \ldots , r_k(x)) \text{ if } t_k \le t < t_{k+1}.$$

Then we can define an information filtration $\mathcal{G} = (\mathcal{G}_t)_{t \le 1}$ as before: for each t let $\mathcal{G}_t = \sigma(\{r(t, \cdot)\})$. The class of admissible controls consists of $\mathcal{G}$ - adapted functions $u : [0,1] \times \mathcal{C}^n \to M$ (or $\mathcal{M}(M)$).

Conditions on the coefficients are again minimal ones to guarantee solutions unique in distribution. Proof of the main results is similar to that in §§ 4,5 except that analysis of the Girsanov formula is replaced by careful analysis of density of the solution x_t for fixed t, using some estimates from Stroock-Varadhan [10].

REFERENCES

[1] Anderson, R.M., Star-finite representations of measure spaces, Trans.Amer. Math. Soc., 271 (1982), 667-687.

[2] Anderson, R.M. and Rashid, S., A nonstandard characterization of weak convergence, Proc.Amer. Math. Soc. 69 (1978), 327-332.

[3] Cutland, N.J., Internal controls and relaxed controls, J. London Math. Soc., 27 (1983), 130-140.

[4] Cutland, N.J., Optimal controls for partially observed stochastic systems using nonstandard analysis, in Stochastic Differential Systems, Lecture Notes in Control and Information Sciences 43 (Eds. M. Kohlmann and N. Christopeit, Springer, Berlin 1982), 276-284.

[5] Cutland, N.J., Optimal controls for partially observed stochastic systems: an infinitesimal approach, Stochastics 8 (1983), 239-257.

[6] Cutland, N.J., Nonstandard measure theory and its applications, Bull. Lond. Math. Soc., 15(1983), 529-589.

[7] Doob, J.L., Stochastic Processes, Wiley, New York, 1953.

[8] Keisler, H.J., An infinitesimal approach to stochastic analysis, Memoirs of Amer. Math. Soc. 297 (1984).

[9] Loeb, P.A., Weak limits of measures and the standard part map, Proc. Amer. Math. Soc. 77 (1979), 128-135.

[10] Stroock, D.W. and Varadhan, S.R.S., Multidimensional Diffusion Processes, Springer-Verlag, Berlin, 1979.

SOME RESULTS ON BELLMAN EQUATION IN HILBERT SPACES AND APPLICATIONS TO INFINITE DIMENSIONAL CONTROL PROBLEMS

G. DA PRATO

Scuola Normale Superiore
56100 PISA, Italy

1. INTRODUCTION

We are here concerned with the problem

$$\begin{cases} \phi_t = \frac{1}{2}\operatorname{Tr}(S\phi_{xx}) - \frac{1}{2}|S\phi_x|^2 + \langle Ax + SF(x), \phi_x \rangle + g(x) \\ \phi(0,x) = \phi_0(x) \end{cases} \tag{1.1}$$

under the following hypotheses:

(H1) S is a self-adjoint, positive nuclear operator in a separable Hilbert space H. S is given by

$$Sx = \sum_{i=1}^{\infty} \lambda_i \langle x, e_i \rangle e_i \tag{1.2}$$

where $\{e_i\}$ is a complete orthonormal system in H and $\lambda_i > 0$, $i = 1,2,\ldots$

(H2) A is a self-adjoint negative operator in H and

$$Ae_i = \mu_i e_i \qquad \text{with } \mu_i \leqslant 0 \quad , \quad i = 1,2,\ldots \tag{1.3}$$

Remark 1.1

Hypothesis (H1) can be weakened by assuming $\lambda_i \leqslant 0$. Hypothesis (H2) implies that A and B commute; following [3] we could only assume $\{e_i\} \subset D(A)$ plus a suitable condition.

Before stating assumptions on F, g and ϕ_0 let us give some notations and definitions.

We shall denote by $C_b(H)$ the set of all mappings $\psi : H \to \mathbb{R}$ uniformly continuous and bounded. $C_b(H)$, endowed with the norm:

$$\|\psi\|_\infty = \sup_{x \in H} |\psi(x)| \tag{1.4}$$

is a Banach space; $C_b^k(H)$, $k = 1,2,\ldots$, is defined similarly.

Definition 1.2

The mapping $\psi \in C_b(H)$ is said <u>S-differentiable</u> in H if :

i) <u>For any</u> $x,y \in H$ <u>there exists the limit</u>

$$\lim_{h \to 0} \frac{1}{h} (\psi(x + hSy) - \psi(x)) \overset{\text{def}}{=} L_x(y) \tag{1.5}$$

ii) L_x is linear continuous in H .

If ψ is S-differentiable in H we shall denote by $S\psi_x(x)$ the element of H defined by

$$< S\psi_x(x) , z > = L_x(z) \qquad\qquad \forall z \in H \qquad . \tag{1.6}$$

We set finally

$$C_S^1(H) = \{\psi \in C_b(H) ; S\psi_x \in C_b(H)\} \tag{1.7}$$

$C_S^1(H)$, endowed with the norm

$$\|\psi\|_S = \|\psi\|_\infty + \|S\psi_x\|_\infty \qquad\qquad , \tag{1.8}$$

is a Banach space. We shall assume:

(H3) $\phi_0 \in C_S^1(H)$, $F \in C_b(H,H)$, $g \in C_b(H)$.

Eq. (1.1) is connected with the following optimal control problem:

minimize

$$J(t,x,u) = E \left\{ \int_t^T [g(y(s)) + \frac{1}{2} |u(s)|^2] ds + \phi_0(y(t)) \right\} \tag{1.9}$$

over all $u \in M_w^2(0,T;H)$ subject to the state equation

$$\begin{cases} dy(s) = (Ay(s) + SF(y(s)) + Su(s))ds + dw_s \\ \\ y(t) = x \qquad\qquad t \leqslant s \end{cases} \tag{1.10}$$

where w_s is a H-valued Brownian motion whose covariance is S and $M_w^2(0,T;H)$ is the set of all adapted processes u such that:

$$E \int_0^T |u(s)|^2 ds < +\infty \quad . \tag{1.11}$$

If H is finite dimensional, Eq. (1.1) and problem (1.9) are studied in [4].

In Section 2 we study the linear problem:

$$\begin{cases} \phi_t = \frac{1}{2} \operatorname{Tr}(S\phi_{xx}) + \langle Ax, \phi_x \rangle \\ \phi(0,x) = \phi_0(x) \end{cases} \tag{1.12}$$

and smoothing properties of the semi-group

$$(T_t\phi_0)(x) = \phi(t,x) \quad , \quad \phi_0 \in C_b(H) \quad . \tag{1.13}$$

Then, in Section 3 we write Eq. (1.1) in the mild form:

$$\phi(t,x) = (T_t\phi_0)(x) + \int_0^t T_{t-s}\left(-\frac{1}{2}|S\phi_x|^2 + \langle F(x), S\phi_x\rangle + g(x)\right)ds \tag{1.14}$$

that we solve by successive approximations.

Finally, in Section 4 we study Dynamic Programming for the Control Problem (1.9).

We remark that the corresponding infinite horizon problem is studied in [3].

2. THE LINEAR PROBLEM

We are here concerned with Eq. (1.12).

The formal solution of Eq. (1.12) is given by

$$\phi(t,x) = E\phi_0(e^{tA}x + X_t) \tag{2.1}$$

where X_t is the Gaussian process defined by

$$X_t = \int_0^t e^{(t-s)A}\, dw_s \quad . \tag{2.2}$$

The following proposition is proved in [3].

Proposition 2.1

Assume (H1) *and* (H2) , *let* T_t *be the semi-group defined by* (1.13). *Then we have:*

$$\|T_t\psi\|_\infty \leq \|\psi\|_\infty \qquad \forall\, \psi \in C_b(H) \tag{2.3}$$

$$\|S(T_t\psi)_x\|_\infty \leq \|S\psi_x\|_\infty \qquad \forall\, \psi \in C_S(H) \quad . \tag{2.4}$$

Moreover if $\psi \in C_b(H)$ *and* $t > 0$ *we have* $T_t\psi \in C_S(H)$ *and*

$$S(T_t\psi)_x(x) = \Lambda(t)E(X_t\psi(e^{tA}x + X_t)) \tag{2.5}$$

where

$$\Lambda(t)e_i \begin{cases} = 2\mu_i\, e^{t\mu_i}(1 - e^{2t\mu_i})^{-1} & \text{if } \mu_i < 0 \\ = -\frac{1}{t} & \text{if } \mu_i = 0 \end{cases}$$

Finally

$$\|S(T_t\psi)_x\|_\infty \leq \frac{\gamma}{\sqrt{t}} \tag{2.7}$$

where

$$\gamma = \operatorname*{Sup}_{\alpha>0} \frac{\alpha e^{-\alpha/2}}{1-e^{-\alpha}} \quad . \tag{2.8}$$

We consider now a finite dimensional approximation of T_t. For any $h \in \mathbb{N}$ we set

$$\begin{cases} P_h x = \sum_{i=1}^{h} < x,e_i > e_i \quad , \quad A_h = AP_h \quad , \quad S_h = SP_h \\ w_h(t) = P_h w_t \quad . \end{cases} \tag{2.9}$$

Moreover

$$(T_t^h\psi)(x) = E\psi(e^{tA_h}P_h x + X_t^h) \quad , \qquad \forall \psi \in C_b(H) \tag{2.10}$$

$$X_t^h = \int_0^t e^{(t-s)A_h}\, dw_h(s) \quad . \tag{2.11}$$

Proposition 2.2

Assume (H1) and (H2), let $x \in H$, $\psi \in C_b(H)$. Then we have:

$$\lim_{h\to\infty} (T_t^h\psi)(x) = (T_t\psi)(x) \quad \text{uniformly in} \quad t \in [0,T] \tag{2.12}$$

If moreover $\psi \in C_S(H)$, then

$$\lim_{h\to\infty} (S(T_t^h\psi)_x)(x) = (S(T_t\psi)_x)(x) \quad \text{uniformly in} \quad t \in [0,T]. \tag{2.13}$$

Proof

We have

$$X_t - X_t^h = \int_0^t e^{(t-s)B_h}\, dZ_h(s)$$

where $B_h = (1 - P_h)A$ and $Z_h = (1 - P_h)w_h$.
By a Kotelenez inequality ([7]) it follows:

$$P(\sup_{t\in[0,T]} |X_t - X_t^h|^2 \geq \varepsilon) \leq \frac{T}{\varepsilon} \sum_{i=h+1}^{\infty} \lambda_i \quad .$$

Thus X_t^h converges to X_t in probability, uniformly in $t \in [0,T]$. Now all statements follow easily since ψ is uniformly continuous #

3. THE NON LINEAR PROBLEM

We assume here (H1), (H2), (H3) and write Eq. (1.14) in the form:

$$\phi = \Gamma\phi \tag{3.1}$$

where

$$(\Gamma\phi)(t,x) = (T_t\phi_0)(x) + \int_0^t T_{t-s}[-\tfrac{1}{2}|S\phi_x(s,\cdot)|^2 + \\ + \langle F(\cdot), S\phi_x(x,\cdot)\rangle + g(\cdot)](x)ds . \tag{3.2}$$

We shall solve Eq. (3.1) by the contractions principle in Z_r, where

$$Z_r = \Big\{\zeta \in C([0,T]\times H;\mathbb{R})\,;\zeta(t,\cdot) \in C_S^1(H) \text{ and } \\ \|\zeta\|_{Z_r} = \sup_{t\in[0,T]} (\|\zeta(t,\cdot)\|_\infty + \|S\zeta_x(t,\cdot)\|_\infty) \leq r\Big\} \tag{3.3}$$

Proposition 3.1 (Local existence)

Assume (H1), (H2) and (H3); then there exist $T_0 > 0$ and a unique solution ϕ of (3.1) continuous in $[0,T_0] \times H$ and such that

$$\sup_{t \in [0,T_0]} \|\phi(t,\cdot)\|_S < +\infty \quad . \tag{3.4}$$

Proof

Choose $r > 0$ such that

$$\|\phi_0\|_\infty + \|S\phi_{0x}\|_\infty \leq \frac{r}{2} \tag{3.5}$$

We firstly prove that if T is small then Γ maps Z_r into itself. Setting $\psi = \Gamma\phi$ we have

$$\|\psi(t,\cdot)\| \leq \frac{r}{2} + tM_r \quad .$$

where

$$M_r = \frac{r^2}{8} + \frac{r\|F\|_\infty}{2} + \|g\|_\infty \quad . \tag{3.6}$$

Moreover, since

$$S\psi_x(t,\cdot) = S(T_t\phi_0)_x + \int_0^t S[T_{t-s}(-\frac{1}{2}|S\phi_x|^2 + <F,S\phi_x> + g)]_x ds \tag{3.7}$$

we have, by (2.4) and (2.7)

$$\|S\psi_x(t,\cdot)\|_\infty \leq \frac{r}{2} + 2\sqrt{t}\,\gamma M_r$$

so that, choosing T' such that

$$T' = \min\left\{\frac{r}{2M_r} \;,\; \frac{r^2}{16\gamma^2 M_r^2}\right\} \tag{3.8}$$

Γ maps Z_r (with $T = T'$) into itself.

Let us prove now that Γ is a contraction. To this purpose set $\psi = \Gamma\phi$, $\overline{\psi} = \Gamma\overline{\phi}$, then by

$$\psi(t,\cdot) - \overline{\psi}(t,\cdot) = \int_0^t T_{t-s}[<F - \frac{1}{2}(S\phi_x + S\overline{\phi}_x), S\phi_x - S\overline{\phi}_x> ds \tag{3.9}$$

it follows that

$$\|\psi(t,\cdot) - \overline{\psi}(t,\cdot)\|_\infty \leq t(r + \|F\|_\infty)\|\phi - \overline{\phi}\|_{Z_r}$$

$$\|(S\psi_x)(t,\cdot) - (S\overline{\psi}_x)(t,\cdot)\|_\infty \leq 2\gamma\sqrt{t}(r + \|F\|_\infty)\|\phi - \overline{\phi}\|_{Z_r}$$

thus Γ is a $\frac{1}{2}$-contraction provided $T \leq T''$ where

$$T'' = \min\left\{ \frac{1}{3}(r + \|F\|_\infty)^{-1} , \frac{1}{36}\gamma^{-2}(r + \|F\|_\infty)^{-2} \right\} \quad ; \tag{3.10}$$

choosing finally

$$T_0 = \min\{ T', T'' \} \tag{3.11}$$

the conclusion follows #

We consider now, for further use, the finite dimensional approximating problem:

$$\begin{cases} \zeta_t^h = \frac{1}{2}\operatorname{Tr}(S^h\zeta_{xx}^h) - \frac{1}{2}|S^h\zeta_x^h|^2 + \langle A^h x + P^h F(x), S^h\zeta_x^h\rangle + g^h(x) \\ \zeta^h(0,x) = \phi_0^h(x) \quad , \quad \forall x \in P_h H \end{cases} \tag{3.12}$$

where ϕ_0^h and g^h are the restrictions respectively of ϕ_0 and g to $P_h H$. Setting $\phi^h(t,x) = \zeta^h(t,P_h x)$, ϕ^h verifies the integral equation:

$$\phi^h = \Gamma^h\phi^h \tag{3.13}$$

where

$$(\Gamma^h\phi^h)(t,x) = (T_t^h\phi_0)(x) + \int_0^t T_{t-s}^h\left[-\frac{1}{2}|S^h\phi_x^h|^2 + \langle P^h F, S^h\phi^h\rangle + g^h\right] ds \, . \tag{3.14}$$

<u>Proposition 3.2</u> (Approximation)

<u>Assume</u> (H1), (H2), (H3) <u>and let</u> T_0 <u>be defined by</u> (3.11). <u>Then</u> <u>Eq.</u> (3.15) <u>has a unique solution</u> ϕ^h <u>continuous in</u> $[0,T_0] \times H$ <u>and for any</u> $x \in H$ <u>we have</u>

$$\phi^h(t,x) \to \phi(t,x) \quad , \quad S\phi_x^h(t,x) \to S\phi(t,x) \tag{3.15}$$

<u>uniformly for</u> $t \in [0,T_0]$.

<u>Proof</u>

It is easy to check that all constants involved in the proof of Proposition 3.1 can be chosen independent of h. Thus existence and uniqueness of ϕ^h in $[0,T_0]$ follow. Let us prove (3.15). We have

$$\phi^h = \lim_{n\to\infty}\phi_n^h \quad , \quad \phi = \lim_{n\to\infty}\phi_n \quad \text{in } Z_r \tag{3.16}$$

where

$$\phi^h_{n+1} = \Gamma^h \phi^h_n \quad , \quad \phi_{n+1} = \Gamma\phi_n \tag{3.17}$$

and the convergence in (3.16) is uniform in h. Moreover

$$|\phi(t,x) - \phi^h(t,x)| \leqslant |\phi(t,x) - \phi_n(t,x)| + |\phi_n(t,x) - \phi^h_n(t,x)| + \\ + |\phi^h_n(t,x) - \phi^h(t,x)|$$

For any $\varepsilon > 0$ let $n_\varepsilon \in \mathbb{N}$ such that $\|\phi^h - \phi^h_n\|_{Z_r} < \varepsilon$ for $n > n_\varepsilon$ and $h \in \mathbb{N}$; by a recurrence argument and by Proposition 2.2, (3.15) follows #

<u>Theorem 3.3</u> (Global existence)

<u>Assume</u> (H1), (H2) <u>and</u>:

(H3)' $\phi_0, g \in C_S(H)$, $F \in C^1_b(H,H)$

<u>Then, for any</u> $T > 0$, Eq. (3.1) <u>has a unique global solution</u> ϕ <u>continuous in</u> $[0,T] \times H$ <u>and such that</u>

$$\sup_{t\in[0,T]} \|\phi(t,\cdot)\|_S < +\infty \quad . \tag{3.18}$$

<u>Moreover</u> (3.15) <u>holds for any</u> $x \in H$ <u>uniformly for</u> $t \in [0,T]$.

<u>Proof</u>

By standard arguments we only need to prove an a priori estimate for $\|\phi(t,\cdot)\|_S$; for this, by virtue of Proposition 3.2, it suffices to estimate $\|\zeta^h(t,\cdot)\|_S$ where ζ^h is the solution of (3.12). Now $\zeta = \zeta^h$ is the solution of the finite dimensional problem

$$\begin{cases} \zeta_t = \frac{1}{2}\sum_{i=1}^h \lambda_i \zeta_{x_i x_i} - \frac{1}{2}\sum_{i=1}^h \lambda_i^2 \zeta_{x_i}^2 + \sum_{i=1}^h (\mu_i x_i + \lambda_i F_i)\zeta_{x_i} + g(x) \\ \zeta(0,x) = \phi_0(x) \quad , \quad x \in H_h = P_h H \end{cases} \tag{3.19}$$

where $\zeta_x = (\zeta_{x_1},\ldots,\zeta_{x_h})$, $\zeta_{x_i} = \langle\zeta_x, e_i\rangle$, $F_i(x) = \langle F(x), e_i\rangle$.
By a maximum argument we get easily

$$\|\zeta(t,\cdot)\|_\infty \leqslant \|\phi_0\|_\infty + t\|g\|_\infty \quad . \tag{3.20}$$

Setting $Z = S^h\zeta_x = (Z_1, Z_2, \ldots, Z_h)$ we obtain, by differentiating (3.19),

$$Z_t^j = \frac{1}{2}\sum_{i=1}^{h} \lambda_i Z^j_{x_i x_i} - \sum_{i=1}^{h} Z^i Z^j_{x_i} + \mu_j Z^j +$$

$$+ \sum_{i=1}^{h} \lambda_j \frac{\partial F_i}{\partial x_j} Z_i + \sum_{i=1}^{h} \lambda_i F_i Z^j_{x_i} + \lambda_j g_{x_j} =$$

$$= \mathcal{L}_j(Z^j) + (M(Z))_j + \lambda_j g_{x_j} \tag{3.21}$$

where

$$\mathcal{L}_j(v) = \frac{1}{2}\sum_{i=1}^{h} \lambda_i v_{x_i x_i} - \sum_{i=1}^{h} Z_i v_{x_i} + \mu_j v \tag{3.22}$$

and

$$(M(Z))_j = \sum_{i=1}^{h} \lambda_j \frac{\partial F_i}{\partial x_j} Z^i \quad . \tag{3.23}$$

We have clearly, by maximum arguments

$$\|e^{t\mathcal{L}} v\|_\infty \leq \|v\|_\infty \qquad \forall v \in C_b(H,H) \quad . \tag{3.24}$$

Moreover, by (3.23) it follows, setting $u = M(v)$,

$$|u_j(x)|^2 \leq \lambda_j^2 \sum_{i=1}^{h} \left|\frac{\partial F_i(x)}{\partial x_j}\right|^2 \sum_{i=1}^{h} |v_i(x)|^2$$

$$\leq \lambda_j^2 \left|\frac{\partial F(x)}{\partial x_j}\right|^2_{\mathbb{R}^h} |v(x)|^2_{\mathbb{R}^h}$$

so that

$$\|u\|^2_\infty \leq \mathrm{Tr}(S^2)\ \|F'_x\|_\infty\ \|v\|^2_\infty$$

and this implies

$$\|e^{tM} v\|_\infty \leq \mathrm{Tr}(S)\ \|F'_x\|_\infty\ \|v\|_\infty \qquad \forall v \in C_b(H,H) \quad . \tag{3.25}$$

Collecting (3.24) and (3.25) we find the required a priori estimate, namely

$$\|S\zeta_x(t,\cdot)\|_\infty \leq \exp(\mathrm{Tr}(S)\ \|F'_x\|_\infty t)\ \|S\phi_{0x}\|_\infty +$$

$$+ \int_0^t \exp(\mathrm{Tr}(S)\ \|F'_x\|_\infty s)\ \|Sg_x\|_\infty ds \quad \#$$

4. DYNAMIC PROGRAMMING

We assume here (H1), (H2) and (H3)'. We denote by $J(t,x)$ the value function of the control problem (1.9), (1.10)

$$J(t,x) = \inf \{J(t,x,u); u \in M_w^2(0,T;H)\} \tag{4.1}$$

Lemma 4.1

Let $u \in M_w^2(0,T;H)$ and let y be the solution of Eq. (1.10). Then the following identity holds:

$$\psi(t,x) + \frac{1}{2} E \int_t^T |S\phi_x(s,y(s)) + u(s)|^2 ds = E(\int_t^T [g(y(s)) + \frac{1}{2} |u(s)|^2 ds + \phi_0(y(T)) \tag{4.2}$$

where

$$\psi(t,x) = \phi(T-t,x) \tag{4.3}$$

and ϕ is the solution of Eq. (3.1)

Proof

Let $\psi^h(t,x) = \phi^h(T-t,x)$, where ϕ^h is the solution of Eq. (3.13). Since ϕ^h is the solution of a finite-dimensional problem, (4.2) is standard for ψ^h; now the conclusion follows from Theorem 3.3 #

Consider now the closed loop equation

$$\begin{cases} dy = (Ay + SF(y) - S^2\psi_x(y))ds + dw_s \\ y(t) = x \qquad s \geqslant t \end{cases} \tag{4.4}$$

that has a unique mild solution y^* since $\psi \in C_b^1(H,H)$.

We can prove the result

Proposition 4.2

Assume (H1), (H2), (H3)' and let ϕ be the solution of Eq. (3.1). Then we have:

$$J(t,x) = \phi(T-t,x) \tag{4.5}$$

Moreover there exists a unique optimal control u^* given by

$$u^*(s) = -S\phi_x(T-s,y^*(s)) \qquad . \tag{4.6}$$

Proof

The proof relies on identity (4.2) (see for instance [1]) #

REFERENCES

[1] V. BARBU, G. DA PRATO, Hamilton Jacobi Equations in Hilbert Spaces, PITMAN, London (1983).

[2] V. BARBU, G. DA PRATO, Solution of the Bellman Equation Associated with an Infinite Dimensional Stochastic Control Problem and Synthesis of Optimal Control, SIAM J. Control and Optimization, 21, 4 (1983 531-550.

[3] G. DA PRATO, Some Results on Bellman Equation in Hilbert Spaces, SIAM Journal on Control and Optimization (to appear).

[4] W. H. FLEMING, R.W. RISHEL, Deterministic and Stochastic Optimal Control, Springer, Berlin, (1975).

[5] L. GROSS, Potential Theory on Hilbert Space, J. Funct. Anal., 1 (1967) 123-181.

[6] T. HAVÂRNEANU, Existence for the Dynamic Programming Equation of Control Diffusion Processes in Hilbert Space Nonlinear Analysis, T.M.A. (to appear).

[7] P. KOTELENEZ, A Submartingale Type Inequality with Applications to Stochastic Evolution Equations, STOCHASTICS 8 (1982) 139-151.

A PDE APPROACH TO ASYMPTOTIC ESTIMATES FOR OPTIMAL EXIT PROBABILITIES

Wendell H. Fleming[1] and Panagiotis E. Souganidis[2]
Lefschetz Center for Dynamical Systems
Division of Applied Mathematics
Brown University
Providence, RI 02912

1. Introduction.

The theory of nonlinear, first order PDE of Hamilton-Jacobi type has been substantially developed with the introduction by M. G. Crandall and P.-L. Lions [1] of the class of viscosity solutions. This turns out to be the correct class of generalized solutions for such equations. M. G. Crandall, L. C. Evans and P.-L. Lions [2] provide a simpler introduction to the subject while the book by P.-L. Lions [12] and the review paper by M. G. Crandall and P. E. Souganidis [4] provide a view of the scope of the theory and references to much of the recent literature.

Recently, L. C. Evans and H. Ishii [7] illustrated the usefulness of the viscosity solution methods in studying various asymptotic problems concerning stochastic differential equations with small noise intensities. They gave new proofs based on PDE methods for results of Ventcel-Freidlin type which were previously treated by quite different (probabilistic) techniques. In the present work we use a similar PDE-viscosity solution method to give a new proof and extend a result of W. H. Fleming and C.-P. Tsai [11] concerning optimal exit probabilities and differential games. In the asymptotic formula for the minimum exit probability there is a crucial quantity I, which turns out to be the lower value of an associated differential game. As a function of an initial time s and initial state $x \in \Omega$ $(\Omega \subset \mathbb{R}^N)$ the lower value $I = I(s,x)$ is a viscosity solution of a boundary-value problem

$$I_s + H(x,I_x) = 0 \quad \text{for} \quad 0 < s < T,\ x \in \Omega, \tag{1.1}$$

with H of the form (3.3) below and with boundary conditions $0, +\infty$ as described in (3.5).

Only indications of the proofs are given here, with details to appear elsewhere. The proofs in [11] are based on rather involved differential game theoretic and probabilistic arguments. We think that the method indicated below is considerably simpler and more appealing.

1) Partially supported by NSF under Grant No. MCS 8121940, by ONR under Grant No. N00014-83-K-0542 and by AFOSR under Grant No. AF-AFOSR 81-0116.
2) Partially supported by ONR Grant No. N00014-83-K-0542 and by NSF under Grant No. DMS-8401725.

2. Minimum exit probabilities.

Following [11] let $\xi(\cdot)$ be an N-dimensional stochastic process with continuous sample paths defined for times $t \geq s$. Let $D \subset \mathbb{R}^N$ be open and bounded with smooth boundary ∂D. For initial time s and state $x = \xi(s) \in D$, let τ_{sx} denote the exit time from D (i.e., the first t such that $\xi(t) \in \partial D$). For fixed $T > s \geq 0$, $P(\tau_{sx} \leq T)$ is the exit probability.

We assume that $\xi(\cdot)$ is a controlled Markov diffusion process, satisfying in the Ito-sense the stochastic differential equation

$$d\xi(t) = b[\xi(t),y(t)]dt + \varepsilon^{\frac{1}{2}}\sigma[\xi(t)]dw(t), \tag{2.1}$$

where $y(t)$ is a control applied at time t, $\varepsilon > 0$ is a parameter, and $w(\cdot)$ is an N-dimensional brownian motion. In [11], it was assumed that σ = identity matrix. We assume that $y(t) \in Y$, where $Y \subset \mathbb{R}^M$ is compact. Moreover, $b(\cdot,\cdot)$, $\sigma(\cdot)$ are Lipschitz; and the matrices $a(x) = \sigma(x)\sigma'(x)$ have all eigenvalues $\geq c > 0$.

The control process $y(\cdot)$ is assumed to have the feedback form

$$y(t) = \underline{y}(t,\xi(t))$$

where $\underline{y}: [s,T] \times \mathbb{R}^N \to Y$ is a Borel measurable function. Let

$$q_{\underline{y}}^{\varepsilon}(s,x) = P(\tau_{s,x} \leq T).$$

Of course, the exit probability depends on ε and $\underline{y}$ in view of (2.1), namely $\tau_{sx} = \tau_{sx}^{\varepsilon,\underline{y}}$. The minimum exit probability is

$$q^{\varepsilon} = \min_{\underline{y}} q_{\underline{y}}^{\varepsilon}. \tag{2.2}$$

The function $q^{\varepsilon}(s,x)$ satisfies the dynamic programming equation

$$q_s^{\varepsilon} + \frac{\varepsilon}{2} \operatorname{tr} a(x) q_{xx}^{\varepsilon} + \min_{y\in Y}\{b(x,y)\cdot q_x^{\varepsilon}\} = 0, \tag{2.3}$$

in the cylinder $[0,T] \times D$, where q_x denotes the gradient vector and $\operatorname{tr} a q_{xx} = \sum_{i,j} a_{ij} q_{x_i x_j}$. The boundary conditions are

$$\begin{aligned} q^{\varepsilon}(s,x) &= 1 \quad \text{for } s < T,\ x \in \partial D \\ q^{\varepsilon}(T,x) &= 0 \quad \text{for } x \in D. \end{aligned} \tag{2.4}$$

In general, it is difficult to get effective information about q^{ε} and the optimal control law in this way. Instead, we seek an asymptotic formula for q^{ε}, valid for small $\varepsilon > 0$, of the form

$$-\lim_{\varepsilon\downarrow 0} \varepsilon \log q^{\varepsilon} = I, \tag{2.5}$$

where I turns out to be the lower value of a certain differential game. Equation (2.5) can be written as

$$q^{\varepsilon} = \exp\{-\frac{I+o(1)}{\varepsilon}\} ,$$

which is a weaker result than a WKB expansion

$$q^{\varepsilon} = \exp\{-\tfrac{I}{\varepsilon}\}\cdot\text{asymptotic series in powers of } \varepsilon. \tag{2.6}$$

The expansion (2.6) cannot be expected to hold except in certain regions where $I(s,x)$ is a smooth function of (s,x). We have no results concerning (2.6), although an expansion up to terms involving $\varepsilon,\varepsilon^2$ for a similar problem arising in stochastic control was given in [9, Section 6].

3. The differential game.

A formal description of the game is as follows. There are two players, a maximizing player who chooses $y(t) \in Y$ and a minimizing player who chooses $z(t) \in \mathbb{R}^N$. The state $\phi(t)$ of the game at time t satisfies

$$\phi(t) = x + \int_s^t z(r)dr. \tag{3.1}$$

Let θ denote the exit time of $\phi(t)$ from D, and $\theta \wedge T = \min(\theta,T)$. Let

$$L(x,y,z) = \tfrac{1}{2}(b(x,y)-z)'a(x)^{-1}(b(x,y)-z),$$

$$\chi(x) = \begin{cases} 0, & x \in \partial D \\ +\infty, & x \in D. \end{cases}$$

The game payoff is

$$\int_s^{\theta\wedge T} L(\phi(t),y(t),z(t))dt + \chi(\phi(\theta\wedge T)). \tag{3.2}$$

We consider the "lower" game in which (formally speaking) the minimizing player has the information advantage of knowing both $y(t)$ and $\phi(t)$ before $z(t)$ is chosen, while his opponent knows only $\phi(t)$ before choosing $y(t)$.

This formal description can be made precise in one of several possible ways, each of which involves concepts of game strategy. The Elliott-Kalton formulation [5], [6], [8], [13] is convenient here. _Let_ $I = I(s,x)$ _denote the lower value of the game_, in the Elliott-Kalton sense. Let

$$H(x,p) = \max_{y\in Y} \min_{z\in\mathbb{R}^N} [L(x,y,z) + p\cdot z]. \tag{3.3}$$

The Isaacs (or dynamic programming) equation associated with this lower game is

$$I_s + H(x,I_x) = 0 \tag{3.4}$$

Theorem. Let $I^{\varepsilon} = -\varepsilon \log q^{\varepsilon}$. Then:

(a) $I(s,x)$ is the unique viscosity solution to (3.4) in the cylinder $(0,T) \times D$ with the boundary conditions

$$\begin{aligned} &I(s,x) = 0 \quad \text{for} \quad 0 < s < T, \quad x \in \partial D \\ &I(s,x) \to +\infty \quad \text{as} \quad s \uparrow T, \text{ for } x \in D \end{aligned} \tag{3.5}$$

(b) $\lim_{\varepsilon \to 0} I^{\varepsilon} = I$;

As candidates for viscosity solutions we admit functions

$$I \in C([0,T'] \times \bar{D}), \ \forall\, T' < T, \tag{3.6}$$

where $C(\mathcal{O})$ is the space of continuous functions defined in $\mathcal{O}$, which satisfy the boundary condition (3.5).

4. Indications of the proof.

Let us indicate the proof of (b), with details to appear elsewhere. The fact that I is a viscosity solution follows by standard arguments concerning the Elliott-Kalton formulation of the value of differential games [8] and some rescaling [11]. The uniqueness part of (a) was pointed out to us by P.-L. Lions and a somewhat simpler argument to get uniqueness was given by M. G. Crandall. It is because of the infinity boundary condition for $s = T$ in (3.5) that uniqueness is not a standard result. For more details see M. G. Crandall, P.-L. Lions and P. E. Souganidis [3].

From (3.3)

$$H(x,p) = -\frac{1}{2} p'a(x)p + \max_{y \in Y}\{b(x,y)\cdot p\}.$$

From this and (2.3), (2.4),

$$I_s^{\varepsilon} + \frac{\varepsilon}{2} \operatorname{tr} a(x) I_{xx}^{\varepsilon} + H(x, I_x^{\varepsilon}) = 0 \tag{4.1}$$

and I^{ε} satisfies the same boundary conditions (3.5) as I. Since equation (3.4) is obtained by putting $\varepsilon = 0$ (the "vanishing viscosity" method) this suggests that $I^{\varepsilon} \to I$ as $\varepsilon \to 0$, but of course does not prove it.

The main steps of the proof are as follows.

1. $0 \le I^{\varepsilon}(s,x) \le C_1(s), \quad 0 \le s < T.$

This has a very simple probabilistic proof [11, Lemma 2.1] and also can be proved nonprobabilistically by choosing suitable comparison functions.

2. $|I_x^{\varepsilon}(s,x)| \le C_2(s), \quad 0 \le s < T.$

For $x \in \partial D$, such an estimate is obtained using suitable barrier functions

In the interior of D, it is then obtianed from the "Bernstein trick" applying the maximum principle to

$$z^{\varepsilon} = \zeta^2 |I_x^{\varepsilon}|^2 - kI$$

for an appropriate cut-off function ζ and k a large enough constant.

3. As $\varepsilon \to 0$, I^{ε} converges uniformly on compact subsets of $[0,T) \times \overline{D}$ to a limit which is the unique viscosity solution to (3.4), (3.5). This is immediate from 1 and 2.

References

1. M. G. Crandall and P.-L. Lions, "Viscosity solutions of Hamilton-Jacobi equations", Trans. Amer. Math. Soc. 277 (1983), 1-42.
2. M. G. Crandall, L. C. Evans, and P.-L. Lions, "Some properties of viscosity solutions of Hamilton-Jacobi equations", Trans. Amer. Math. Soc. 282 (1984), 487-502.
3. M. G. Crandall, P.-L. Lions and P. E. Souganidis, in preparation.
4. M. G. Crandall and P. E. Souganidis, "Developments in the theory of nonlinear first-order partial differential equations", Proceedings of International Symposium on Differential Equations, Birmingham, Alabama (1983), Knowles and Lewis, eds.; North Holland.
5. R. J. Elliott and N. J. Kalton, "The existence of value in differential games", Memoirs Amer. Math. Soc. 126 (1972).
6. L. C. Evans and H. Ishii, "Nonlinear first order PDE on bounded domains", to appear.
7. L. C. Evans and H. Ishii, "A PDE approach to some asymptotic problems concerning random differential equations with small noise parameters", to appear.
8. L. C. Evans and P. E. Souganidis, "Differential games and representation formulas for solutions of Hamilton-Jacobi-Isaacs equations", to appear in Indiana Univ. Math. Journal.
9. W. H. Fleming, "Stochastic control for small noise intensities", SIAM J. Control 9 (1971), 473-517.
10. W. H. Fleming, "Exit probabilities and optimal stochastic control", Appl. Math. Optim. 7 (1978), 329-346.
11. W. H. Fleming and C.-P. Tsai, "Optimal exit probabilities and differential games", Appl. Math. Optim. 7 (1981), 253-282.
12. P.-L. Lions, Generalized solutions of Hamilton-Jacobi equations, Pitman, 1982.
13. P.-L. Lions and P. E. Souganidis, "Differential games, optimal control and directional derivatives of viscosity solutions of Bellman's and Isaacs' equations", to appear in SIAM Journal of Control and Optimization.

OPTIMAL STOCHASTIC CONTROL WITH

STATE CONSTRAINTS

P.L. Lions
Ceremade
University Paris-Dauphine
Place de Lattre de Tassigny
75775 Paris Cedex 16

I . Introduction :

We consider here the control of stochastic systems described by stochastic differential equations and we investigate the various ways of imposing state constraints. Roughly speaking there are three possibilities : i) exit problems, in that case one stops the process when it leaves the domain defining the constraints ; ii) reflecting boundary conditions, here one assumes that the system automatically behaves at the boundary of the domain in such a way the state remains inside (a typical example being of course the reflexion of the processes) ; iii) forced state constraints, here we restrict a priori our attention to controls such that the state of the system does not leave the constraints domain.

Since we will always deal with complete observations it is well-known that the value function (or minimum cost function) is related to the solution of a fully nonlinear second-order elliptic equation (called the Hamilton-Jacobi-Bellman equation, HJB in short) possibly degenerate. For more details on these standard facts, we refer to W.H. Fleming and R. Rishel [10], A. Bensoussan and J.L. Lions [1], [2], N.V. Krylov [13], P.L. Lions [20], [21].

In the first case (exit problems) the boundary conditions - which combined with the HJB equation should determine the value function - are of Dirichlet type. In general, the value function is not smooth and this prevents the standard application of dynamic programming optimality principle to derive HJB equations ; however this difficulty is circumvented in P.L. Lions [22], [23] by the use of viscosity solutions introduced by M.G. Crandall and P.L. Lions in the context of the classical Hamilton-Jacobi equations [5] - see also M.G. Crandall, L.C. Evans and P.L. Lions [4], P.L. Lions [24]. In addition regularity results are available; we refer to P.L. Lions [25], [21] for regularity results in degenerate situations; and in the uniformly elliptic case, $C^{1,1}$ regularity was first obtained by L.C. Evans and P.L. Lions [9] (see also L.C. Evans and A. Friedman [8], P.L. Lions [26]), then $C^{2,\alpha}_{loc}$ regularity was obtained by L.C. Evans [6], [7], N.V. Krylov [14] and the global $C^{2,\alpha}$ regularity is due to N.V. Krylov [15], [16], [17], L. Cafarelli, J.J. Kohn, L. Nirenberg and J. Spruck [3].

We will report here on some recent results concerning cases ii) and iii) ; and we will also consider the associated ergodic control problems. Let us also warn the reader that we will not consider the most general situation iii) but we will restrict ourselves to the simplest case.

II . Reflecting boundary conditions.

Let $\mathcal{O}$ be a smooth (bounded for example) open set in $\mathbb{R}^N$, let $\mathcal{A}$ be a separable metric space (values of the control) and let (Ω,F,F^t,P,B_t) be a probability space with a m-dimensionnal Brownian motion B_t satisfying the usual conditions. The state of the system is given by a stochastic differential equation with reflecting boundary conditions :

$$(1)\quad \begin{cases} dX_t = \sigma(X_t,\alpha_t)dB_t + b(X_t,\alpha_t)dt - \gamma(X_t,\alpha_t)dA_t \\ X_t, A_t \text{ are continuous, } A_t \text{ has bounded variations} \\ X_o = x \in \mathcal{O} \quad , \quad X_t \in \overline{\mathcal{O}} \text{ for all } t \geqslant 0 \\ A_t = \int_o^t 1_{\partial\mathcal{O}}(X_s)\, d|A|_s \end{cases}$$

where $|A|_s$ denotes the total variation of A_t on $[0,s]$, and α_t - the control process - is an arbitrary progressively measurable process with values in $\mathcal{A}$. We will assume that γ satisfies :

$$(2)\qquad \exists\, \nu > 0\ ,\ \forall\ (x,\alpha) \in \partial\mathcal{O} \times \mathcal{A}\ ,\qquad (\gamma(x,\alpha)n(x)) \geqslant \nu > 0$$

where n is the unit outward normal and that σ, b, γ are smooth in x, uniformly for $\alpha \in A$ (say $C^{1,1}(\bar{\mathcal{O}})$) and continuous in α so that (1) admits a unique solution (X_t, A_t) - we refer to N. Ikeda and S. Watanabe [12], A. Bensoussan and J.L. Lions [2], P.L. Lions and A.S. Sznitman [30] for the solution of such problems.

We next introduce the cost function :

$$(3) \qquad J(x,\alpha_t) = E \int_0^\infty f(X_t,\alpha_t)\, e^{-\lambda t}\, dt$$

where f is smooth in x, uniformly in α (say $C^{1,1}(\bar{\mathcal{O}})$) and continuous in α and where $\lambda > 0$. Finally the value function is given by

$$(4) \qquad u(x) = \inf_{\alpha_t} J(x,\alpha_t) \quad , \qquad \forall\, x \in \bar{\mathcal{O}}$$

where the infimum is taken over all possible controls α_t.

In the results which follow, we use the assumptions

$$(5) \qquad \partial\mathcal{O} = \Gamma_+ \cup \Gamma_- \text{ where } \Gamma_+, \Gamma_- \text{ are closed, disjoint and possibly empty ;}$$

$$(6) \qquad \exists \mu > 0 \,,\; \forall\, (x,\gamma) \in \Gamma_+ \times A \,, \qquad (a_{ij}) \geq \mu I_N$$

in the sense of symmetric matrices, where $a = \frac{1}{2} \sigma\sigma^T$;

$$(7) \qquad \forall\, (x,\gamma) \in \Gamma_- \times A \,, \quad a_{ij}(x,\alpha) n_i n_j = 0 \,, \quad b_i(x,\alpha) n_i - a_{ij}(x,\alpha) \partial_{ij} d(x) \leq 0$$

where $d(x) = \mathrm{dist}(x, \partial\mathcal{O})$;

$$(8) \qquad \lambda > \lambda_0 = \sup \Big\{ 2\partial_\xi b(x,\alpha)\cdot\xi + 3 \,\mathrm{Tr}\, [\partial_\xi \sigma(x,\alpha)\cdot\partial_\xi \sigma^T(x,\alpha)] \;/\; x \in \bar{\mathcal{O}} \,,\; \alpha \in A \,,\; \xi \in S^{N-1} \Big\} .$$

<u>Theorem 1</u> : We assume (5)-(8). Then the value function is the unique function satisfying :

i) $\quad u \in C^{0,1}(\bar{\mathcal{O}})$, $u \in C^{1,1}(\bar{V}) \cap C^{2,\alpha}(V)$ for some open neighborhood V of Γ_+ and for some $\alpha \in \,]0,1[$

ii) u is semi-concave i.e. : $\exists C > 0$, $\forall \xi \in S^{N-1}$, $\partial_\xi^2 u \leq C$ in $\mathcal{D}'(\mathcal{O})$.

iii) $A_\alpha u \in L^\infty(\mathcal{O})$ for all $\alpha \in A$ and $\sup_{\alpha \in A} \|A_\alpha u\|_{L^\infty(\mathcal{O})} < \infty$.

iv) u satisfies the HJB equation :

(HJB) $$\sup_{\alpha \in A} [A_\alpha u - f_\alpha] + \lambda u = 0 \qquad \text{a.e. in } \mathcal{O}$$

with $$A_\alpha = -a_{ij}(x,\alpha)\partial_{ij} - b_i(x,\alpha)\partial_i \quad , \quad f_\alpha(x) = f(x,\alpha)$$

v) u satisfies the boundary condition :

(9) $$\sup_{\alpha \in A} [\gamma(x,\alpha) \cdot Du(x)] = 0 \quad , \quad \forall x \in \Gamma_+ .$$

Theorem 2 : If we assume :

(10) $$\exists \mu > 0 , \forall (x,\alpha) \in \bar{\mathcal{O}} \times A , \qquad (a_{ij}) \geq \mu I_N$$

then the value function is the unique solution in $C^{1,1}(\bar{\mathcal{O}})$ of (HJB) satisfying the boundary condition :

(9') $$\sup_{\alpha \in A} [\gamma(x,\alpha) \cdot Du(x)] = 0 \quad , \quad \forall x \in \partial\mathcal{O} .$$

In addition $u \in C^{2,\alpha}(\mathcal{O})$ for some $\alpha \in]0,1[$.

Remarks : i) Theorem 2 is due to P.L. Lions and N. Trudinger [31], and Theorem 1 is deduced from Theorem 2 (and [31]) by the method of P.L. Lions [25] (see also P.L. Lions [27]).

ii) Some partial results in the uniformly elliptic case were first obtained by J.L. Menaldi [32], P.L. Lions [25].

iii) If γ is independent of x and (9) holds, there exists an optimal Markovian control.

We next consider the case of ergodic problems $(\lambda \to 0)$ and we assume (for simplicity) that $\mathcal{O}$ is connected. We denote by u_λ the value function corresponding to a positive discount factor λ.

Theorem 3 : We assume (10). Let x_o be any point in $\mathcal{O}$. Then, $(\lambda u_\lambda \, , \, u_\lambda - u_\lambda(x_o))$ converge uniformly to $(\bar{u}, v_o) \in \mathbb{R} \times C^{1,1}(\bar{\mathcal{O}})$. In addition, $v_o \in C^{2,\alpha}(\mathcal{O})$ for some $\alpha \in [0,1]$ and $(\bar{u}, v_o)$ is the unique solution in $\mathbb{R} \times C^{1,1}(\bar{\mathcal{O}})$ of

$$
(11) \qquad \begin{cases} \sup\limits_{\alpha \in A} \, [A_\alpha v_o - f_\alpha] + \bar{u} = 0 & \text{a.e. in } \mathcal{O} \, , \quad v_o(x_o) = 0 \\ \sup\limits_{\alpha \in A} \, [\gamma(x,\alpha) \cdot Dv_o(x_o)] = 0 & \text{on } \Gamma_+ \, . \end{cases}
$$

Finally, $\bar{u}$ is given by :

$$
(12) \qquad \bar{u} = \lim_{T \to \infty} \frac{1}{T} \inf_{\alpha_t} E \int_0^T f(X_s, \alpha_s) \, ds \quad , \qquad \forall \, x \in \bar{\mathcal{O}} \, .
$$

Remarks : i) Additional characterizations of $\bar{u}, v_o$ are possible ; ii) if γ is independent of α , the ergodic problem (12) admits an optimal Markovian control ; iii) some related ergodic problems appear in J.M. Lasry [18], F. Gimbert [11], P.L. Lions and B. Perthame [29].

III . Forced state constraints.

We want to consider here control problems where we restrict the class of controls to those such that the state of the system (i.e. X_t) never leaves $\bar{\mathcal{O}}$. It is clear enough that if $\mathcal{O}$ is bounded (say) and σ is nondegenerate (i.e. $\sigma\sigma^T$ definite positive), unbounded drifts are to be used to constrain X_t . Hence the simplest situation is :

$$
(13) \qquad dX_t = dB_t + \alpha(X_t) \, dt \quad , \qquad X_o = x \in \mathcal{O}
$$

where we immediately took feedback controls that is functions $\alpha(x) \in C^{0,1}(\mathcal{O})$ and we will say that α is admissible if the solution of (13) defined up to time $\tau = \inf \, (t \geq 0 \, , \, X_t \notin \mathcal{O})$ satisfies :

$$
(14) \qquad X_t \in \mathcal{O} \, , \, \forall \, t \geq 0 \quad \text{or} \quad \tau = +\infty \quad \text{a.s.} \, , \quad \forall \, x \in \mathcal{O} \, .
$$

And again to simplify, we will consider the following example of cost functions :

(15) $$J(x,\alpha) = E\int_0^\infty \left\{f(X_t) + \frac{1}{q}\,|\alpha(X_t)|^q\right\} e^{-\lambda t}\,dt$$

where $1 < q < \infty$, $\lambda > 0$, $f \in C^{0,1}(\mathcal{O})$ and f is bounded from below ; so that $J(x,\alpha) \leqslant +\infty$. And the value function is given by

(16) $$u(x) = \inf\{J(x,\alpha) \;/\; \alpha \in C^{0,1}(\mathcal{O}) \;,\; \alpha \text{ admissible}\} \;.$$

<u>Remark</u> : If σ presents some degeneracy, the state constraints may be realized in a simpler way and the results concerning the value function are then quite different from those below. A typical example is the deterministic situation and we refer to P.L. Lions [28] for an analysis of that case.

The results concerning the value function u depend on q and on the behavior of f near $\partial\mathcal{O}$ (recall that we denote by $d(x) = \mathrm{dist}(x,\partial\mathcal{O})$). Let us mention also that in the above case the HJB equation reduces to :

(17) $$-\frac{1}{2}\,\Delta u + \frac{1}{p}\,|\nabla u|^p + \lambda u = f \qquad \text{in } \mathcal{O}$$

where p is the conjugate exponent of q : $p = q/(q-1)$. The results which follow are due to J.M. Lasry and P.L. Lions [18].

<u>Theorem 4</u> : Assume that $q \geqslant 2$ and that f is bounded. Then $u \in C^2(\mathcal{O})$ is the unique solution of (17) satisfying :

(18) $$u(x) \to +\infty \qquad \text{if} \qquad d(x) \to 0_+ \;.$$

In addition, u is the maximum solution of (17) in $C^2(\mathcal{O})$ (or in $W^{1,p}_{loc}(\mathcal{O})$) and u satisfies : $u(x)d(x)^{\beta} \to C_o$ as $d(x) \to 0_+$, with $\beta = q-2$, $\beta C_o = \left(\frac{p}{2}\,(\beta+1)\right)^{-1/(p-1)}$ if $q > 2$, $C_o d^{-\beta}$ is replaced by $(-\mathrm{Log}\, d)$ if $q = 2$. Finally, there exists a unique optimal control given by $\alpha = -\nabla u$.

<u>Remark</u> : The result still holds if f satisfies :

$$f(x)\, d(x)^q \to 0 \qquad \text{as} \qquad d(x) \to 0_+ \;.$$

Theorem 5 : Assume that $q < 2$ and that f is bounded. Then, any solution of (17) in $C^2(\mathcal{O})$ bounded from below is bounded from above and belongs to $C^{0,\theta}(\bar{\mathcal{O}})$ with $\theta = 2-q$. Furthermore, u is the maximum solution of (17) in $C^2(\mathcal{O})$, and $u \in C^{0,\theta}(\bar{\mathcal{O}})$. Finally, there exists a unique optimal control given by $\alpha = -\nabla u$.

Roughly speaking the two preceding results correspond to an exit problem with an infinite exit cost (Th. 4) or at least the largest possible exit cost (Th. 5) : these huge exit costs force the system to remain inside $\mathcal{O}$. Another way to realize the constraint is by the application of a cost f which blows up near $\partial\mathcal{O}$:

Theorem 6 : Assume that $q \geqslant 2$ and that f satisfies

(19) $$f(x)d(x)^{\gamma} \to C_1 > 0 \quad \text{as} \quad d(x) \to 0_+$$

with $\gamma > q$. Then $u \in C^2(\mathcal{O})$ is the unique solution of (17) bounded from below. Furthermore, u satisfies

(20) $$u(x)d(x)^{\beta} \to C_2 \quad \text{as} \quad d(x) \to 0_+$$

with $\beta = (\gamma-p)/p$, $\beta C_2 = (C_1 p)^{-1/p}$.

Finally, there exists a unique optimal control given by $\alpha = -\nabla u$.

We conclude with the ergodic analogue of Theorem 4 ;

Theorem 7 : We denote by u_λ the value function given by (16) and we assume that $q \geqslant 2$ and that f is bounded. Then if x_o is a fixed point in $\mathcal{O}$, $(\lambda u_\lambda \, , \, u_\lambda - u_\lambda(x_o))$ converge uniformly on compact subsets of $\mathcal{O}$ to $(\bar{u}, v_o)$ $\in \mathbb{R} \times C^2(\mathcal{O})$ where $(\bar{u}, v_o)$ is the unique solution in $\mathbb{R} \times C^2(\mathcal{O})$ of

$$\begin{cases} -\frac{1}{2}\,\Delta v_o + \frac{1}{p}\,|\nabla v_o|^p + \bar{u} = f \quad \text{in } \mathcal{O}, \qquad v_o(x_o) = 0 \\ v_o(x) \to +\infty \quad \text{if} \quad d(x) \to +\infty \quad . \end{cases}$$

Furthermore, v_o satisfies :

$$v_o(x)d(x)^{\beta} \to C_o \quad \text{if} \quad q > 2 \quad , \quad v_o(x)\left(\text{Log } \frac{1}{d}\right)^{-1} \to 1 \quad \text{if} \quad q = 2 \ ;$$

and $\bar{u}$ is given by

$$(21) \qquad \bar{u} = \lim_{T \to \infty} \frac{1}{T} \inf \left\{ E \int_o^T f(X_t) + \frac{1}{q} |\alpha(X_t)|^q \, dx \ / \ \alpha \ \text{admissible} \right\}$$

for all $x \in \bar{\mathcal{O}}$. Finally, there exists a unique optimal control of the ergodic problem (21) given by $\alpha = -\nabla v_o$.

Bibliography :

[1] A. Bensoussan and J.L. Lions : Applications des inéquations variationnelles en contrôle stochastique. Dunod, Paris, 1978.

[2] A. Bensoussan and J.L. Lions : Contrôle impulsionnel et inéquations quasi-variationnelles. Dunod, Paris, 1982.

[3] L. Cafarelli, J.J. Kohn, L. Nirenberg and J. Spruck : The Dirichlet problem for nonlinear second-order elliptic equations. II-Complex Monge-Ampère and uniformly elliptic equations. To appear.

[4] M.G. Crandall, L.C. Evans and P.L. Lions : Some properties of viscosity solutions of Hamilton-Jacobi equations. Trans. Amer. Math. Soc., 1984.

[5] M.G. Crandall and P.L. Lions : Viscosity solutions of Hamilton-Jacobi equations. Trans. Amer. Math. Soc., 277 (1983), p. 1-42 ; announced in C.R. Acad. Sci. Paris, 292 (1981), p. 183-186.

[6] L.C. Evans : Classical solutions of fully nonlinear, convex, second-order elliptic equations. Comm. Pure Appl. Math., 35 (1982), p. 333-363.

[7] L.C. Evans : Classical solutions of the Hamilton-Jacobi-Bellman equation for uniformly elliptic operators. Trans. Amer. Math. Soc., 275 (1983), p. 245-255.

[8] L.C. Evans and A. Friedman : Optimal stochastic switching on the Dirichlet problem for the Bellman equation. Trans. Amer. Math. Soc., 253 (1979), p. 365-389.

[9] L.C. Evans and P.L. Lions : Résolution des équations de Hamilton-Jacobi-Bellman pour des opérateurs uniformément elliptiques. C.R. Acad. Sci. Paris, 290 (1980), p. 1049-1052.

[10] W.H. Fleming and R. Rishel : Deterministic and stochastic optimal control. Springer, Berlin, 1975.

[11] F. Gimbert : Problème ergodique pour les équations quasilinéaires avec conditions de Neumann. Thèse de 3e cycle, Université Paris-Dauphine, 1984, and to appear in J. Funct. Anal..

[12] N. Ikeda and S. Watanabe : Stochastic equations and diffusion processes. North-Holland, Amsterdam, 1981.

[13] N.V. Krylov : Controlled diffusion processes. Springer, Berlin 1980.

[14] N.V. Krylov : Boundedly nonhomogeneous elliptic and parabolic equations. Izv. Mat. Ser., 46 (1982), p. 487-523.

[15] N.V. Krylov : On degenerate nonlinear elliptic equations. Mat. Sb., 120 (1983), p. 311-330.

[16] N.V. Krylov : Boundedly nonhomogeneous elliptic and parabolic equations in a domain. Izv. Mat. Ser., 47 (1983), p. 75-108.

[17] N.V. Krylov : On degenerate nonlinear elliptic equations. II. Mat. Sb., 121 (1983), p. 211-232.

[18] J.M. Lasry : Contrôle stochastique ergodique. Thèse d'Etat, Université Paris-Dauphine, 1974.

[19] J.M. Lasry and P.L. Lions : Infinite boundary conditions for nonlinear elliptic equations and optimal stochastic control with state constraints. C.R. Acad. Sci. Paris, 1984, and detailed paper to appear.

[20] P.L. Lions : On the Hamilton-Jacobi-Bellman equations. Acta Applic., 1 (1983), p. 17-41.

[21] P.L. Lions : Hamilton-Jacobi-Bellman equations and the optimal control

of stochastic systems. In Proceedings International Congress of Mathematicians, Warsaw, 1983.

[22] P.L. Lions : Optimal control of diffusion processes and Hamilton-Jacobi-Bellman equations. Parts 1,2. Comm. P.D.E., 8 (1983), p. 1101-1174, p. 1229-1276.

[23] P.L. Lions : Some recent results in the optimal control of diffusion processes. In Stochastic Analysis, Proceedings of the Taniguchi International Symposium on Stochastic Analysis, Katata and Kyoto, 1982. Kinokuniya, Tokyo, 1984.

[24] P.L. Lions : Generalized solutions of Hamilton-Jacobi equations. Pitman, London, 1982.

[25] P.L. Lions : Optimal control of diffusion processes and Hamilton-Jacobi-Bellman equations. Part 3. In Nonlinear Partial Differential Equations and their applications, Collège de France Seminar, Vol. V. Pitman, London, 1983.

[26] P.L. Lions : Résolution analytique des problèmes de Bellman-Dirichlet. Acta Math., 146 (1981), p. 151-166 ; announced in C.R. Acad. Sci. Paris, 287 (1978), p. 747-750.

[27] P.L. Lions : Optimal control of diffusion processes. Colloque ENST-CNET. Lecture Notes in Math., Springer, Berlin, 1984.

[28] P.L. Lions : Optimal control and viscosity solutions. In Proceedings of the Conference on Dynamic Programming, Roma 1984 ; eds. Capuzzo-Dolcetta, Fleming and Zolezzi. Springer, Berlin, 1984-85.

[29] P.L. Lions and B. Perthame : Quasi-variational inequalities and ergodic impulse control. To appear.

[30] P.L. Lions and A.S. Sznitman : Stochastic differential equations with reflecting boundary conditions. Comm. Pure Appl. Math., 1984.

[31] P.L. Lions and N.S. Trudinger : Hamilton-Jacobi-Bellman equations and oblique derivative boundary conditions. To appear.

[32] J.L. Menaldi : In Advances in Filtering and Optimal Stochastic Control, Ed. Fleming, Gorostiza. Springer, Berlin, 1983.

ON IMPULSE CONTROL WITH PARTIAL OBSERVATION

G. Mazziotto *

L. Stettner **

J. Szpirglas *

J. Zabczyk ***

Abstract:

This paper presents an existence result for an impulse control problem with partial observation. The unobserved process evolves between any two successive impulse times as a Feller Markov process on a locally compact separable state space, and the observation process is of a "signal + white noise" type.

* PAA/TIM/MTI - Centre National d'Etudes des Télécommunications, Issy les Moulineaux, France.

** Institute of Mathematics, Polish Academy of Sciences, Warsaw, Poland.

*** Institute of Mathematics, Polish Academy of Sciences, and Control Theory Centre, University of Warwick, England.

1 - INTRODUCTION

The paper is concerned with the existence of a separated optimal strategy for an impulse control problem with partial observation. The complete information case was solved by M. Robin ([10]) and by J.P. Lepeltier and B. Marchal ([6]). An extension to the incomplete information case was obtained by G. Mazziotto and J. Szpirglas ([9]). In ([9]) the following conditions have been imposed on the model: compactness of the state space and finiteness of the life time of the unobserved Feller proces. In the present paper we dispense with these two assumptions. The main difficulty encountered here, comes from the fact that, as soon as the state space of the unobserved process is no longer compact, the associated filtering process takes its values in a non-locally compact space.

The paper is organized as follows. In Section 2 we formulate an impulse control problem with partial observation and define the model with which we will work in next chapters. The model is in the spirit of the paper by G. Mazziotto and J. Szpirglas ([9]). However the way the solution is presented can be adapted to different modelling techniques, see report ([14]). Section 3 gathers results on filtering processes and filtering equations. In particular results on continuous dependence of the filter on the initial condition due to L. Stettner ([11]) and L. Stettner and J. Zabczyk ([12]) are presented. The section 4 introduces stopping time results needed for the proof of the separation principle. The main existence theorem is formulated in Section 5, and the proof of the result is sketched. The proof uses ideas from discrete time dynamic programming, and is similar to the one given in ([10]) or ([9]).

The full version of the presented results, with complete proofs, will appear in a subsequent paper by the authors.

2 - FORMULATION OF THE CONTROL PROBLEM

The model describing the evolutions of the unobserved process X^v submitted to the impulse control v, and of the noisy observation Y, is constructed as in ([10]), ([6]) or ([9]) for the control theory part. For the filtering theory part, we use the reference probability method in order to get an observation filtration clearly independent of each control, as in ([9]).

The observation is represented by the coordinate process $Y = (Y_t \,;\, t \geq 0)$ on the space $\Omega = C([0,\infty), \mathbb{R}^d)$ of the continuous functions from $[0,\infty)$ into $\mathbb{R}^d$. Let $°\underline{\underline{G}} = (°\underline{\underline{G}}_t \,;\, t \geq 0)$ be the natural filtration generated by Y, and $°\underline{\underline{A}} = °\underline{\underline{G}}$. As it will be seen further, the impulse control problem leads to consider various probabilities which are not equivalent on $(\Omega, {}^\circ\underline{\underline{A}})$. Nevertheless we need a proper extension of filtration ${}^\circ\underline{\underline{G}}$ which does not depend on those probabilities. Namely, for each bounded ${}^\circ\underline{\underline{G}}$-predictable process H, let $\mathbb{P}^H$ be the probability on $(\Omega, {}^\circ\underline{\underline{A}})$ such that the process $(Y_t - \int_0^t H_s\, ds\ ;\ t \geq 0)$ is a Brownian motion. If $\underline{\underline{G}}^H$ denotes the $\mathbb{P}^H$-augmentation of ${}^\circ\underline{\underline{G}}$ (i.e., for any t, $\underline{\underline{G}}_t^H$ is the σ-field generated by all the $\mathbb{P}^H$-negligible sets of ${}^\circ\underline{\underline{A}}$ and by the σ-field ${}^\circ\underline{\underline{G}}_{t+} = \bigcap_{s>0} {}^\circ\underline{\underline{G}}_{t+s}$, see (2)), we define $\underline{\underline{G}}$ by

$$\forall\, t \geq 0 \ :\ \underline{\underline{G}}_t = \bigcap_H \underline{\underline{G}}_t^H \quad \text{and} \quad \underline{\underline{A}} = \underline{\underline{G}}_\infty,$$

where H runs over the set of all the ${}^\circ\underline{\underline{G}}$-predictable bounded processes. Note that $\mathbb{P}^0$ is simply the Wiener measure on $(\Omega, \underline{\underline{A}})$.

A strategy is a sequence $v = (T_n, \xi_n\ ;\ n > 0)$ of random variables where the T_n 's represent the successive instants of impulse, and the ξ_n 's are the corresponding impulsions assumed to take their values in a compact set U. Not all sequences v will be allowed as legitimate strategies, their choice will be limited by the amount of available information. The proper strategies, called admissible controls, must depend on the observation Y, in a progressive way. Namely, an admissible control is a sequence $v = (T_n, \xi_n\ ;\ n > 0)$ of random variables on $(\Omega, \underline{\underline{A}})$ such that $(T_n\ ;\ n > 0)$ is an increasing sequence of $\underline{\underline{G}}$-stopping times such that $T_{n+1} > T_n$ on $\{T_n < \infty\}$ and $\lim_n T_n = \infty$, and such that $\forall\, n$, ξ_n is a $\underline{\underline{G}}_{T_n}$-measurable random variable with values in the compact set U. The set of admissible controls is denoted by $\underline{\underline{V}}$.

The unobserved process X^v submitted to the admissible control $v = (T_n, \xi_n\ ;\ n > 0)$ is assumed to have free markovian evolutions in a state space E, according to a given Feller transition semi-group $P = (P_t\ ;\ t \geq 0)$, and to have a forced jump at each time T_n of amount $\phi(X_{T_n}^v, \xi_n)$, where ϕ is a given transformation from $E \times U$ into E, called the operational function.

In order to represent the evolution of the unobserved process between two successive impulsions, we first consider a Markov process X in a canonical form $X = (\overline{\Omega}, \overline{\underline{\underline{A}}}, \overline{\underline{\underline{F}}}_t, \overline{\theta}_t, X_t, (\overline{\mathbb{P}}_x\ ;\ x \in E))$ where $\overline{\Omega}$ is the set of all the cad-lag (i.e., right-continuous and with left-limits, see (2)) functions from $[0,\infty)$ into E, $\overline{\underline{\underline{F}}} = (\overline{\underline{\underline{F}}}_t\ ;\ t \geq 0)$ is the filtration generated by the coordinate process $X = (X_t\ ;\ t \geq 0)$ with $\overline{\underline{\underline{A}}} = \overline{\underline{\underline{F}}}_\infty$ and properly completed with respect to the family of probabilities $(\overline{\mathbb{P}}_x\ ;\ x \in E)$, $(\overline{\theta}_t\ ;\ t \geq 0)$ is the translation semi-group on $\overline{\Omega}$. We suppose that the state space E is separable locally compact, equipped with the Borel σ-field $\underline{\underline{E}}$. A point at infinity is denoted by δ and the

Banach space of all continuous functions on E vanishing at δ as $C_0(E)$. We assume that the transition semi-group of X, $P = (P_t\ ;\ t \geq 0)$, enjoys the following Feller 's conditions:

i) if $f \in C_0(E)$, then $P_t f \in C_0(E)$, $\forall\ t \geq 0$

ii) if $f \in C_0(E)$, then $P_t f(x) \to f(x)$ as $t \to 0$, $\forall\ x \in E$.

Let $\hat{E}$ be the set of all the probability laws on $(E,\underline{\underline{E}})$, equipped with the topology of weak convergence, and endowed with its Borel σ-field $\hat{\underline{\underline{E}}}$. For any $\mu \in \hat{E}$, $\mathbb{P}_\mu$ denotes the probability defined on $(\Omega,\underline{\underline{A}})$ by $\mathbb{P}_\mu = \mu.\mathbb{P}_.$. The infinitesimal generator of semi-group P is denoted by $\underline{\underline{L}}$, and its domain by $\underline{\underline{D}}(\underline{\underline{L}})$. Finally $b(E)$ and $b(\hat{E})$ denote the set of all the Borel bounded functions on E and $\hat{E}$, respectively, and $C(\hat{E})$ stands for the set of all the continuous bounded functions on $\hat{E}$.

We now construct the model describing the evolution of the unobserved process X^v submitted to the control $v = (T_n,\xi_n\ ;\ n > 0)$ by the method developed in ([6]) and ([10]).

Set $\tilde{\Omega} = \Omega \times (\bar{\Omega})^{\mathbb{N}}$, $\tilde{\underline{\underline{A}}} = \underline{\underline{A}} \otimes (\bar{\underline{\underline{A}}})^{\otimes\mathbb{N}}$, with the coordinate processes $Y = (Y_t\ ;\ t \geq 0)$ and $(X^n = (X^n_t\ ;\ t \geq 0)\ ;\ n \in \mathbb{N})$ respectively. To any admissible control $v = (T_n,\xi_n\ ;\ n \geq 0)$, we associate the process $X^v = (X^v_t\ ;\ t \geq 0)$ defined on $(\tilde{\Omega},\tilde{\underline{\underline{A}}})$ as follows:

$$\forall\ t \geq 0 : X^v_t = \sum_{n=0}^{\infty} X^n_{t-T_n}\ 1\!\!1_{\{T_n < t \leq T_{n+1}\}} \quad , \text{ where } T_0 \text{ is } 0.$$

The process X^v is right- and left-limited by construction.

The observation filtration generated by Y on $(\tilde{\Omega},\tilde{\underline{\underline{A}}})$ is again denoted by $°\underline{\underline{G}}$. The filtration on $(\tilde{\Omega},\tilde{\underline{\underline{A}}})$ describing the whole evolution of the system (Y,X^v) under control v is defined as in ([6]), and denoted by $\underline{\underline{F}}^v$. For any initial law μ, the reference probability $\mathbb{P}^v_\mu$ is defined on the space $(\Omega,\underline{\underline{F}}^v_\infty)$ as follows.

$$\mathbb{P}^v_\mu(d\omega,d\omega_0,d\omega_1,\ldots) = \mathbb{P}^0(d\omega)\ K^v_\mu(\omega;d\omega_0,d\omega_1,\ldots)$$

where the kernel K^v_μ is constructed thanks to Ionescu-Tulcea theorem by

$$\forall\ n \in \mathbb{N} : K^v_\mu(\omega,d\omega_0,\ldots,d\omega_n) = \mathbb{P}_\mu(d\omega_0)\ \ldots\ \mathbb{P}_{\phi(X^v_{T_n}(\omega,\ldots,\omega_{n-1}),\xi_n)}(d\omega_n)$$

We assume that the operational function ϕ is bounded continuous from $E \times U$ into E, such that two simultaneous impulses are equivalent to a single impulse i.e., there exists ψ from $U \times U$ into U such that

$$\forall\ x \in E,\ \forall\ \xi_1,\ \xi_2 \in U : \phi(\phi(x,\xi_1),\xi_2) = \phi(x,\psi(\xi_1,\xi_2))\ .$$

The definition of $\mathbb{P}^v_\mu$ implies that μ is the initial law of X^v, and $X^v_{T_n^+} = \phi(X^v_{T_n},\xi_n)$ $\mathbb{P}^v_\mu$-a.s. , $\forall\ n \in \mathbb{N}$.

We denote by $\underline{\underline{F}}^{v,\mu,0}$ (resp. $\underline{\underline{G}}^{v,\mu,0}$) the $\mathbb{P}^v_\mu$-augmentation of $\underline{\underline{F}}^v$ (resp. $°\underline{\underline{G}}$).

Now, let us define a change of probability on $(\tilde{\Omega},\tilde{A})$ such that under this new probability, the system (Y,X^v) is a classical filtering model, of "signal+white noise" type.
For h a given bounded continuous function from E into $\mathbb{R}^d$, let L^v be the $(\underline{\underline{F}}^{v,\mu,0}, \mathbb{P}^v_\mu)$-martingale defined by:

$$\forall\ t \geq 0 : L^v_t = \exp\{\int_0^t h(X^v_s).dY_s - \frac{1}{2}\int_0^t |h(X^v_s)|^2\, ds\} \quad .$$

From the boundedness of h, it is well known (see ([7])) that, for any finite t, L^v_t is the Radon-Nikodym derivative of a probability $Q^{v,t}_\mu$ on $(\tilde{\Omega},\underline{\underline{F}}^v_t)$ which is equivalent to $\mathbb{P}^v_\mu$. Moreover, using the functional structure of the space $(\tilde{\Omega},\tilde{\underline{\underline{A}}})$, one can prove that these $Q^{v,t}_\mu$'s define by projectivity a probability Q^v_μ on $(\tilde{\Omega},\underline{\underline{F}}^v_\infty)$. This probability has no reason to be equivalent to $\mathbb{P}^v_\mu$. We denote by $\underline{\underline{F}}^{v,\mu}$ (resp. $\underline{\underline{G}}^{v,\mu}$) the Q^v_μ-augmentation of $\underline{\underline{F}}^v$ (resp. $°\underline{\underline{G}}$).

It can be checked as in ([9]) that under probability Q^v_μ, the evolution of process X^v between any two successive impulse times remains markovian, with the same semi-group P. Moreover a simple application of Girsanov theorem (see ([7])) shows that the process W^v defined by

$$\forall\ t \geq 0 : W^v_t = Y_t - \int_0^t h(X^v_s)\, ds$$

is a Brownian motion.

Finally, let us state the impulse control problem in partial observation to be solved. The average cost function associated to each admissible control $v = (T_n,\xi_n\, ; n > 0)$ and to each initial law μ, is defined as follows.

$$J(v,\mu) = E_{Q^v_\mu}\left(\int_0^\infty e^{-\alpha s} f(X^v_s)\, ds + \sum_{m=1}^{\infty} e^{-\alpha T_n} C(X^v_{T_n},\xi_n) \right)$$

where f is a positive bounded continuous function on E - the cost by unit of time-, C is a continuous bounded function on $E \times U$ greater than a given positive constant - the impulse cost-, and α is a positive number - the actualization factor. To avoid simultaneous impulses, we suppose the impulse cost such that :

$$\forall\ x \in E,\ \forall\ \xi_1,\ \xi_2 \in U : C(x,\xi_1) + C(\phi(x,\xi_1),\xi_2) > C(x,\psi(\xi_1,\xi_2)).$$

Then, the problem is to find an admissible control v* such that

$$J(v^*,\mu) = \inf_{v \in \underline{\underline{V}}} J(v,\mu) \quad , \text{ for } \mu \text{ fixed.}$$

3 - THE FILTERING PROCESSES

The filter of X^v knowing the observation Y is a $\hat{E}$-valued process defined by:

$$\forall t \geq 0, \forall f \in b(E) : \Pi_t^v(f) = E_{Q_\mu^v}(f(X_t^v) / \underline{\underline{G}}_t^{v,\mu}) \quad \text{a.s.} ,$$

if $\mu \in \hat{E}$ is the initial law of X^v, supposed fixed, and v any admissible control. From the construction of the model by reference probability method, we obtain the following Kallianpur-Striebel formula:

$$\forall t \geq 0, \forall f \in b(E) : \Pi_t^v(f) = K_\mu^v(L_t^v\, f(X_t^v)) / K_\mu^v(L_t^v) \quad .$$

The initial law μ is supposed to be fixed, therefore we often omit to mention μ in the sequel. For instance the filter Π^v defined above does depend on μ.

From these formulas, it is easily deduced that the process Π^v is right-limited and left-continuous for the weak topology on $\hat{E}$, and such that

$$\forall n > 0 : \Pi_{T_n^+}^v = \hat{\phi}(\Pi_{T_n}^v, \xi_n)$$

where $\hat{\phi}$ is the function from $\hat{E} \times U$ into $\hat{E}$ defined by

$$\forall \mu \in \hat{E}, \forall \xi \in U : \hat{\phi}(\mu,\xi) = \mu(\phi(.,\xi)) \quad .$$

In this paragraph, we first prove that the filter Π^v can be expressed in terms of a battery of filters $(\Pi^{v,n} ; n \in \mathbb{N})$ simply related to the processes $(X^n ; n \in \mathbb{N})$ of the definition of X^v, and to the control v. Then, we show that all these filters $\Pi^{v,n}$'s are solution in law of a unique filtering equation, with various appropriate initial conditions. Finally, results on the continuous dependence, with respect to the initial law, of the solutions of this equation are quoted.

Let $v \in \underline{\underline{V}}$. For each $n \in \mathbb{N}$, we define an $\hat{E}$-valued process $\Pi^{v,n} = (\Pi_t^{v,n} ; t \geq 0)$ verifying:

$$\forall t \geq 0, \forall f \in b(E) : \Pi_t^{v,n}(f) = E_{Q_\mu^v}(f(X_t^n) / \underline{\underline{G}}_{T_n+t}^{v,\mu})$$

It can be shown that we can choose a version of process $\Pi^{v,n}$ which is defined on $(\Omega,\underline{\underline{A}})$ and $\underline{\underline{G}}$-adapted, and with all its trajectories continuous for the weak topology on E.

For any n, we denote by $\underline{\underline{G}}^n$ the filtration $(\underline{\underline{G}}_{T_n+t} ; t \geq 0)$. Clearly any random variable S is a $\underline{\underline{G}}^n$-stopping time if and only if T_n+S is a $\underline{\underline{G}}$-stopping time.

Then, from the definition of $\Pi^{\nu,n}$, we have the following identity for any $\underline{\underline{G}}^n$-stopping time S such that $0 < S \leq T_{n+1} - T_n$:

$$\forall\ f \in b(E)\ :\ \Pi_S^{\nu,n}(f) = \Pi_{T_n+S}^{\nu}(f) \qquad \Omega^{\nu}\text{-a.s.}\ .$$

In the sequel we make use of the following notion of solution of the filtering equation. Let $(\Omega,\underline{\underline{A}},\mathbb{P})$ be an arbitrary probability space endowed with a right-continuous filtration $\underline{\underline{H}} = (\underline{\underline{H}}_t\ ;\ t \geq 0)$, and with a d-dimensional $\underline{\underline{H}}$-Brownian motion $B = (B_t\ ;\ t \geq 0)$. On this space, let μ be a $\underline{\underline{H}}_0$-measurable $\widehat{E}$-valued random variable, and let $q = (q_t\ ;\ t \geq 0)$ be an $\underline{\underline{H}}$-adapted $\widehat{E}$-valued stochastic process which is continuous for the weak topology. We say that the term $(\Omega,\underline{\underline{A}},\underline{\underline{H}},\mu,q,B,\mathbb{P})$ is a solution of equation (I) iff:

$$\text{(I)}\begin{cases} \forall\ f \in \underline{\underline{D}}(\underline{\underline{L}}),\ \forall\ t \geq 0\ ,\quad \mathbb{P}\text{-a.s.}\ : \\ q_t(f) = \mu(f) + \int_0^t q_s(\underline{\underline{L}}f)\ ds + \int_0^t (q_s(hf) - q_s(h)q_s(f)).dB_s \end{cases}$$

Let denote by $C(\widehat{E})$ the space of all bounded continuous functions on $\widehat{E}$. Recall that if E is non compact, then the set $\widehat{E}$ has no reason to be locally compact. However the weak topology on $\widehat{E}$ can be generated by the metric ρ defined as follows:

$$\forall\ \mu,\ \nu \in \widehat{E}\ :\ \rho(\mu,\nu) = \sum_{i=0}^{\infty} 2^{-i}\ |\mu(f_i) - \nu(f_i)|\ /\ 1 + \|f_i\|$$

where $(f_i\ ;\ i \in \mathbb{N})$ is a countable family of functions of $\underline{\underline{D}}(\underline{\underline{L}})$ dense in $C_0(E)$ (the existence of which comes from the Feller property of semigroup P).

The following theorem is due basially to H. Kunita ([5]), who proved it for a compact space E. The extension to locally compact spaces is contained in ([12]) for the Feller property, and the uniqueness is proved as in ([13]).

<u>THEOREM 1</u> : For an arbitrary $\underline{\underline{H}}_0$-measurable initial condition μ, there exists a unique solution $q^{\mu} = (q_t^{\mu}\ ;\ t \geq 0)$ of (I), a.s. continuous and $\underline{\underline{H}}$-adapted. Moreover, the solution q^{μ} is a Markov process with respect to $\underline{\underline{H}}$ and $\mathbb{P}$, with a transition semi-group $\widehat{P} = (\widehat{P}_t\ ;\ t \geq 0)$ satisfying the following Feller property.

i) $\forall\ g \in C(\widehat{E})\ :\ \widehat{P}_t g \in C(\widehat{E})$ for any $t \geq 0$,

ii) $\forall\ g \in C(\widehat{E}),\ \forall\ \nu \in \widehat{E}\ :\ \lim_{t \downarrow 0} \widehat{P}_t g(\nu) = g(\nu)$.

With some computation, we can verify that, for any n, the

term $(\Omega, \underline{\underline{A}}, \underline{\underline{G}}^n, \Pi^v_{T_n+}, \Pi^{v,n}, I^n, Q^v)$ is a solution of (I), where I^n is the innovation process defined by

$$\forall\ t \geq 0 : I^n_t = Y_{T_n+t} - Y_{T_n} - \int_0^t \Pi^{v,n}_s(h)\ ds$$

Consequently, each $\Pi^{v,n}$ is a Markov process with the Feller semi-group $\hat{P}$.

The dependence of the solution q^μ of (I) on the initial condition $q^\mu_0 = \mu$ is studied by Stettner in ([11]). The following results will be developed in a subsequent version of this paper.

THEOREM 2 : Let $(\mu_n ; n \in \mathbb{N})$ be a sequence in $\hat{E}$ converging weakly toward μ_∞. The following assertions hold.

i) $\forall\ \varepsilon > 0,\ \forall\ T > 0 : \exists$ compact $\Gamma \subset \hat{E}$ such that
$\forall\ n \leq \infty : \mathbb{P}(\{q^{\mu_n}_t \in \Gamma\ ,\ \forall\ t \leq T\}) \geq 1 - \varepsilon$

ii) $\forall\ \eta > 0,\ \forall\ \varepsilon > 0,\ \forall\ T > 0 : \exists$ N such that
$\forall\ n \geq N : \mathbb{P}(\{\sup_{t \leq T} \rho(q^{\mu_n}_t, q^{\mu_\infty}_t) < \eta\}) \geq 1 - \varepsilon$

iii) $\forall\ \eta > 0,\ \forall\ \varepsilon > 0 : \exists\ \delta > 0$ such that
$\forall\ \mu \in \hat{E} : \mathbb{P}(\{\sup_{t \leq \delta} \rho(\mu, q^\mu_t) < \eta\}) \geq 1 - \varepsilon$

The assertion i) is classical for a $\mathbb{R}^d$-valued Feller process ([10]), but for the filter which is valued in the non-locally compact space $\hat{E}$ it needs a particular proof. This is done in ([12]) and ([11]). The assertions ii) and iii) are proved in ([11]), mainly by using Eq. (I).

4 - SOME OPTIMAL STOPPING RESULTS :

In this section we present results on optimal stopping needed in the sequel. The conditions imposed allow to use in the proofs discrete-time approximation technique which goes back to Maskevicius ([8]), see also ([12]). However most of them could be deduced of the more general approaches of Bismut and Skalli ([1]), or El Karoui ([3]). We mainly work on a Feller process taking its values in a metric but non-locally compact space $\hat{E}$. The main result concerns the continuity of the value function. Our formulation well suited for impulse control application is taken from ([14]).

Let $\hat{P} = (\hat{P}_t ; t \geq 0)$ be a fixed semi-group acting on $b(\hat{E})$, let be given $g \in C(\hat{E})$, and let $\alpha > 0$. We start from the following set of inequalities.

For $v \in b(\hat{E})$: $v \geq g$

$$v \geq e^{-\alpha t} \hat{P}_t v \quad , \forall t \geq 0 .$$

When the subset in $b(\hat{E})$ of the functions v which satisfy the above inequalities has a unique minimal element, this solution is called reduite of g, and denoted by Rg. General existence results can be found in ([3]). The following theorem states conditions for a better regularity of the reduite Rg.

<u>THEOREM 3</u> : Assume that for arbitrary $g \in C(\hat{E})$, the function $(t,x) \to \hat{P}_t g(x)$ is continuous on $[0,\infty) \times \hat{E}$ with respect to each variable separately. Then Rg exists, and is lower semi-continuous.

Next theorem interpretes Rg as the value function of an optimal stopping problem.

<u>THEOREM 4</u> : Let $\Pi = (\Pi_t \,;\, t \geq 0)$ be a right-continuous Markov process with transition semi-group $\hat{P} = (\hat{P}_t \,;\, t \geq 0)$, defined on a probability space $(\Omega, \underline{\underline{A}}, \mathbb{P})$ endowed with a right-continuous filtration $\underline{\underline{F}} = (\underline{\underline{F}}_t \,;\, t \geq 0)$. If $\underline{\underline{T}}(\underline{\underline{F}})$ denotes the set of all the $\underline{\underline{F}}$-stopping times, then for arbitrary $\underline{\underline{F}}_0$-measurable random variable $\xi \geq 0$, we have:

$$E(\xi \,.\, Rg(\Pi_0)) = \sup_{T \in \underline{\underline{T}}(F)} E(\xi \,.\, e^{-\alpha T} g(\Pi_T)) \,.$$

To insure continuity of the reduite Rg for all $g \in C(\hat{E})$ as well as existence of an optimal stopping time, we have to impose the following condition (C).

(C) For arbitrary convergent sequence $(\mu_n \,;\, n \in \mathbb{N})$ in $\hat{E}$, and any number $\varepsilon > 0$, there exists a compact set L containing the μ_n 's and a number $\delta > 0$ such that, if $g \in C(\hat{E})$ with $0 \leq g \leq 1$ and $g \leq \delta$ on L, then $Rg(\mu_n) \leq \varepsilon$ for $n = 1,2,\ldots$.

This condition (C) is always satisfied for Feller semi-groups on a locally compact space E, see ([12]), ([14]). But it can be verified that this condition (C) is also fullfilled for the filters considered in Section 3. This is an easy consequence of the Assertion i) of the Theorem 2, and of the interpretation of the reduite from Theorem 4.

The following theorem is the main result of this section.

THEOREM 5 : Assume that for arbitrary $g \in C(\hat{E})$ transformation $(t,\mu) \to \hat{P}_t g(\mu)$ is continuous on $[0,\infty) \times \hat{E}$, and Condition (C) holds. Then, Rg is continuous on $\hat{E}$.
Moreover, if $\Pi = (\Pi_t \,;\, t \geq 0)$ is a right-continuous quasi-left-continuous Markov process of transition semi-group $\hat{P}$, defined on $(\Omega, \underline{\underline{A}}, \mathbb{P})$ with respect to a right-continuous filtration $\underline{\underline{F}}$ then, there exists an optimal stopping time T^* for the problem

$$E(e^{-\alpha T^*} g(\Pi_{T^*})) = \sup_{T \in \underline{\underline{T}}(\underline{\underline{F}})} E(e^{-\alpha T} g(\Pi_T))$$

such that $\mathbb{P}$-a.s.,

$$T^* = \inf\{t \geq 0 : Rg(\Pi_t) = g(\Pi_t)\} \quad .$$

The second part of the theorem follows from classical results on optimal stopping, see ([1]) or ([3]). However the first part seems to be new, and extends the result of Robin ([10]).

We finish this section with an abstract version of a Bellman equation for impulse control problems. For this purpose, let be given a monotonic transformation M on $C(\hat{E})$ i.e., such that $g_1 \leq g_2$ implies $Mg_1 \leq Mg_2$ on $\hat{E}$. Let $f \in C(\hat{E})$ be given, and define

$$h = \int_0^\infty e^{-\alpha t} \hat{P}_t f \, dt \quad \text{and} \quad g = M(0) \ .$$

THEOREM 6 : Assume that for a positive $\gamma > 0$, $\gamma h \leq g$, and that assumptions of Theorem 5 are satisfied. Then, there exists a unique function $u \in C(\hat{E})$ such that, for arbitrary $\underline{\underline{F}}_0$-measurable variable $\xi \geq 0$,

$$E(\xi . u(\Pi_0)) = \inf_{T \in \underline{\underline{T}}(\underline{\underline{F}})} E(\xi . (\int_0^T e^{-\alpha t} f(\Pi_t) \, dt + e^{-\alpha T} Mu(\Pi_T))) .$$

Moreover, there exists an optimal stopping time T^* for the problem above such that, $\mathbb{P}$-a.s.,

$$T^* = \inf\{t \geq 0 : Mu(\Pi_t) = u(\Pi_t)\} \quad .$$

This result follows from the previous theorem, and from a fixed point theorem of Hanouzet and Joly ([4]) type. The main property needed is the concavity of the reduite operation. For details, see ([14]).

5 - THE SEPARATED IMPULSE CONTROL PROBLEM :

Let us come back to the impulse control problem stated in Section 2. The average cost to be minimized can be easily expressed by means of the filtering process. For any $\mu \in \hat{E}$ and $v \in \underline{\underline{V}}$, we get

$$J(v,\mu) = E_{Q_\mu^v}\left(\int_0^\infty e^{-\alpha t}\,\hat{f}(\Pi_t^v)\,dt + \sum_{m=1}^\infty e^{-\alpha T_m}\,\hat{C}(\Pi_{T_m}^v,\xi_m)\right),$$

where $\hat{f}$ and $\hat{C}$ are the positive bounded continuous functions respectively defined on $\hat{E}$ and $\hat{E} \times U$ by :

$$\forall\, \mu \in \hat{E},\ \forall\, \xi \in U\ :\ \hat{f}(\mu) = \mu(f) \text{ and } \hat{C}(\mu,\xi) = \mu(C(.,\xi)) \quad .$$

Then, the separated control problem consists in finding an admissible optimal control v^* in terms of the filter Π^{v^*}.

Let consider the transformation M defined on $C(\hat{E})$ by

$$\forall\, g \in C(\hat{E})\ :\ Mg(\mu) = \inf_{\xi\in U}\{\hat{C}(\mu,\xi) + g(\hat{\phi}(\mu,\xi))\} \quad , \forall\, \mu \in \hat{E}.$$

We also consider the function $u \in C(\hat{E})$ defined according to Theorem 6 with M given above, and with any term $(\Omega,\underline{\underline{A}},\underline{\underline{H}},\mu,q^\mu,B,\mathbb{P})$ solution of Eq. (I). Namely, for any $\underline{\underline{H}}_0$-measurable $\Theta \geq 0$, we have

$$E(\Theta.u(q_0^\mu)) = \inf_{T\in\underline{\underline{T}}(\underline{\underline{H}})} E(\Theta.(\int_0^T e^{-\alpha t}\,\hat{f}(q_t^\mu)\,dt + e^{-\alpha T}\,Mu(q_T^\mu))) \quad .$$

Now, let us state our main result.

THEOREM 7 : The function u defined above gives the optimal cost cost of the separated problem i.e., for any $\mu \in \hat{E}$:

$$u(\mu) = \inf_{v\in\underline{\underline{V}}} E_{Q_\mu^v}\left(\int_0^\infty e^{-\alpha t}\,f(\Pi_t^v)\,dt + \sum_{m=1}^\infty e^{-\alpha T_m}\,C(\Pi_{T_m}^v,\xi_m)\right).$$

Moreover, there exists an optimal admissible control for the separated problem, as well as for the initial problem.

Proof: Let us prove by induction on n that, for arbitrary admissible contro $v = (T_k,\xi_k\ ;\ k > 0)$, we have for any $\mu \in \hat{E}$ the following inequality.

$$u(\mu) \leq E_{Q_\mu^v}\left(\int_0^{T_n} e^{-\alpha t}\,f(\Pi_t^v)\,dt + \sum_{k=1}^n e^{-\alpha T_k}\,C(\Pi_{T_k}^v,\xi_k) + e^{-\alpha T_n}\,u(\Pi_{T_n^+}^v)\right).$$

From Theorem 1, we obtain that the term $(\Omega,\underline{\underline{A}},\underline{\underline{G}}^n,\Pi_{T_n^+}^v,\Pi^{v,n},I^n,Q_\mu^v)$ is the unique solution of Eq. (I), with transition semi-group $\hat{P}$.

Therefore, from the definition of u, we get for any $\underline{\underline{G}}_0^n$-measurable variable $\Theta \geq 0$:

$$E(\Theta.u(\Pi^v_{T_n^+})) = \inf_{S\in \underline{\underline{T}}(\underline{\underline{G}}^n)} E_{Q^v_\mu}(\Theta.(\int_0^S e^{-\alpha t} f(\Pi_t^{v,n})\, dt + e^{-\alpha S} Mu(\Pi_S^{v,n}))) .$$

Particularly, $S = T_{n+1} - T_n$ is a $\underline{\underline{G}}^n$-stopping time, and $e^{-\alpha T_n}$ is $\underline{\underline{G}}_{T_n}$-measurable, thus

$$E(e^{-\alpha T_n} u(\Pi^v_{T_n^+})) \leq E_{Q^v_\mu}(\int_0^{T_{n+1}-T_n} e^{-\alpha t} f(\Pi_t^{v,n})\, dt +$$

$$+ e^{-\alpha T_{n+1}}(C(\Pi^{v,n}_{T_{n+1}-T_n},\xi_{n+1}) + u(\phi(\Pi^{v,n}_{T_{n+1}-T_n},\xi_{n+1}))))$$

Using the fact that $\Pi_S^{v,n} = \Pi^v_{T_n+S}$ a.s. for any $\underline{\underline{G}}^n$-stopping time S such that $S \leq T_{n+1} - T_n$, we get :

$$E(e^{-\alpha T_n} u(\Pi^v_{T_n^+})) \leq E_{Q^v_\mu}(\int_{T_n}^{T_{n+1}} e^{-\alpha t} f(\Pi_t^v)\, dt +$$

$$+ e^{-\alpha T_{n+1}} (C(\Pi^v_{T_{n+1}},\xi_{n+1}) + u(\Pi^v_{T^+_{n+1}}))) .$$

The initial inequality at order n+1 is easily deduced of what preceds. Since u is bounded and $\lim_n T_n = \infty$, we obtain by taking the limit in n,

$$u(\mu) \leq J(v,\mu) \quad \forall \mu \in E \text{ and } \forall v \in \underline{\underline{V}} .$$

Now, let us construct by induction on n, an impulse control which achieves the infimum.

First notice that from the continuity of function u and from the compactness of set U, one can find a measurable selection $\mu \to \xi^*(\mu)$ such that

$$Mu(\mu) = C(\mu,\xi^*(\mu)) + u(\phi(\mu,\xi^*(\mu))) \quad , \quad \forall \mu \in E .$$

Suppose we have found n impulses $(T_1^*,\xi_1^*;\ldots;T_n^*,\xi_n^*)$ such that for $v^n = (T_1^*,\xi_1^*;\ldots;T_n^*,\xi_n^*;\infty,0)$:

$$u(\mu) = E_{Q^v_\mu}(\int_0^{T_n} e^{-\alpha t} f(\Pi_t^{v^n})\, dt + \sum_{m=1}^n e^{-\alpha T_m} C(\Pi^{v^n}_{T^*_m},\xi^*_m)$$

$$+ e^{-\alpha T_n^*} u(\Pi^{v^n}_{T_n^{*+}})) .$$

Let define

$$S_n^* = \inf\{t \geq 0 : Mu(\Pi_t^{v^n,n}) = u(\Pi_t^{v^n,n})\}$$

$$\xi^*_{n+1} = \xi^*(\Pi^{v^n,n}_{S^*_n}) \quad \text{and} \quad T^*_{n+1} = T^*_n + S^*_n \quad .$$

We easily check by using Theorem 6, that the preceding equality also holds for the control $v^{n+1} = (T^*_1,\xi^*_1;\ldots;T^*_n,\xi^*_n;T^*_{n+1};\xi^*_{n+1};\infty,0)$. Then, by induction on n, we construct the optimal impulse control v* and the corresponding filtering process Π^{v^*}. This achieves the proof.

REFERENCES :

(1) J.M. BISMUT, B. SKALLI: "Temps d'arrêt optimal, théorie générale des processus et processus de Markov". Z.f.Wahr.V.Geb. 39 N°1(1977), 301-314.

(2) C. DELLACHERIE, P.A. MEYER: "Probabilités et Potentiel" Tomes 1 & 2 Hermann, Paris (1975) & (1980).

(3) N. EL KAROUI : "Les aspects probabilistes du controle stochastique". in Ecole d'été de Saint-Flour VIII. Lect. N. Maths N° 876, Springer Verlag, Berlin (1981).

(4) B. HANOUZET, J.L. JOLY: "Convergence uniforme des itérés définissant la solution d'une inéquation quasi-variationnelle abstraite". C. R. Acad. Sc. Paris 286 (1978), 735-738.

(5) H. KUNITA : "Asymptotic Behaviour of the Non-linear Filtering Errors of Markov processes". J. Mult. Anal. 1, N°4 (1971), 365-393.

(6) J.P. LEPELTIER, B. MARCHAL : "Théorie générale du controle impulsionnel". Thèse Univ. du Maine & Thèse Univ. Paris 6 (1980).

(7) R.S. LIPSTER, A.N. SHIRYAYEV : "Statistics of Random Processes". Appl. of Math. N°5, Springer Verlag, Berlin (1977).

(8) V. MASKEVICIUS : "Convergence of the value of the game in optimal stopping problems of Markov processes". Liet. Mat. Rink. 14 (1974), 113-127.

(9) G. MAZZIOTTO, J. SZPIRGLAS : "Separation Principle for Impulse Control with Partial Information". Stochastics 10 (1983), 47-73.

(10) M. ROBIN : "Controle impulsionnel des processus de Markov". Thèse Univ. Paris 9 (1978).

(11) L. STETTNER : "On Optimal Stopping of Feller Markov Processes with Incomplete Information in Locally Compact State Space". Preprint(1983).

(12) L. STETTNER, J. ZABCZYK : "Optimal Stopping for Feller Processes". Preprint IM Polish Academy of Sciences N° 284 (1983).

(13) J. SZPIRGLAS : "Sur l'équivalence d'équations différentielles stochastiques à valeurs mesures intervenant dans le filtrage markovien non-linéaire". Ann. Inst. H. Poincaré Vol XIV, N°1 (1978), 33-59.

(14) J. ZABCZYK : "On the Synthesis Problem in Impulse Control". Control Theory Centre Report, Univ. of Warwick (1984).

CONSTRUCTION AND CONTROL OF REFLECTED DIFFUSION WITH JUMPS

José Luis MENALDI (*)
Department of Mathematics
Wayne State University
Detroit, Michigan 48202
USA

and

Maurice ROBIN
INRIA Rocquencourt
Boîte Postale 105
78153 le Chesnay Cedex
France

ABSTRACT

By means of a penalization argument on the domain, we construct a diffusion process with jumps and with instantaneous reflection at the boundary. This approach enables us to give a stochastic interpretation of the solution of Hamilton-Jacobi-Bellman equations for second order integro-differential operators with Neumann boundary conditions.

INTRODUCTION

In this paper we consider some optimal control problems where the state of the system obeys a stochastic differential equation of Wiener-Poisson type in a bounded domain with a reflection condition on the boundary.

First we give a method for constructing the reflected process, using a penalization argument on the domain. Roughly speaking, we start with a jump diffusion processes $(y^{\varepsilon}(t),\ t > 0)$, indexed by $\varepsilon > 0$ and defined in the whole space $\mathbb{R}^d$, with drift coefficient $g(y^{\varepsilon}(t), v(t)) - \frac{1}{\varepsilon}\beta(y^{\varepsilon}(t))$, diffusion term $\sigma(y^{\varepsilon}(t), v(t))$, and jump intensity $\gamma(y^{\varepsilon}(t), \zeta, v(t))$. The domain $\bar{\mathcal{O}}$ is either convex or smooth in $\mathbb{R}^d$ and the function β represents a penalization factor, which measure the distance to the open set $\mathcal{O}$ in an appropriate way and according to the reflected direction ν prescribed on the boundary $\partial\mathcal{O}$. The functions g,σ,γ are given and $v(.)$ stands for the control which is an adapted process taking values in a compact metric space V. By letting ε go to zero, we construct a controlled process $(y(t),\ t \geq 0)$ modelling the state of the dynamic system. This process is refered to as the reflected diffusion with jumps on the domain $\bar{\mathcal{O}}$. By means of this construction, we obtain a stochastic interpretation of variational inequalities and Hamilton-Jacobi-Bellman equations for integro-differential operator with Neumann boundary conditions. Related works are contained in Anulova [1],[2], Chaleyat-Maurel et al.[6] Bensoussan et Lions [3],[4], P.L. Lions et al. [11] and Shalaumov [17].

(*) This research has been completed during a visit to the University of Paris IX - -Dauphine.

1. CONSTRUCTION OF THE PROCESS

Let $(\Omega,F,P,F_t,w_t,\mu_t,\ t \geq 0)$ be a complete Wiener-Poisson space in $\mathbb{R}^n \times \mathbb{R}^m_*$, $\mathbb{R}^m_* = \mathbb{R}^m - \{0\}$, with Levy measure π, i.e. (Ω,F,P) is a complete probability space, $(F_t,\ t \geq 0)$ is a right continuous increasing family of complete sub σ-algebras of F, $(w_t,\ t \geq 0)$ is a standard Wiener process in $\mathbb{R}^n$ with respect to $(F_t,\ t \geq 0)$, $(\mu_t,\ t \geq 0)$ is a martingale measure in $\mathbb{R}^m_*$ w.r.t. $(F_t,\ t \geq 0)$, independent of $(w_t,\ t \geq 0)$ and corresponding to a standard Poisson random measure $p(t,A)$, namely, for any Borel measurable subset of $\mathbb{R}^m_*$, we have $\mu_t(A) = p(t,A) - t\ \pi(A)$, $E\{p(t,A)\} = t\ \ \pi(A)$. For a detailed study of diffusion processes with jumps we refer to Bensoussan and Lions [4], for instance.

Suppose that V is a compact metric space and that $g = (g_i, i=1,\ldots,d)$, $\sigma = (\sigma_{ik}$, $i=1,\ldots,d$, $k=1,\ldots,n)$, $\gamma=(\gamma_i, i=1,\ldots,d)$ are Borel measurable functions, $g : \mathbb{R}^d \times V \to \mathbb{R}^d$, $\sigma : \mathbb{R}^d \times V \to \mathbb{R}^d \times \mathbb{R}^n$, $\gamma : \mathbb{R}^d \times \mathbb{R}^m_* \times V \to \mathbb{R}^d$, satisfying

$$\left.\begin{aligned} & |g(x,v)|^p + |\sigma(x,v)|^p + \int_{\mathbb{R}^m_*} |\gamma(x,\zeta,v)|^p\ \pi(d\zeta) \leq C_p, \\ & |g(x,v) - g(x',v)|^p + |\sigma(x,v) - \ (x',v)|^p + \\ & + \int_{\mathbb{R}^m_*} |\gamma(x,\zeta,v) - \gamma(x',\zeta,v)|^p\ \pi(d\zeta) \leq C_p |x-x'|^p , \end{aligned}\right\} \tag{1.1}$$

for every $p \geq 2$, x,x' in $\mathbb{R}^d$, v in V and some constant C_p independent of x, x' and v, and with $|.|$ denoting the appropriate Euclidian norm.

The variable v is to be replaced by a progressively measurable stochastic process $(v(t),\ t \geq 0)$ with values into V.

1.1. Convex domains

For a given open subset $\mathcal{O}$ of $\mathbb{R}^d$ such that

$$\mathcal{O} \text{ is convex and bounded,} \tag{1.2}$$

we assume that

$$x + \gamma(x,\zeta,v) \in \bar{\mathcal{O}}\ , \quad \forall x \in \bar{\mathcal{O}}\ , \quad \forall \zeta \in \mathbb{R}^m_*\ , \quad \forall v \in V, \tag{1.3}$$

where $\bar{\mathcal{O}}$ denotes the closure of $\mathcal{O}$. This means that all jumps origined at $\bar{\mathcal{O}}$ are always inside $\bar{\mathcal{O}}$.

A reflected diffusion process with jump $(y(t),\ t \geq 0)$ and its associated reflecting process $(\eta(t),\ t \geq 0)$ is a pair of progressively measurable stochastic processes which are right continuous having left-hand limits such that

$$\left.\begin{array}{ll}
\text{(i)} & y(t) \text{ takes values into the closure } \bar{O} \text{ and } \eta(t) \text{ is continuous with locally bounded variation ;} \\
\text{(ii)} & dy(t) = g(y(t),\ v(t))dt + \sigma(y(t),\ v(t))dw_t + \\
& + \int_{\mathbb{R}^m_*} \gamma(y(t),\zeta,v(t))\ d\mu_t(\zeta) - d\eta(t),\ t \geq 0, \\
& y(0) = x, \quad \eta(0) = 0; \\
\text{(iii)} & \text{for every } (z(t),\ t \geq 0) \text{, progressively measurable stochastic process which is right continuous having left-hand limits and takes values into the closure } \bar{O}, \\
& \text{we have} \\
& \int_0^T (y(t) - z(t))d\eta(t) \geq 0,\ \forall T \geq 0.
\end{array}\right\} \qquad (1.4)$$

This problem is refered to as a stochastic variational inequality (SVI) for reflected diffusion processes with jumps in convex domains. We approximate the SVI (1.4) by means of a classical penalty argument on a diffusion process with jumps in the whole space $\mathbb{R}^d$.

Without loss of generality, we may assume that

$$\gamma(x,\zeta,v) = \gamma(pr(x),\zeta,v)\ ,\ \forall x \in \mathbb{R}^d,\ \forall \zeta \in \mathbb{R}^m_*\ ,\ \forall v \in V\ , \qquad (1.5)$$

where pr(.) denotes the orthogonal projection on the closure $\bar{O}$.

Consider the stochastic differential equation

$$\left.\begin{array}{l}
dy^\varepsilon(t) = g(y^\varepsilon(t),\ v(t))dt + \sigma(y^\varepsilon(t),\ v(t))dw_t + \\
+ \int_{\mathbb{R}^m_*} \gamma(y^\varepsilon(t),\zeta,v(t))d\mu_t(\zeta) - \frac{1}{\varepsilon}\beta(y^\varepsilon(t))dt\ ,\ t \geq 0, \\
y^\varepsilon(0) = x,
\end{array}\right\} \qquad (1.6)$$

with $\varepsilon > 0$, x in $\bar{O}$ and $\beta = (id - pr)^*$, i.e.

$$\beta(x) = \frac{1}{2}\,\mathrm{grad}(\min\,\{|x-y|^2 : y \in \bar{O}\}),\quad \forall x \in \mathbb{R}^d. \qquad (1.7)$$

The problem (1.4) has been introduced by Bensoussan and Lions [3] for the deterministic case, i.e. $\sigma = 0$, $\gamma = 0$, where $\eta(t)$ becomes Lipschitzian and existence and uniqueness of the solution $(y(t),\ \eta(t),\ t \geq 0)$ were proved. The penalization (1.6) was used by Shalaumov [17] in a half-space and in P.L. Lions et al. [11] for diffusion processes, i.e. $\gamma = 0$. In both cases, only a weak convergence of the image measures of the processes was obtained.

THEOREM 1.1. Let the assumptions (1.1), (1.2), (1.3) and (1.5) hold. Then the SVI (1.4) has one and only one solution $(y(t), \eta(t), t \geq 0)$. Moreover, for each $p \geq 1$, $T > 0$, we have

$$E\{\sup_{0\le t\le T} |y^{\varepsilon}(t) - y(t)|^p\} \to 0 \text{ as } \varepsilon \to 0, \tag{1.8}$$

$$E\{\sup_{0\le t\le T} |\frac{1}{\varepsilon}\int_0^t \beta^*(y^{\varepsilon}(s))ds - \eta(t)|^p\} \to 0 \text{ as } \varepsilon \to 0, \tag{1.9}$$

where β^* denotes the transpose of β and the limits are uniform with respect to x in $\bar{\mathcal{O}}$ and any v(.) progressively measurable process valued into V. □

The method of the proof is very similar to [13]. A crucial difference is to show that assumptions (1.3), (1.5) imply

$$E\{\sup_{0\le t\le T} |\beta(y^{\varepsilon}(t))|^p\} \le C\,\varepsilon^{p/2-1}\quad,\quad \forall\varepsilon > 0, \tag{1.10}$$

for some constant C. Details of the proof, in the case without control v(.), can be found in [14] . Note that the uniqueness of the solutions for the SVI (1.4) is a direct consequence of the convexity of the domain $\mathcal{O}$. However, the existence of solutions, even for $\gamma = 0$ and $\mathcal{O}$ smooth, is more complicated to be established, cfr. Bensoussan and Lions [4].

1.2. Smooth Domains

Instead of (1.2) and (1.3), we assume that

$$\left.\begin{array}{l} \mathcal{O} \text{ is bounded and there exists a function } \rho(x) \text{ from } \mathbb{R}^d \text{ into } \mathbb{R} \\ \text{which is twice continuously differentiable and satisfies} \\ \mathcal{O} = \{x \in \mathbb{R}^d : \rho(x) < 0\},\ \partial\mathcal{O} = \{x \in \mathbb{R}^d : \rho(x) = 0\}, |\nabla\rho| \ge 1 \text{ on } \partial\mathcal{O}, \end{array}\right\} \tag{1.11}$$

where ∇ stands for the grandient operator, and

$$\left.\begin{array}{l} x + t\gamma(x,\zeta,v) \in \bar{\mathcal{O}}\quad,\ \forall x \in \bar{\mathcal{O}},\ \forall\zeta \in \mathbb{R}^m_*\ , \\ \qquad\qquad \forall v \in V\ ,\ \forall t \in [0,1]\ . \end{array}\right\} \tag{1.12}$$

Suppose we are given a vector field ν defined in neighbourhood of the closure $\bar{\mathcal{O}}$, which is twice continuously differentiable and

$$\nu(x).\nabla\rho(x) \ge \delta > 0 \qquad \forall x \in \partial\,\mathcal{O}\ . \tag{1.13}$$

Note that $|\nabla\rho(x)|^{-1}\nabla\rho(x)$ is the outward unit normal to $\mathcal{O}$ at the point x.
A jump diffusion process $(y(t),\ t \geq 0)$, with instantaneous reflection at the boundary $\partial\mathcal{O}$ in the direction ν, is a progressively measurable stochastic process, which is right continuous having left-hand limits and

$$\left.\begin{array}{ll}
\text{(i)} & y(t) \text{ takes values into the closure } \bar{\mathcal{O}} \text{ and there exists} \\
& \text{an increasing continuous real process } \xi(t) \text{ such that} \\
\text{(ii)} & dy(t) = g(y(t),\ v(t))dt + \sigma(y(t),\ v(t))dw_t + \\
& + \int_{\mathbb{R}^m_*} \gamma(y(t),\zeta,v(t))d\mu_t(\zeta) - \nu(y(t))d\xi(t),\ t \geq 0 \\
\text{(iii)} & d\xi(t) = \chi(y(t) \in \partial\mathcal{O})d\xi(t),\ t \geq 0\ ,
\end{array}\right\} \tag{1.14}$$

where $\chi(y \in \partial\mathcal{O})$ denotes the characteristic function of the boundary $\partial\mathcal{O}$. The stochastic process $(\xi(t),\ t \geq 0)$ is called the increasing process associated to the reflected diffusion process with jumps $(y(t),\ t \geq 0)$.
A stochastic equation similar to (1.14) has been considered in Anulova [1,2] for a half-space and in Chaleyat-Maurel et al. [6] for smooth domains. Based on the hypothesis (1.11) and without loss of generality, we can construct a Lipschitzian function pr(.) from $\mathbb{R}^d$ into itself such that

$$\left.\begin{array}{l}
|x - pr(x)| \leq dist(x,\bar{\mathcal{O}}) \\
pr(x) \in \bar{\mathcal{O}}\ ,\ \forall x \in \mathbb{R}^d\ ,\ dist(x,\bar{\mathcal{O}}) < \delta_o\ ,
\end{array}\right\} \tag{1.15}$$

where $dist(x,\bar{\mathcal{O}})$ denotes the distance between the point x and the set $\bar{\mathcal{O}}$ and δ_o is some positive constant.
Since the functions γ and ρ be modified outside of $\bar{\mathcal{O}}$ without changing (1.13), we can assume that

$$\left.\begin{array}{ll}
\gamma(x,\zeta,v) = \gamma(pr(x),\zeta,v), & \forall x \in \mathbb{R}^d\ ,\ dist(x,\bar{\mathcal{O}}) < \delta_o\ , \\
& \forall\zeta \in \mathbb{R}^m\ ,\ \forall v \in V,
\end{array}\right\} \tag{1.16}$$

$$\left.\begin{array}{l}
|\nabla\rho(x)| \geq 1\ ,\ \forall x \notin \mathcal{O}. \\
\rho(x) = |x-a|\ ,\ \forall x \notin \bar{\mathcal{O}},\ dist(x,\bar{\mathcal{O}}) > \delta_o,
\end{array}\right\} \tag{1.17}$$

for some point a in the interior $\mathcal{O}$ and the same δ_o, pr(.) of (1.15). It is convenient to have the representation

$$\nu(x) = M(x)\nabla\rho\ (x)\quad ,\quad \forall x \in \partial\mathcal{O} \tag{1.18}$$

where M(x) is a twice continuously differentiable function from a neighbourhood of the boundary $\partial\mathcal{O}$ into the set of symmetric matrices $d \times d$, satisfying

$$z\ M(x)z^* \geq \delta|z|^2\ ,\ \forall z \in \mathbb{R}^d\ , \tag{1.19}$$

with δ being the constant in (1.13). Note that in view of the assumption (1.11), this representation can be obtained.

We construct a vector field $\beta(x)$, defined and Lipschitzian on the whole space $\mathbb{R}^d$, as follows :

$$\beta(x) = \rho^+(x)\chi(x)M(x)\nabla\rho(x) + (1-\chi(x))\rho^+(x)\nabla\rho(x), \tag{1.20}$$

where ρ^+ denotes the positive part of the function ρ given in (1.11) and $\chi(x)$ is a smooth function satisfying

$$\left.\begin{aligned} &0 \le \chi(x) \le 1 \ , \ \forall x \in \mathbb{R}^d, \\ &\chi(x) = 1 \text{ in a neighbourhood of the closure } \bar{O}, \\ &\chi(x) = 0 \text{ if either } \operatorname{dist}(x,\bar{O}) > \delta_o \text{ or } M(x) \text{ is not defined.} \end{aligned}\right\} \tag{1.21}$$

Consider the stochastic differential equation

$$\left.\begin{aligned} &dy^\varepsilon(t) = g(y^\varepsilon(t), v(t))dt + \sigma(y^\varepsilon(t), v(t))dw_t + \\ &+ \int_{\mathbb{R}^m_*} \gamma(y^\varepsilon(t),\zeta,v(t))d\mu_t(\zeta) - \frac{1}{\varepsilon}\beta(y^\varepsilon(t))dt, \ t \ge 0 \\ &y^\varepsilon(0) = x \end{aligned}\right\} \tag{1.22}$$

i.e. the same equation as (1.6), but with a different β, and denote by

$$\xi^\varepsilon(t) = \frac{1}{\varepsilon} \int_0^t \rho^+(y^\varepsilon(s))ds \ , \ t \ge 0 \tag{1.23}$$

<u>THEOREM 1.2</u>. Assume the conditions (1.1), (1.11), (1.12), (1.13), (1.15),..., (1.21) hold. Then the stochastic equation (1.14) has one and only one solution $(y(t), \xi(t), t \ge 0)$. Moreover, for each $p \ge 1$, $T > 0$, we have

$$E\{ \sup_{0\le t\le T} |y^\varepsilon(t)-y(t)|^p \} \to 0 \quad \text{as } \varepsilon \to 0, \tag{1.24}$$

$$E\{ \sup_{0\le t\le T} \{\xi^\varepsilon(t)-\xi(t)|^p \} \to 0 \quad \text{as } \varepsilon \to 0, \tag{1.25}$$

where the limits are uniform with respect to x in $\bar{O}$ and any $v(.)$ progressively measurable process valued into V. □

The methods to be used are essentially those of Theorem 1.1. We just need to establish some key before applying the technique presented in [13] . For instance, if K is a Lipschitz constant of $\nabla\rho$ we have

$$\rho^+(x)(x'-x)\nabla\rho(x) \le \rho^+(x)\rho^+(x') + K\rho^+(x)\ |x'-x|^2 \ , \ \forall x \in \mathbb{R}^d . \tag{1.26}$$

This inequality replaces the monotonicity property of the function β in (1.7). Details of the proof, in the case without control $v(.)$, can be found in [14].

1.3. Some remarks

If in lieu of (1.11), we suppose that

$$\left.\begin{array}{l}\text{there exists a one-to-one map } \psi \text{ from a neighbourhood of closure } \bar{\mathcal{O}} \\ \text{into a neighbourhood of the closed unit ball } \bar{B} \text{ in } \mathbb{R}^d\text{, such that } \psi \\ \text{and its inverse } \psi^{-1} \text{ are three times continuously differentiable} \\ \text{satisfying } \psi(\mathcal{O}) = B \text{ , } \varphi(\partial\mathcal{O}) = \partial B,\end{array}\right\} \quad (1.27)$$

then the proof of Theorem 1.2. is reduced to an application of Theorem 1.1. on the domain $\psi_1(\bar{\mathcal{O}})$ for a suitable diffeomorphism ψ_1. Note that (1.27) is satisfied for any bounded simply connected set $\mathcal{O}$ in $\mathbb{R}^d$ having a smooth, connected and orientable boundary $\partial\mathcal{O}$. Moreover, the hypothesis (1.27) is not really needed. It suffices to know that the domain $\mathcal{O}$ is such that it can be transformed into a convex set $\mathcal{O}'$, via a diffeomorphism of class C^2 with the property of mapping the direction ν into the normal n' of $\mathcal{O}'$. In this case, $\mathcal{O}$ could have only piecewise smooth boundary and still satisfy the above property. As well as in Theorem 1.1. or Theorem 1.2. we have the following estimates :

$$E \{ \sup_{0\le t\le T} |y^{\varepsilon}(t)-y(t)|^p \} \le C \varepsilon^q \quad , \quad \forall\varepsilon > 0 \qquad (1.28)$$

$$E \{ \sup_{0\le t\le T} |\frac{1}{\varepsilon} \int_0^t \beta^*(y^{\varepsilon}(s))ds - \eta(t)|^p \le C \varepsilon^q \; , \; \forall\varepsilon > 0 \qquad (1.29)$$

$$E \{ \sup_{0\le t\le T} |\xi^{\varepsilon}(t)-\xi(t)|^p \} \le C \varepsilon^q \quad , \quad \forall\varepsilon > 0, \qquad (1.30)$$

for any numbers $p > 2q > 2$ and some constant C independent of $\varepsilon > 0$, x in $\bar{\mathcal{O}}$ and the control v(.). Also, the processes y(t), $\eta(t)$ and $\xi(t)$ are Lipschitz continuous w.r.t. the initial data x in $\bar{\mathcal{O}}$. For instance, given any $p \ge 2$ there exists a constant α_p, independent of x, x' in $\bar{\mathcal{O}}$ and the control v(.), such that

$$E \{|y_x(\theta) - y_{x'}(\theta)|^p e^{-\alpha_p\theta} \} \le |x-x'|^p \quad , \quad \forall x,x' \in \bar{\mathcal{O}} \qquad (1.31)$$

for any finite stopping time θ.

Most of the results in Theorem 1.1. and 1.2. can be extended to more general situation under suitable assumptions. For example, we can replace the Lipschitzian character of coefficients g, σ, γ by a non-degeneracy hypothesis on σ, and thus make use of the so-called martingale formulation of the problem. Unbounded domains $\mathcal{O}$, stopping times and random variables as initial data and an evolution version of (1.4), (1.14) can be considered.

Under some conditions, we may allow the matrix M(x) in (1.18), (1.19) to depend on the control v, i.e. M(x,v). This is the reflecting direction ν is also controlled, i.e. $\nu(x,v)$.

2. CONTROL PROBLEMS FOR REFLECTED DIFFUSION WITH JUMPS

2.1. Continuous control problem

2.1.1. Drift control case : we consider in this section the simplest case where the control appears only in the drift term $g(x,v)$, extensions will be mentioned below. More precisely, we take the framework of § 1.2 i.e. the assumptions (1.1), (1.11), (1.12), (1.13) with the following additional hypothesis :

$$\left.\begin{array}{l}\sigma,\ \gamma \text{ does not depend on } v, \text{ and } m = d \text{ for the sake of simplicity,}\\ \sigma\sigma^* \text{ is strictly elliptic, } V \text{ is a compact metric space,}\\ g(x,v) \text{ is continuous w.r.t. } v.\end{array}\right\} \quad (2.1)$$

Moreover, we are given

$$f : \mathbb{R}^d \times V \to \mathbb{R}, \text{ measurable and bounded, continuous w.r.t. } v. \qquad (2.2)$$

In order to define the control problem, we use the method of Bensoussan and Lions [4] (cfr. also Bensoussan and Menaldi [5] .

Let $\mathbf{V}$ be the class of progressively measurable processes v with values in V.

$$\left.\begin{array}{l}\text{Denote } y_x(t) \text{ by the solution of (1.14) corresponding to}\\ (\sigma,\ \gamma,\ \nu), \text{ with } g \equiv 0\ .\end{array}\right\} \quad (2.3)$$

Then define, for $v \subset \mathbf{V}$,

$$q_v(t) = \sigma^{-1}(y_x(t))g(y_x(t),v(t))$$

and consider the probability measure P^v on (Ω,F)

$$\frac{dP^v}{dP}\Big|_{F_t} = \exp\Big(\int_0^t q_v(s)dw_s - \frac{1}{2}\int_0^t |q_v(s)|^2\, ds\Big).$$

If we define P_x^v as the image of P^v on $D(0,\infty;\mathbb{R}^d)$ via the process $y_x(t)$, one can show that P_x^v is the solution of the martingale problem associated to the operator

$$\left.\begin{array}{l} L_v\,\varphi = -\frac{1}{2}\sum\limits_{i,j=1}^{d}(\sigma\sigma^*)_{ij}\frac{\partial^2\varphi}{\partial x_i \partial x_j} - \sum\limits_{i=1}^{d} g_i(x,v)\frac{\partial\varphi}{x_i} - \\ \qquad - \int [\varphi\,(x+\gamma) - \varphi(x) - \gamma.\nabla\varphi].\pi(dz)\end{array}\right\} \quad (2.4)$$

(cfr. Bensoussan and Lions [4] , El Karoui [7]).

We can, thus, define the cost functional as

$$J_x(v) = E^v\Big\{\int_0^\infty e^{-\alpha t} f(y_x(s),\ v(s))ds\Big\} \qquad (2.5)$$

where α is a given discount factor. The problem is to minimize $J_x(v)$ over $\mathbf{V}$. We will need the following additional notations and assumptions :

$$M(x,B) = \pi(\zeta|\gamma\ (x,\zeta)\in\ B),$$

for any Borel subset B of $\mathbb{R}_*^d$ and M is such that

$$M(x,dz) = c(x,z)m(dz) \qquad (2.6)$$

with $0 \le c(x,z) \le 1$ and since $x + \gamma(x,\xi) \in \bar{\mathcal{O}}$, $\forall x \in \bar{\mathcal{O}}$, $\forall \xi$, we must have

$$c(x,z) = 0 \text{ if } x + z \notin \bar{\mathcal{O}}.$$

Notice that, because of (1.1), we have

$$\int_{\mathbb{R}^d_*} |z|^2 \, m(dz) < +\infty .$$

Let also

$$\left.\begin{aligned}
A\varphi(x) &= - \sum_{ij=1}^{d} a_{ij}(x) \frac{\partial^2 \varphi}{\partial x_i \partial x_j} , \quad a = \sigma\sigma^* \\
B\varphi(x) &= \int_{\mathbb{R}^d_*} [\varphi(x+z) - \varphi(x) - z.\nabla\varphi \chi_{|z|<1}] \, c(x,z) m(dz) \\
h(x,v) &= g(x,v) + \int_{\mathbb{R}^d_*} z \, \chi_{|z| \ge 1} \, . \, c(x,z) m(dz) \\
H(\varphi) &= \inf_{v \in V} \{ h(x,v) \nabla\varphi + f(x,v) \}.
\end{aligned}\right\} \quad (2.7)$$

One can then write, at least formally, the following Hamilton-Jacobi-Equation (HJB) for the optimal cost

$$u(x) = \inf_{v \in \mathbf{V}} J_x(v) \quad (2.8)$$

$$\left.\begin{aligned}
&Au - Bu - H(u) + \alpha u = 0, \\
&\nu . \frac{\partial u}{\partial x} \Big|_{\partial \mathcal{O}} = 0
\end{aligned}\right\} \quad (2.9)$$

for which we can use a simplified version the following result of Bensoussan and Lions[4].

<u>THEOREM 2.1</u>. Under the assumptions (1.1), (1.11), (1.12), (1.13), (2.1), (2.2), (2.6) [or $M(x, \mathbb{R}^d - \{0\}) \le \bar{M}$ instead of (2.6)], there exists a unique solution $u \in W^{2,p}(\mathcal{O})$, $2 \le p < +\infty$ of (2.9). □

We can now state :

<u>THEOREM 2.2</u>. Under the assumptions (1.1), (1.11), (1.12), (1.13), (2.1), (2.2), (2.6), [or $M(x, \mathbb{R}^d - \{0\}) \le \bar{M}$ instead of (2.6)], the solution of (2.9) satisfies $u(x) = \inf\{ J_x(v) \; ; \; v \in \mathbf{V} \}$, and there exists an optimal control.

<u>Outline of the Proof</u>

First, we notice that we have the Itô's formula : $\forall \varphi \in C_b^2(\bar{\mathcal{O}})$,

$$\left.\begin{aligned}
E\{ \varphi(y_x^v(t))\} = \varphi(x) &+ E\{ \int_0^t (-L_v \varphi)(y_x^v(s), v(x)) ds \} \\
&- E\{ \int_0^t (\nu.\nabla\varphi)(y_x^v(s)) d\xi_s ,
\end{aligned}\right\} \quad (2.10)$$

where we use the notation

$$\left.\begin{aligned}
(L_v \varphi)(x,v) = A\varphi &- \sum_{i=1}^{d} h_i(x,v) \frac{\partial \varphi}{\partial x_i} \\
&- \int_{\mathbb{R}^d_*} [\varphi(x,z) - \varphi(x) - z\nabla\varphi\chi_{|z|<1}] \, c(x,z) m(dz).
\end{aligned}\right\} \quad (2.11)$$

The equality (2.10) comes from the general Itô's formula for semi-martingale and the fact that the process $y_x^v(t)$ stays in $\bar{O}$. Then, as in Bensoussan and Lions [4], we have

$$|E\int_0^T \varphi(y_x^v(t))dt| \leq C|\varphi|_{L^p} \qquad (2.12)$$

for $\varphi \in L^p(\mathbb{R}^d)$, and therefore, we can extend (2.10) to $\varphi \in W^{2,p}(O)$, $p > n+1$. Since we have, from (2.9) :

$$-(L_v u)(x,v) + \alpha u \leq f(x,v),\ \forall v \in V,$$

applying the Itô's formula for any $v \in \mathbf{V}$, we get

$$u(x) \leq J_v(v). \qquad (2.13)$$

Next, since we can find a Borel function $\hat{v}(x)$ realizing the infimum in $H(u)$, we have

$$-(L_v u)(x,\hat{v}(x)) + \alpha u = f(x,\hat{v}(x)) \qquad (2.14)$$

and thus, taking $\hat{v}(t) = \hat{v}(y_x(t))$ with $y_x(t)$ defined in (2.3), we obtain

$$u(x) = J_x(\hat{v}) \qquad (2.15)$$

which complete the proof. □

REMARK 2.1. One can, under suitable assumptions, consider cases where γ depends on v, for instance

$$\left.\begin{aligned}&\pi(\xi|\gamma(x,v,\xi) \in B) = \int_B c_0(x,z)(1 + c_1(x,v,z))m(dz),\\ &|c_1(x,v,z)| \leq C|z|,\end{aligned}\right\} \qquad (2.16)$$

which allows to construct the controlled process like in Bensoussan and Menaldi [5]. This would give the stochastic interpretation of the HJB equation studied in Bensoussan and Lions [4] Chapitre 3 § 2. □

2.1.2. General Case : for more general cases, namely, when γ, σ depend on v, and when O is a convex non smooth domain, we can state some results using the semigroup formulation. For that purpose, we consider as an admissible system $S = \{ WP, y(t), v(t), \eta(t), t \geq 0 \}$ where WP stands for the Wiener-Poisson space $WP = \{\Omega, F, P, F_t, w_t, \mu_t, t\}$ and where y, v, η are related by (1.4).

Defining, for f and h measurable bounded functions ,

$$J_x(S,h,t) = E\{ \int_0^t e^{-\alpha s} f(y(s),v(s))ds + h(y(t))e^{-\alpha t} \}$$

$$\left.\begin{aligned}&Q(t)h(x) = \inf \{J_x(S,h,t),\ S \text{ admissible system}\} \quad \text{and}\\ &u(x) = \inf (J_x(S,0,\infty),\ S \text{ admissible system}).\end{aligned}\right\} \qquad (2.17)$$

Let B_s the set of upper semicontinuous bounded functions on $\bar{O}$. We will also assume

$$\left.\begin{aligned}&|f(x,v) - f(x',v')| \leq k(|x-x'| + |v-v'|)\\ &|g(x,v) - g(x',v')|^2 + |\sigma(x,v) - \sigma(x',v')|^2\\ &+ \int_{\mathbb{R}^d_*} |\gamma(x,z,v) - \gamma(x',z,v')|^2\, \pi(dz) \leq c_2[|x-x'|^2 + |v-v'|^2].\end{aligned}\right\} \qquad (2.18)$$

We then have

Lemma 2.1. Under the assumptions of § 1.2, and (2.18), the operators $(Q(t),\ t \geq 0)$ is a non linear semigroup on B_s that is

(i) $Q(t) : B_s \to B_s$, $Q(0) = I$, $Q(t+s) = Q(t)Q(s)$,

(ii) $\| Q(t)h - Q(s)h\| \to 0$ $t \to s$, if $h \in C(\overline{\mathcal{O}})$,

$$\text{(iii)}\begin{cases} \| Q(t)h\| \leq \frac{1}{\alpha}(1-e^{-\alpha t})\| f\| + e^{-\alpha t} \| h \| \quad \forall t \geq 0 \\ \| Q(t)h_1 - Q(t)h_2 \| \leq e^{-\alpha t} \| h_1 - h_2 \| \end{cases}$$

$$\text{(iv)}\begin{cases} \text{and if } \alpha > \mu = \frac{\alpha_2}{2} \\ \alpha_2 = \sup \{ c(x,x',v) |x-x'|^2 : x,x' \in \mathbb{R}^d,\ v \in V \} \\ c(x,x',v) = (x-x').(g(x,v) - g(x',v)) + \frac{1}{2} \sum\limits_{i,k=1}^{d} (\sigma_{ik}(x,v)-\sigma_{ik}(x',v))^2 + \\ + \int [|x-x' + \gamma(x,\zeta,v)- \gamma(x',\zeta,v)|^2 - |x-x'|^2-2(x-x')(\gamma(x,\zeta,v)- \gamma(x',\zeta,v)]\pi(d\zeta), \\ \text{then, if } |h(x) - h(x')| \leq K|x-x'| ,\ K > \frac{C_2}{\alpha-\mu}, \text{ where } C_2 \text{ is defined in (1.1),} \end{cases}$$

we have

$$|Q(t)h(x) = Q(t)h(x')| \leq K |x-x'|. \tag{2.19}$$

The proof of (i), (ii), (iii) is identical to those of Krylov [8], Nisio [15], and is therefore omitted. Assume that h is K-Lipschitz continuous, then from (1.31) we have

$$|J_x(S,h,t) - J_{x'}(S,h,t)| \leq \frac{C_2}{\alpha-\mu}(1-e^{-(\alpha-\mu)t}) |x-x'| + K e^{-(\alpha-\mu)t} |x-x'|$$

and therefore, if $K > \frac{k}{\alpha-\mu}$, we get

$$|Q(t)h(x) - Q(t)h(x')| \leq K|x-x'|. \qquad \square$$

THEOREM 2.3. Under the assumptions of § 1.2 and (2.18) and if $\alpha > \mu$, then $u(x)$ is the unique solution of

$$Q(t)u = u,\ u \in W^{1,\infty}(\mathcal{O}) \tag{2.20}$$

and u satisfies the dynamic programming equation in the following sense

$$u(x) = \inf\{ J_x(S,u,\theta),\ S \text{ admissible} \} \tag{2.21}$$

for any stopping time θ.

Outline of the Proof : form the definition of u, it is easy to obtain

$$u = \lim_{t\to\infty} Q(t)h \ , \text{ for any } h \in B_s. \tag{2.22}$$

On the other hand, $Q(t)\bar{u} = \bar{u}$ has a unique solution in B_s, which is $W^{1,\infty}(\mathcal{O})$ if the assumption of (iv) of Lemma 2.1 are satisfied. From (2.22) we conclude that $\bar{u} = u$. The proof of (2.21) is similar to the diffusion processes case. □

REMARK 2.2. The main difficulty is proving that $Q(t+s) = Q(t)Q(s)$ but this is done exactly like in Krylov [8], Bensoussan and Lions [4], P.L. Lions [9], P.L. Lions and Menaldi [10], and therefore we do not reproduce the proof here. □

REMARK 2.3. When σ does not depend on v, and $\mathcal{O}$ is convex but non smooth, then we can still obtain an HJB equation in variational form assuming $\sigma\sigma^* \in W^{1,\infty}$.

2.2. Optimal Stopping

Using the results of § 1 without the intervention of v, one can see that $y_x(t)$ solution of (1.4)(resp (1.14)) define a Feller Markov process with values in a compact metric space. Therefore, the general results on optimal stopping of Feller process (cfr.[16]) apply : let

$$\left.\begin{aligned} & f \in C(\mathcal{O}) \ , \ \psi \in C(\bar{\mathcal{O}}), \\ & J_x(\tau) = E\left\{\int_0^\tau f(y_x(t))dt + e^{-\alpha\tau}\psi(y_x(\tau))\right\} \\ & \Phi(t)\varphi(x) = E\varphi(y_x(t)),\ \forall\varphi \in C(\bar{\mathcal{O}}), \end{aligned}\right\} \tag{2.23}$$

THEOREM 2.4. Under the assumptions of § 1 and (2.23), the function

$$u(x) = \inf\{ (J_x(\tau) : \tau \quad F_t \text{ stopping time} \}$$

is the maximum element of the set of functions satisfying

$$\left.\begin{aligned} & u \le e^{-\alpha t}\ \Phi(t)u + \int_0^t e^{-\alpha s}\ \Phi(s)f\,ds \\ & u \le \psi \\ & u \in C(\bar{\mathcal{O}}) \end{aligned}\right\} \tag{2.24}$$

Various forms of variational inequality can be obtained for that problem depending on the regularity of the data. For instance, we can state the following result : define

$$A\varphi(x) = -\sum_{i,j=1}^{d} a_{ij}\frac{\partial^2\varphi}{\partial x_i\partial x_j} - \sum_{i=1}^{d} g_i\frac{\partial\varphi}{\partial x_i} -$$

$$- \int_{\mathbb{R}^d_*} [\varphi(x+z) - \varphi(x) - z.\nabla\varphi]\, c_o(x,z)m(dz)\,.$$

THEOREM 2.5. Assume that the assumptions of § 1.2 and (2.23) hold, that $\psi \in W^{2,p}(\mathcal{O})$, $\nu.\frac{\partial\psi}{\partial x}|_\Gamma \ge 0$ and that $\sigma\sigma^*$ is strictly elliptic, then u is the unique solution of

$$\left.\begin{aligned} & -Au + \alpha u \le f \ \ \text{, a.e. on } \mathcal{O},\ \nu.\frac{\partial u}{\partial x}|_\Gamma = 0\ ,\quad u \le \varphi \\ & (Au + \alpha u - f)(u - \psi) = 0\ ,\quad u \le \psi\ , \end{aligned}\right\} \tag{2.25}$$

□

The proof can be done by classical argument using the penalized problem which gives an equation for which the results of § 2.1 are relevant. □

REMARK 2.4. Using the techniques of [12], much more general results can be stated, including results for the degenerate case. □

REMARK 2.5. One could also consider, at least in semigroup formulation, the continuous control problem with optimal stopping (cfr. Bensoussan and Lions [4])and the optimal cost is the maximum solution of

$$u \leq Q(t)u, \; u \leq \psi \quad , \; u \in C(\bar{O}) \; .$$

REMARK 2.6. As for optimal stopping, the semigroup formulation of impulse control problems can be used since we have a Feller process. We refer to Bensoussan and Lions [4], and [16] for a general study.

REFERENCE

[1] S.V. Anulova, On processes with Levy Generating Operator in a Half-Space, Math. USSR Izv., 13 (1979), pp. 9-51.

[2] S.V. Anulova, On Stochastic Differential Equations with Boundary Conditions in a Half-Space, Math. USSR Izv., 18 (1982), pp. 423-437.

[3] A. Bensoussan and J.L. Lions, Diffusion Processes in Bounded Domains and Singular Perturbations Problems for Variational Inequalities with Neumann Boundary Conditions, Lectures Notes in Math., 451 (1975), Springer Verlag, pp. 8-25.

[4] A. Bensoussan and J.L. Lions, Contrôle Impulsionnel et Inéquation Quasi-Variationnelles, Dunod, Paris, 1982.

[5] A. Bensoussan and J.L. Menaldi, Optimal Stochastic Control of Diffusion Processes with Jumps Stopped at the Exit of a Domain, in Stochastic Analysis and Applications, Ed. M.A. Pinsky, M. Dekker, New York, 1983, pp. 81-104.

[6] M. Chaleyat-Maurel, N. EL Karoui and B. Marchal, Réflexion Discontinue et Systèmes Stochastiques, Ann. Prob., 8, (1980), pp. 1049-1067.

[7] N. EL Karoui, Méthodes Probabilistes en Contrôle Stochastique, Lectures Notes in Math., 876 (1980), pp. 74-238.

[8] N.V. Krylov, Controlled Diffusion Processes, Springer Verlag, Berlin, 1980.

[9] P.L. Lions, Generalized solutions of Hamilton Jacobi equations, Pitman, London, 1982.

[10] P.L. Lions and J.L. Menaldi, Optimal Control of Stochastic Integrals and Hamilton-Jacobi-Bellmann equations, Parts I and II, SIAM J. Control Optimization, 20 (1982), pp. 58-81 and pp. 82-95.

[11] P.L. Lions, J.L. Menaldi and S.A. Sznitman, Contruction de Processus de Diffusion Réfléchis par Pénalisation du Domaine, C.R. Acad. Sc. Paris, I-292 (1981) pp. 559-562.

[12] J.L. Menaldi, On the Optimal Stopping Time Problem for Degenerate Diffusions, SIAM J. Control Optimization, 18 (1980), pp. 697-721.

[13] J.L. Menaldi, Stochastic Variational Inequality for Reflected Diffusions, Indiana Univ. Math. J., 32 (1983), pp. 733-744.

[14] J.L. Menaldi and M. Robin, Reflected Diffusion Processes with Jumps. Ann. Prob., to appear. See also C.R.Acad.Sc.Paris, I-297 (1983), pp. 533-536.

[15] M. Nisio, On a Nonlinear Semigroup Attached to Stochastic Optimal Control, Publ. R.I.M.S., Kyoto University, 12 (1976), pp. 513-537.

[16] M. Robin, Contrôle Impulsionnel des Processus de Markov, Thèse d'Etat, Université de Paris IX-Dauphine, 1978.

[17] V.A. Shalaumov, On the Behaviour of a Diffusion Process with Large Drift Coefficient in a Half-Space, Theory Prob. Appl., 6 (1980), pp. 592-598.

Lecture Notes in Control and Information Sciences

Edited by M. Thoma

Vol. 43: Stochastic Differential Systems
Proceedings of the 2nd Bad Honnef Conference
of the SFB 72 of the DFG at the University of Bonn
June 28 – July 2, 1982
Edited by M. Kohlmann and N. Christopeit
XII, 377 pages. 1982.

Vol. 44: Analysis and Optimization of Systems
Proceedings of the Fifth International
Conference on Analysis and Optimization of Systems
Versailles. December 14–17, 1982
Edited by A. Bensoussan and J. L. Lions
XV, 987 pages, 1982

Vol. 45: M. Arató
Linear Stochastic Systems
with Constant Coefficients
A Statistical Approach
IX, 309 pages. 1982

Vol. 46: Time-Scale Modeling of Dynamic Networks
with Applications to Power Systems
Edited by J. H. Chow
X, 218 pages. 1982

Vol. 47: P. A. Ioannou, P. V. Kokotovic
Adaptive Systems with Reduced Models
V, 162 pages. 1983

Vol. 48: Yaakov Yavin
Feedback Strategies for Partially
Observable Stochastic Systems
VI, 233 pages, 1983

Vol. 49: Theory and Application of Random Fields
Proceedings of the IFIP-WG 7/1
Working Conference
held under the joint auspices of the
Indian Statistical Institute
Bangalore, India, January 1982
Edited by G. Kallianpur
VI. 290 pages. 1983

Vol. 50: M. Papageorgiou
Applications of Automatic Control Concepts
to Traffic Flow Modeling and Control
IX, 186 pages. 1983

Vol. 51: Z. Nahorski, H.F. Ravn, R.V.V. Vidal
Optimization of Discrete Time Systems
The Upper Boundary Approach
V, 137 pages 1983

Vol. 52: A. L. Dontchev
Perturbations, Approximations and Sensitivity Analysis
of Optimal Control Systems
IV, 158 pages. 1983

Vol. 53: Liu Chen Hui
General Decoupling Theory of Multivariable
Process Control Systems
XI, 474 pages. 1983

Vol. 54: Control Theory for Distributed
Parameter Systems and Applications
Edited by F. Kappel, K. Kunisch,
W. Schappacher
VII, 245 pages. 1983.

Vol. 55: Ganti Prasada Rao
Piecewise Constant Orthogonal Functions
and Their Application to Systems and Control
VII, 254 pages. 1983.

Vol. 56: Dines Chandra Saha, Ganti Prasada Rao
Identification of Continuous
Dynamical Systems
The Poisson Moment Functional
(PMF) Approach
IX, 158 pages. 1983.

Vol. 57: T. Söderström, P. G. Stoica
Instrumental Variable Methods
for System Identification
VII, 243 pages. 1983.

Vol. 58: Mathematical Theory of
Networks and Systems
Proceedings of the MTNS-83 International
Symposium
Beer Sheva, Israel, June 20–24, 1983
Edited by P. A. Fuhrmann
X, 906 pages. 1984

Vol. 59: System Modelling and Optimization
Proceedings of the 11th IFIP Conference
Copenhagen, Denmark, July 25-29, 1983
Edited by P. Thoft-Christensen
IX, 892 pages. 1984

Vol. 60: Modelling and Performance
Evaluation Methodology
Proceedings of the International Seminar
Paris, France, January 24–26, 1983
Edited by F. Bacelli and G. Fayolle
VII, 655 pages. 1984

Vol. 61: Filtering and Control of Random
Processes
Proceedings of the E.N.S.T.-C.N.E.T. Colloquium
Paris, France, February 23–24, 1983
Edited by H. Korezlioglu, G. Mazziotto, and
J. Szpirglas
V, 325 pages. 1984